Supercapacitor

Maintaining a balanced blend of theory and application, ***Supercapacitor: Fundamentals, Materials, and Devices*** offers a complete guide to next-generation energy storage technology. It introduces supercapacitors as a powerful alternative to conventional batteries, emphasizing their unique electric field–based energy storage mechanism that enables rapid charging, long cycle life, and high power density.

Drawing from both academic and industry-driven research, the book lays a strong foundation by covering supercapacitor classifications, electrode/electrolyte materials, and fabrication techniques. With recent advancements in materials science and device engineering, the book presents novel approaches to supercapacitor design utilizing cutting-edge materials, including MXenes, MOFs, and ionic liquids. It also addresses key challenges and explores promising directions in scalable, sustainable, and high-performance energy storage solutions.

Features

- In-depth overview of supercapacitor types: EDLCs, pseudocapacitors, and hybrid systems.
- Coverage of advanced electrode materials, including MXenes, conducting polymers, and MOFs.
- Detailed fabrication protocols with emphasis on scalable and eco-friendly methodologies.
- Analytical approaches to assess performance: cyclic voltammetry, impedance spectroscopy, and more.

In this edition, chapters on flexible and wearable supercapacitors, ionic liquid–based electrolytes, and sustainable synthesis techniques are included.

This book is a vital resource for graduate students, academic researchers, materials scientists, and professionals in electronics and clean energy fields. By combining fundamental principles with emerging innovations, it offers readers both the theoretical understanding and practical insights needed to succeed in this fast-changing field. Whether designing devices or analyzing new materials, this book provides the tools and context to transform ideas into meaningful energy storage solutions.

Supercapacitor

Fundamentals, Materials, and Devices

Edited by
Dinesh Kumar and Sapna Nehra

CRC Press
Taylor & Francis Group
Boca Raton London New York

CRC Press is an imprint of the
Taylor & Francis Group, an **informa** business

Designed cover image: Shutterstock #2550253679

First edition published 2026
by CRC Press
2385 NW Executive Center Drive, Suite 320, Boca Raton FL 33431

and by CRC Press
4 Park Square, Milton Park, Abingdon, Oxon, OX14 4RN

CRC Press is an imprint of Taylor & Francis Group, LLC

ISBN: 978-1-032-88009-9 (hbk)
ISBN: 978-1-032-88011-2 (pbk)
ISBN: 978-1-003-53583-6 (ebk)

DOI: 10.1201/9781003535836

Typeset in Times
by KnowledgeWorks Global Ltd.

Contents

Preface

Technology is not just about the gadgets, it's about how we use them.

Elon Musk

Supercapacitors are a promising energy storage technology that has the potential to revolutionize the way we store and use energy. Unlike batteries, which store energy in chemical form, supercapacitors store energy in an electric field, making them capable of delivering high-power output and fast charging.

This book aims to provide a comprehensive overview of the fundamentals of supercapacitors, including the types of supercapacitors, materials for supercapacitor electrodes and electrolytes, fabrication techniques, characterization techniques, and applications in energy storage and power electronics. The book also highlights the challenges and future directions in supercapacitors.

This book, divided into 13 meticulously structured chapters, covers a broad spectrum of topics. It begins with an introduction to supercapacitors and their underlying fundamentals. It then explores different types of supercapacitors, electrode materials, electrolytes, and fabrication techniques essential for their development. Key characterization methods are discussed in detail, enabling readers to assess performance metrics effectively. Beyond the core technologies, the book delves into emerging trends and innovations, including the synthesis of metal–organic framework (MOF)–based supercapacitors, high-performance MXene supercapacitors, and the role of ionic liquid electrolytes in achieving superior energy storage capabilities. It also addresses the field's challenges and outlines promising future directions, guiding researchers and engineers toward impactful advancements.

This book is a valuable resource for students, researchers, and professionals seeking in-depth knowledge of supercapacitors and their potential applications. We aim to inspire further exploration and innovation in this rapidly evolving domain by combining fundamental concepts and cutting-edge developments. The book offers industry experts' insights, giving readers a comprehensive understanding of the latest supercapacitor research and developments. Its broad coverage of applications makes it an invaluable resource for researchers, students, and professionals interested in supercapacitors. We hope this book inspires new research concepts and contributes to sustainable development. We appreciate the experts' contributions that have enriched our content and thank readers for joining us in exploring supercapacitor technology's limitless possibilities.

Dinesh Kumar
Sapna Nehra

About the Editors

Dinesh Kumar is a distinguished faculty member at the School of Chemical Sciences, Central University of Gujarat, Gandhinagar. He is an alumnus of the University of Rajasthan, Jaipur, where he earned his master's and doctoral degrees in Chemistry. His scholarly achievements have garnered numerous national and international accolades and fellowships, and he has consistently been ranked among the top 2% of scientists globally from 2021 to 2024. His research primarily focuses on creating mixed metal oxide–based hierarchical nanostructures to develop electrochemical sensors and energy storage devices. His research endeavors are also directed toward synthesizing capped magnetic nanoparticles (MNPs), core–shell nanoparticles (NPs), and biopolymers. These include the development of metal oxide–based nanoadsorbents and nanosensors for removing and detecting inorganic water toxicants, such as heavy metal ions and fluoride. His prolific contributions to science are evidenced by his extensive publication record, which includes 146 articles in prestigious international journals, five books, 140 book chapters, and approximately 105 presentations at various national and international forums. Nineteen students have completed their Ph.D. degrees under his guidance. His work is widely recognized, reflected in his h-index of 41, i10 index of 120, and nearly 6500 citations.

Sapna Nehra has been enriching Nirwan University, Jaipur's academic community as Assistant Professor with additional roles of controller of Examination and Research Coordinator since March 2022. A distinguished alumna of Banasthali Vidyapith, Rajasthan, India, she earned her M.Sc. degree in 2014, followed by a Ph.D. in 2020. Her doctoral research, guided by Dr. Dinesh Kumar at the School of Chemical Sciences, Central University of Gujarat, Gandhinagar, focused on innovative chemical applications. Before her tenure at Nirwan University, she imparted knowledge at Dr. K. N. Modi University, Newai, Tonk, Rajasthan, from 2020 to 2022. She is a prolific author with 13 research papers and 47 book chapters published by prestigious international publishers. She also published three books with national as well as international publishers. In addition, she contributes her expertise as a reviewer to esteemed journals under ScienceDirect, ACS, and Springer Nature. She regularly participates in national and international conferences, showcasing her research, which primarily revolves around developing advanced materials, including polymers, bionanomaterials, and metal oxides, to enhance water purification technologies.

Contributors

Mohamed Magdy Mohamed Abdo
Physics and Mathematics Engineering Department
Faculty of Engineering
Ain Shams University
Cairo, Egypt

A. J. Saleh Ahammad
Department of Chemistry
Jagannath University
Dhaka, Bangladesh

Ahmed M. Bayoumy
Biophysics Group
Physics Department
Faculty of Science
Ain Shams University
Cairo, Egypt

Mohamed Ahmed Belal
Graphene Center of Excellence for Energy and Electronic Applications
Egypt-Japan University of Science and Technology
Alexandria, Egypt

and

Chemistry Department
Faculty of Science
Alexandria University
Alexandria, Egypt

Krishnendu Biswas
Department of Chemistry
School of Advanced Sciences
Vellore Institute of Technology
Chennai, India

Sreeja P C
School of Basic and Applied Sciences
Nirwan University
Jaipur, Rajasthan, India

and

Innovation Centre, Mane Kancor Ingredients Private Limited
Kerala, India

Vishal Choudhary
ICAR-Central Institute of Agricultural Engineering
Madhya Pradesh, India

and

The Graduate School, IARI
New Delhi, India

Nipa Debnath
Department of Physics
Jagannath University
Dhaka, Bangladesh

Ahmed Abd El-Moneim
Graphene Center of Excellence for Energy and Electronic Applications
Egypt-Japan University of Science and Technology
Alexandria, Egypt

and

School of Basic and Applied Science
Egypt-Japan University of Science and Technology
Alexandria, Egypt

and

Physical Chemistry Department
National Research Centre
Cairo, Egypt

Sandip Gangil
ICAR-Central Institute of Agricultural Engineering
Madhya Pradesh, India

Bhupendra Ghritalahre
ICAR-National Rice Research Institute
Cuttack, India

Suresh Jayakumar
Energy and Innovative Materials Laboratory (EIML)
Department of Physics and Nanotechnology
College of Engineering and Technology
SRM Institute of Science and Technology (SRMIST)
Tamil Nadu, India

Paricha Jebin
Department of Physics
Jagannath University
Dhaka, Bangladesh

Ramesh Chandran K
School of Basic and Applied Sciences
Nirwan University
Jaipur, Rajasthan, India
and
Innovation Centre, Mane Kancor Ingredients Private Limited
Kerala, India

Kerim Karabacak
Kutahya Technical Sciences Vocational School
Kutahya Dumlupinar University
Kutahya, Turkiye

Md. Rakib Khan
Department of Chemistry
Jagannath University
Dhaka, Bangladesh

Sanjay Khatri
ICAR-Central Institute of Agricultural Engineering
Madhya Pradesh, India
and
The Graduate School, IARI
New Delhi, India

Dinesh Kumar
School of Chemical Sciences
Central University of Gujarat
Gujarat, India

Manish Kumar
ICAR-Central Institute of Agricultural Engineering
Madhya Pradesh, India
and
The Graduate School, IARI
New Delhi, India

Jaidev Kumbhakar
Cambridge Institute of Polytechnic
Jharkhand, India

Sapna Nehra
School of Basic and Applied Sciences
Nirwan University
Jaipur, Rajasthan, India

P. Chinnappan Santhosh
Energy and Innovative Materials Laboratory (EIML)
Department of Physics and Nanotechnology
College of Engineering and Technology
SRM Institute of Science and Technology (SRMIST)
Tamil Nadu, India

Parmanand Sahu
ICAR-Central Institute of Agricultural Engineering
Madhya Pradesh, India

Ravi Kumar Sahu
ICAR-Central Institute of Agricultural Engineering
Madhya Pradesh, India

Ubaid Sidiqi
School of Chemical Sciences
Central University of Gujarat
Gujarat, India

Anusha Srinivas
School of Basic and Applied Sciences
Nirwan University
Jaipur, Rajasthan, India
and
Innovation Centre
Mane Kancor Ingredients Private Limited
Kerala, India

A. V. Radhamani
Energy and Innovative Materials Laboratory (EIML)
Department of Physics and Nanotechnology
College of Engineering and Technology
SRM Institute of Science and Technology (SRMIST)
Tamil Nadu, India

Vyshakh Viswanath N
Department of Chemistry
School of Advanced Sciences
Vellore Institute of Technology
Chennai, India

Abbreviations

AASCs	aqueous electrolyte–based asymmetric supercapacitors
AC	activated carbon
AGVs	automated guided vehicles
AI	artificial intelligence
AMRs	autonomous mobile robots
ASC	asymmetric supercapacitor
BC	carbon black
BDC/BDA	1,4-benzenedicarboxylic acid
BET	Brunauer–Emmett–Teller
ACN	acetonitrile
CDC	carbide-derived carbon
CE	counter electrode
CNTs	carbon nanotubes
COFs	covalent organic frameworks
CP	chronopotentiometry
CPs	conductive polymer
CTAB	cetyltrimethylammonium bromide
CV	cyclic voltammetry
CVD	chemical vapor deposition
DESs	deep eutectic solvents
DLS	dynamic light scattering
DMF	N,N-dimethylformamide
DoE	Department of Energy
DTG	derivative thermogravimetric
EC	ethyl cellulose
EC/DEC	ethylene carbonate/diethyl carbonate
ECI	electrochemical charge injection
ED	energy density
EDL	electrical double layer
EDLCs	electrical double layer capacitors
EDS	energy-dispersive X-ray spectroscopy
EEG	electrochemical exfoliated graphene
EIS	electrochemical impedance spectroscopy
EMS	energy management systems
EPD	electrophoretic deposition
ESN	electrode superficial nucleation
ESR	equivalent series resistance
ESW	electrochemical stability window
EVs	electric vehicles
FCVs	fuel cell vehicles
FESEM	field emission scanning electron microscopy
FTIR	Fourier transform infrared

GCD	galvanostatic charge–discharge
GO	graphene oxide
GPS	Global Positioning System
HASCs	hybrid asymmetric supercapacitors
hBN	hexagonal boron nitride
HCNS	hollow carbon nanospheres
HESS	hybrid energy storage system
HEVs	hybrid electric vehicles
HR-TEM	high-resolution transmission electron microscopy
HSCs	hybrid supercapacitors
IBED	indirect bipolar electrodeposition
IDE	interdigitated supercapacitor
ILs	ionic liquids
IoT	Internet of Things
KFSI	potassium bis(fluorosulfonyl)imide
LAG	liquid-assisted grinding
LCA	life cycle assessment
LCL	lnductor-Capacitor-lnductor
LDH	layered double hydroxide
LED	light-emitting diode
LIBs	lithium-ion batteries
LIG	laser-induced graphene
LSG	laser-scribed graphene
ML	machine learning
MO	metal oxide
MOFs	metal–organic frameworks
MSCs	micro-supercapacitors
MWCNTs	multiwall carbon nanotubes
NG	neat grinding
NLDFT	non-local density functional theory
OLCs	onion-like carbons
OPW	operating potential window
PAN	polyacrylonitrile
PANI	polyaniline
PC	pseudocapacitors
PCPs	porous coordination polymers
PD	power density
PEDOT	poly(3,4-ethylenedioxythiophene)
PEG	polyethylene glycol
PI	polyimide
PLCs	programmable logic controllers
PPEG	liquid phase exfoliated graphene
PPy	polypyrrole
PSSH	polystyrene sulfonic acid
PTA	purified terephthalic acid

PTFE	polytetrafluoroethylene
PV	photovoltaic
PVA	polyvinyl alcohol
PVDF	polyvinylidene fluoride
PVP	polyvinylpyrrolidone
PW	potential window
PWM	pulse width modulation
RAE	redox-active electrolyte
RE	reference electrode
rGO	reduced graphene oxide
RH	relative humidity
RHPC	rice husk–based hierarchical porous carbon
RTILs	room temperature ionic liquids
RuO_2	ruthenium dioxide
SCE	saturated calomel electrode
SCs	supercapacitors
SE	supporting electrode
SECM	scanning electrochemical microscopy instrument
SEI	solid electrolyte interphase
SEM	scanning electron microscopy
SHE	standard hydrogen electrode
SOHIO	Standard Oil Company, Ohio
SSC	symmetric supercapacitor
SWCNTs	single-wall carbon nanotubes
TEM	transmission electron microscopy
TGA	thermogravimetric analysis
TMO	transition metal oxides
UPS	uninterruptible power supplies
UV–Vis	ultraviolet–visible
VG	vertical graphene
VOCs	volatile organic compounds
WE	working electrode
WiS	water-in-salt
XPS	X-ray photoelectron spectroscopy
XRD	X-ray diffraction
ζ-potential	zeta potential

1 Introduction to Supercapacitors

Vyshakh Viswanath N and Krishnendu Biswas

1.1 INTRODUCTION

The scarcity of fossil fuels and their harmful environmental effects have led the world to adopt renewable energy sources, such as solar, wind, and geothermal, as sustainable solutions. However, the intermittent nature of these energy sources presents a significant challenge, particularly under unfavorable conditions [1]. In this context, efficient energy storage devices, such as batteries, capacitors, and fuel cells, play a crucial role in storing energy for future use [2]. For instance, lithium-ion batteries (LIBs), which offer a high energy density of up to 250 Wh/kg, are widely used in various applications. However, their low power density, sluggish ion transfer kinetics across electrodes, and volume expansion during charge and discharge cycles limit their lifespan and operational efficiency.

In contrast, dielectric capacitors can store charge and operate within milliseconds, delivering only a low energy output [3, 4]. As a promising energy storage device, the supercapacitors (SCs) stand out among electrochemical energy storage systems due to their exceptional characteristics, including a rapid charge–discharge rate, low cost, and the ability to retain charge reversibly for millions of cycles [5]. In specific low-energy applications, SCs can replace batteries or operate in conjunction with them without compromising their cyclic stability. They deliver a steady energy supply on demand without requiring a continuous voltage source, making them well-suited for integration with renewable energy systems where power generation is inconsistent [6, 7]. The charge storage and extraction capacity of SCs depend on several key factors, such as electrode materials and manufacturing conditions. Initially, SCs utilized porous carbon electrodes, but later advancements introduced materials with redox charge storage capabilities, aiming to bridge the gap between batteries and capacitors to enhance both power and energy output [8].

1.2 FUNDAMENTALS OF CHARGE STORAGE IN SCs

Capacitors are classified as passive circuit components, consisting of two metal plates separated by a dielectric material. When a potential difference is applied, opposite charges accumulate on the respective electrodes, generating an electric field that stores electrical energy (Figure 1.1) [9]. The following fundamental equations govern them:

$$C = \frac{Q}{V} \tag{1.1}$$

DOI: 10.1201/9781003535836-1

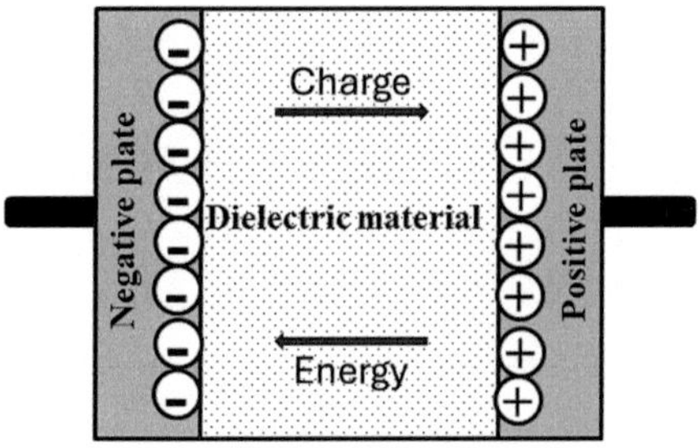

FIGURE 1.1 Schematic representation of a conventional capacitor.

where C represents the capacitance, Q stands for the stored charge, and V denotes the applied potential. In a conventional capacitor, the stored charge is directly proportional to the surface area of the metal plate electrodes and inversely proportional to the distance between them.

$$C = \varepsilon_0 \varepsilon_r \frac{A}{D} \tag{1.2}$$

where A represents the electrode surface area and D is the separation distance, the proportionality constants ε_0 and ε_r correspond to the dielectric constants of free space and the insulating material, respectively. Having discussed the capacitance stored in a capacitor, its performance is typically evaluated in terms of energy and power density [10].

Energy density, which is measured per unit area or unit mass, is expressed by the following equation:

$$E = \frac{1}{2} CV^2 \tag{1.3}$$

The energy density is directly proportional to the capacitance and the voltage window, while power density represents the rate at which energy is dissipated per unit time. Therefore, the resistance associated with the capacitor must be taken into account.

In a conventional circuit, the capacitor is always connected in series with the load resistor, which is analogous to the equivalent series resistance (ESR) of the capacitor. ESR arises from the resistance of various components, including the current collector, electrode, and dielectric material [11]. The following equation gives the power output of the capacitor:

$$P = \frac{V^2}{4\ ESR} \tag{1.4}$$

Capacitors exhibit a high power density and can deliver their full potential within milliseconds, making them an effective power source. However, their low energy density limits their ability to serve as a direct replacement for batteries.

1.3 EVOLUTION OF SUPERCAPACITOR

The concept of electrical charge originated from the well-known experiment in which rubbing amber on wool caused it to attract small objects. Subsequent groundbreaking experiments by Faraday, Millikan, and Thomson offered an in-depth, electron-level understanding of electricity [12]. The fundamental unit of charge is defined as the charge carried by a proton or an electron, with a magnitude of 1.602×10^{19} C. Building on the basic idea of electric charge, the principle of storing electric charges through separation was first demonstrated in the Leyden jar experiment [13, 14]. In Leyden jars, a thin metal foil is separated by glass, allowing charge to be stored electrostatically (Figure 1.2). This innovation laid the foundation for further advancements in capacitor technology.

The rapid electrostatic charge storage in conventional capacitors makes them a viable power source; however, their low capacitance and energy density restrict their use in frequency filtering applications. Enhancing the electrode surface area while reducing the distance between electrodes led to the development of electrolytic capacitors. Aluminum (Al)- or tantalum (Ta)-based electrolytic capacitors are commonly used for air conditioner line filtering, utilizing metals with an insulating oxide surface layer as electrodes and liquid or gel electrolytes for charge storage [15].

Significant research on electrical double-layer capacitors (EDLCs) began in the 1950s at General Electric, pioneered by H. Becker. Initially, EDLCs featured solid porous carbon electrodes immersed in an electrolyte within a container. Further advancements were made by the Standard Oil Company in Cleveland, Ohio (SOHIO), which introduced a separator that allowed for the selective passage of ions, leading

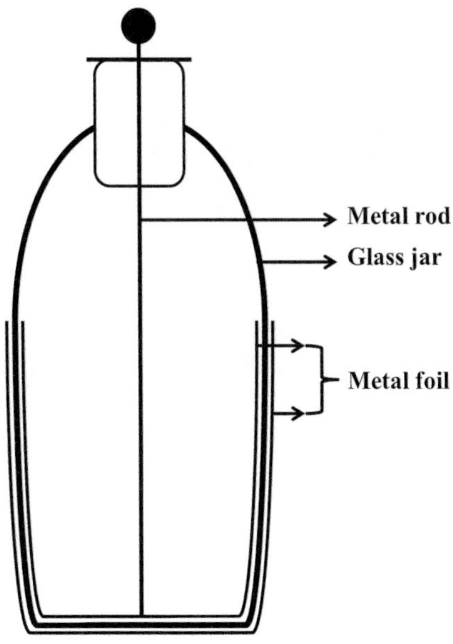

FIGURE 1.2 Cross-sectional representation of Leyden jar.

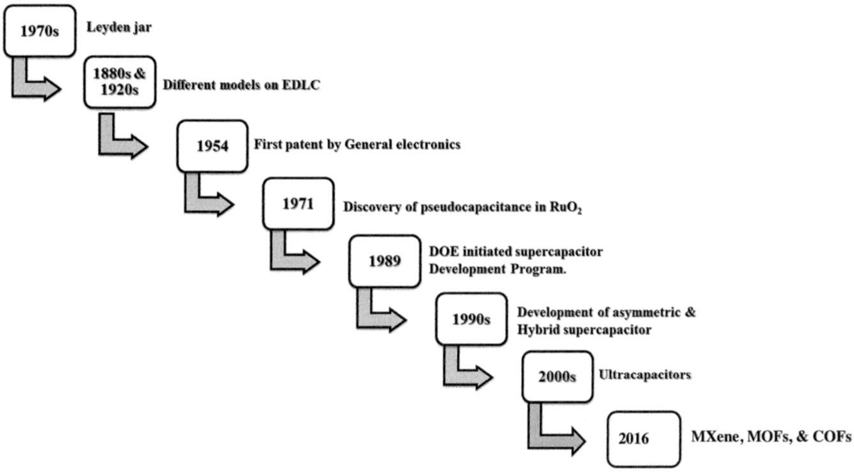

FIGURE 1.3 Timeline for the evolution of SCs.

to extensive research in the development of electrochemical capacitors. Later, the Nippon Electric Company developed a device using porous carbon, termed as "supercapacitor," which was first utilized for memory backup applications [16, 17]. In 1971, ruthenium dioxide (RuO_2) was discovered as a pseudocapacitive electrode material, which significantly departed from traditional carbon-based electrodes. A decade later, the Pinnacle Research Institute pioneered high-power ultracapacitors using ruthenium and tantalum oxides. However, due to the high cost of these materials, commercialization remained limited, and their use was primarily restricted to military applications [18]. By the late 1980s, the U.S. Department of Energy (DoE) launched an ultracapacitor development program at Maxwell Laboratory, facilitating the fabrication of various SC types, including EDLCs, pseudocapacitors (PCs), and asymmetric capacitors, for diverse applications. Numerous companies, such as ELTON, CAP-XX, and Nippon Chemi-con, have conducted extensive research on different SC materials to meet growing energy demands [19]. The rapid advancement in SC research has led to the development of highly efficient and specialized SC variants, including thermal protection SCs, piezoelectric SCs, self-healing SCs, shape-memory SCs, flexible SCs, and flow SCs. A flow diagram illustrating the evolution of SCs is presented in Figure 1.3.

1.4 Supercapacitor

The transition from conventional capacitors to SCs has been driven by significant advancements in electrode materials, fabrication techniques, and charge storage mechanisms. While SCs adhere to the same fundamental equations as capacitors, they utilize high-surface-area electrodes and liquid electrolytes to enhance charge storage. By definition, SCs, also called ultracapacitors or electrochemical capacitors, exhibit significantly higher capacitance and energy density than conventional capacitors, which rely solely on electrostatic charge storage. Due to their rapid charge–discharge

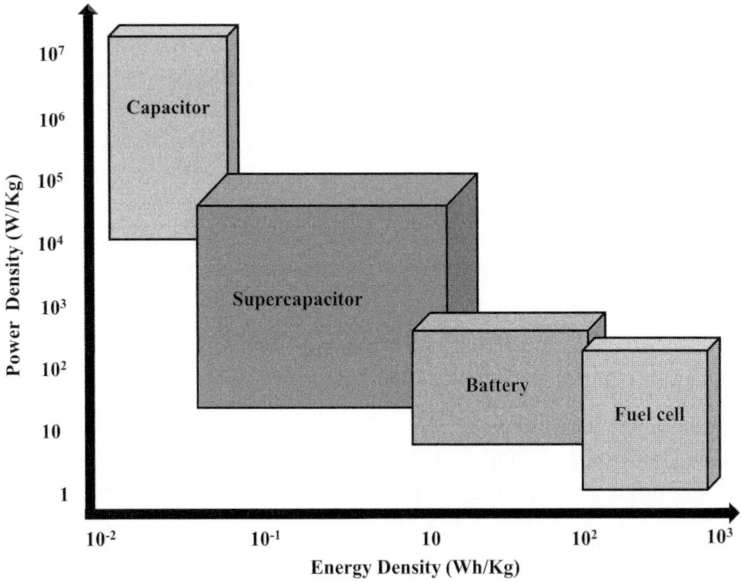

FIGURE 1.4 Ragone plot for various energy storage devices.

kinetics, SCs generate more power than batteries, bridging the gap between batteries and capacitors. The Ragone plot in Figure 1.4 illustrates various energy storage device's energy and power output. As shown, SCs occupy an intermediate position between batteries and conventional capacitors, offering higher energy density than capacitors while delivering greater power density than batteries [20]. It is important to note that the performance of commercial SC cells depends on factors such as gravimetric weight and total volume. In contrast, laboratory-scale evaluations assess active material's performance, making direct comparisons challenging. Therefore, the most reliable metric for comparison is the charge–discharge time and its corresponding ratio.

1.4.1 Components of Supercapacitor

An SC comprises four key components: the electrode, electrolyte, separator, and current collector. The electrode is crucial in determining the device's overall performance [21] among these. It stores charge by interacting with ions under an applied potential, making its properties essential for efficient energy storage. An ideal electrode should have a high surface area with controlled porosity, excellent electrical conductivity, superior electrochemical and thermal stability, environmental friendliness, and cost-effectiveness. The electrolyte is another vital component, as its physical and chemical properties significantly impact the performance and efficiency of the SC. Electrolytes can be categorized into aqueous, ionic, and organic-based types, depending on the operating conditions and materials used. However, no single electrolyte has been identified as perfect, and ongoing research continues to optimize electrolyte formulations to meet diverse performance demands. The separator serves as a barrier to prevent short circuits while allowing the passage of ions without undergoing chemical interactions. An ideal separator should be non-conductive, corrosion-resistant, and have low resistance while ensuring efficient electrolyte wetting. The final component, the current collector, transfers electrons generated during the charge storage process to the external circuit. Given its function, it must exhibit high electrical conductivity and provide mechanical support to the cell. Commonly used current collectors include metal foams, foils, and meshes, as they facilitate uniform electrode material coating and enhance overall performance [22].

1.4.2 Classification of Supercapacitors

SCs are classified into two categories based on their charge storage mechanism and materials: EDLCs and PCs [23]. In EDLCs, charge storage occurs by forming an electrical double layer at the electrode surface, similar to conventional capacitors, without any charge transfer between the electrode and electrolyte. When a potential is applied, oppositely charged ions migrate toward the electrode surface, forming an inner Helmholtz layer and an outer diffusion layer, collectively creating the double layer [21]. The first explanation of double-layer capacitance was proposed by Helmholtz, who considered only a rigid layer of charge at the electrode surface, resulting in a constant capacitance (C_H) that is independent of factors such as the potential window and electrolyte concentration. This model was later refined by the Gouy–Chapman theory, which introduced the concept of a diffuse layer of ions that influences capacitance (C_G). According to this theory, electrolyte ions diffuse over a certain distance, with the ion's kinetic energy determining the layer's thickness. Gouy–Chapman's model suggested that ions are surrounded by oppositely charged solvated molecules, restricting their approach to the electrode surface up to a limited distance. This forms a compact ion layer at the electrode surface, known as the Stern or Helmholtz layer. The Helmholtz model and the Gouy–Chapman diffuse layer theory have developed the modern double-layer capacitance mechanism [24, 25]. The following equation gives the capacitance in the double layer:

$$\frac{1}{C} = \frac{1}{C_H} + \frac{1}{C_G} \tag{1.5}$$

In 1947, Grahame refined this concept by proposing that some ions or charge-neutral particles could penetrate the Stern layer. These penetrating particles, which reside directly on the electrode surface, are specifically adsorbed ions. This advancement introduced the concept of a triple layer [26], consisting of the inner Helmholtz layer, the outer Helmholtz layer, and the diffusion layer (Figure 1.5).

EDLCs utilize high-surface-area porous carbon-based materials as electrodes, including activated carbon, carbon aerogels, and carbon nanotubes. Charge storage in EDLCs follows a non-Faradaic process, where ions are adsorbed onto the electrode surface due to the natural attraction between oppositely charged ions.

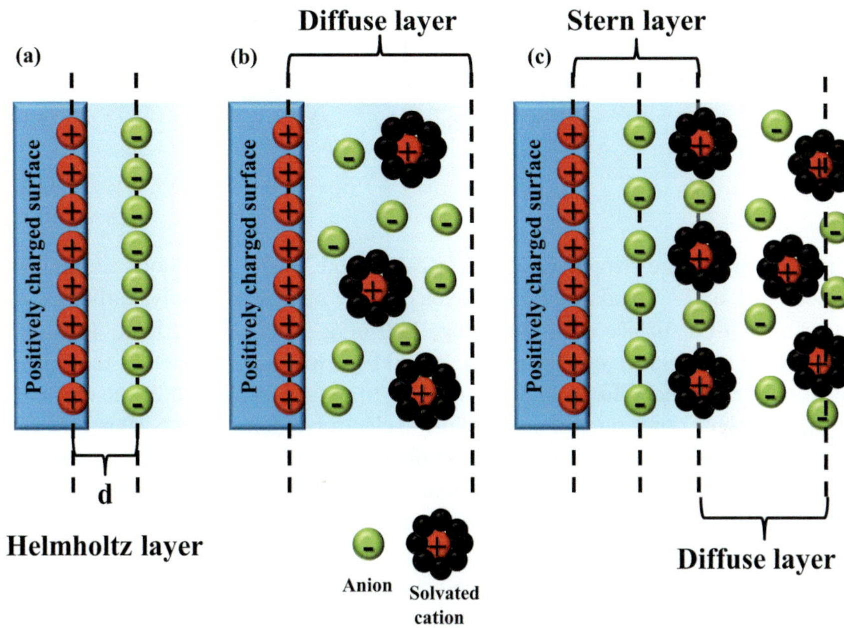

FIGURE 1.5 Schematic representation of (a) Helmholtz model, (b) Gouy-Chapman model and (c) Stern model to explain the electrical double layer mechanism.

(Reprinted with permission from [19]. © Springer 2020.)

During operation, ions from the electrolyte diffuse across the separator and settle within the electrode pores, where they are separated at an atomic or molecular scale by ultrathin dielectric layers. These interfaces function as a double layer, storing charge electrostatically. During the charging process, cations migrate toward the negatively charged electrode, while anions move toward the positively charged electrode. Conversely, these ions return to the electrolyte during discharge, releasing stored energy. The electrochemical charge–discharge process can be represented by Equations 1.6–1.11 [27].

At the positive electrode

$$E_1 + X^- \left(\overrightarrow{Charging}\right) E_1^+ \,//X^- + e^- \tag{1.6}$$

$$E_1^+ \,//X^- + e^- \left(\overrightarrow{Discharging}\right) E_1 + X^- \tag{1.7}$$

At negative electrode

$$E_2 + X^+ + e^- \left(\overrightarrow{Charging}\right) E_2^- \,//X^+ \tag{1.8}$$

$$E_2^- \,//X^+ \left(\overrightarrow{Discharging}\right) E_2 + X^+ + e^- \tag{1.9}$$

The overall charging process can be written as follows:

$$E_1 + E_2 + X^- + X^- \left(\overrightarrow{Charging}\right) E_1^+ //X^- + E_2^- //X^+ \tag{1.10}$$

$$E_1^+ //X^- + E_2^- //X^+ \left(\overrightarrow{Discharging}\right) E_1 + E_2 + X^- + X^- \tag{1.11}$$

where E_1 and E_2 are two electrode surfaces, X^+ and X^- are the cations and anions, respectively, and // stands for electrode/electrolyte interface. Ideally, carbon-based materials with a surface area exceeding 1000 m^2/g should theoretically provide a gravimetric capacitance of approximately 300 F/g. However, in practice, the capacitance is often reduced to around 150 F/g due to partial utilization of the electrode surface, as some pores remain inaccessible to electrolyte ions. By modifying the carbon network, such as incorporating functional groups, dopant atoms, or optimizing the electrolyte composition, it is possible to enhance capacitance performance. Since no charge transfer occurs between the electrode and electrolyte, the charge storage process in EDLCs is highly reversible, ensuring exceptional cyclic stability. EDLC electrodes can endure up to 10^6 continuous charge–discharge cycles, far surpassing the lifespan of conventional batteries. This long cycle life makes EDLC capacitors ideal for applications in remote or extreme environments, such as deep-sea exploration or high-altitude stations, where frequent maintenance is impractical [28, 29].

As previously discussed, EDLCs offer superior cyclic stability and power performance compared to batteries, while also providing higher capacitance and energy storage than conventional capacitors. However, their low energy density and limited capacitance have motivated researchers to explore alternative materials and mechanisms for improved energy storage. PCs have emerged as a promising solution to address this. Unlike EDLCs, PCs utilize Faradaic reactions to store charge, offering a faster charge–discharge rate than batteries and significantly higher capacitance than EDLCs. The presence of surface redox reactions and rapid ion intercalation enables the bulk material to participate in charge storage, resulting in a synergistic energy and power density enhancement, making PCs a strong candidate for next-generation electrode materials. PCs employ metal oxides and conducting polymers as electrode materials [30]. When a potential is applied, the active material undergoes redox reactions, resulting in changes in the valence state of ions and involving electron transfer. While the electrical response of PCs resembles that of EDLCs, where charge storage varies with applied potential, the underlying storage mechanism differs. Here, charge storage occurs via redox reactions within the bilayer, rather than purely electrostatic adsorption. The kinetic behavior of PCs is similar to EDLCs in terms of capacitance, yet their charge transfer process resembles that of batteries due to the involvement of redox reactions. As a result, PCs serve as a bridge between batteries and EDLCs, although they differ from batteries in that their charge storage does not involve a complete chemical phase transformation [31, 32]. The Faradaic process responsible for charge storage can be further classified into three categories: underpotential deposition, redox reaction, and intercalation reaction [33] (Figure 1.6).

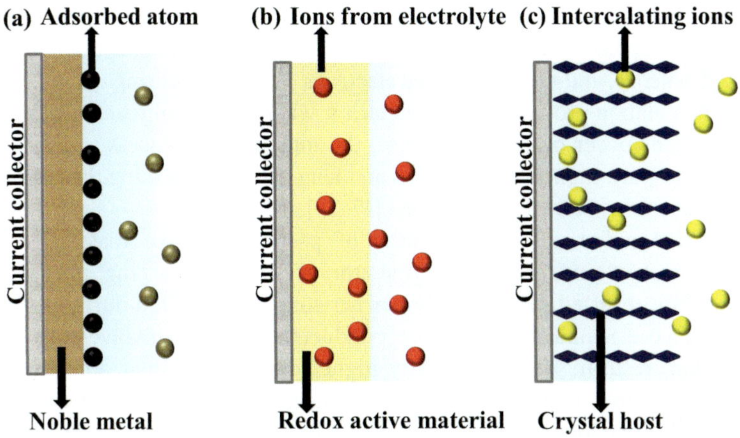

FIGURE 1.6 Schematic representation of different pseudocapacitive charge storage mechanisms: (a) underpotential deposition, (b) redox reaction, and (c) ion intercalation process. (Reprinted with permission from [5]. © Elsevier 2024.)

In underpotential deposition, an ion is deposited onto a metal electrode surface at a potential higher than its standard reduction potential, forming a thin atomic or molecular layer. This process occurs due to strong interactions between the deposited ions and the electrode surface, leading to a more controlled and uniform deposition. Equation 1.12 can be used to describe the process.

$$M + xA^{z+} + xze^{-} \leftrightarrow A * M \tag{1.12}$$

where M represents the metal substrate where the ions are adsorbed, A is the deposited ion with a valency of z, x denotes the number of ions involved, and xze^- signifies the total number of electrons transferred during the process.

During redox reactions, electron transfer occurs at the interface between the electrode and electrolyte, leveraging the multiple oxidation states of metal oxides. Electrolyte ions interact with the electrode, where cations adhere to the surface of oxidized species after undergoing reduction, a process collectively known as a redox reaction. The overall reaction can be expressed as follows:

$$Ox + zC^{+} + ze^{-} \leftrightarrow Rd * C \tag{1.13}$$

In this reaction, Ox represents the oxidized species, C^+ denotes the ions from the electrolyte with a valency of z, Rd refers to the reduced species, and ze^- signifies the number of electrons transferred during the redox process.

The third type of charge storage mechanism in PCs is the intercalation process. In this mechanism, electrolyte ions are reversibly inserted into the interlayer spaces of the electrode material, while electrical neutrality is preserved by modifying the valence state of the electrode material. The process can be expressed as follows:

$$M_l + zC^{+} + ze^{-} \leftrightarrow CM_l \tag{1.14}$$

Where M_l represents the layered electrode material, C^+ denotes the intercalated ion, z is its valency, and ze^- signifies the electrons transferred during the intercalation and de-intercalation process. Unlike batteries, where chemical reactions and phase transformations occur over extended use, the intercalation process in PCs enables rapid ion transport, high-rate capability, and long cycle life, making them more efficient for fast energy storage applications [17, 34].

Conductive polymers, another class of pseudocapacitive materials, utilize the entire electrode material for charge storage. During charging, ions from the electrolyte are inserted into the polymer network and expelled during discharge. Unlike surface-oriented charge storage, conductive polymers involve bulk material participation, resulting in significantly higher capacitance than other materials. Electrical neutrality is maintained during ion insertion by charging the polymer as p-type or n-type, depending on the inserted ions. Despite their superior capacitive performance, challenges such as electrode cracking during charge storage and the limited availability of negatively charged polymers hinder their widespread application [35, 36].

As discussed, PCs offer higher capacitance than EDLCs due to their charge storage mechanism and effective utilization of bulk materials, resulting in greater energy and power density. However, despite these advantages, they suffer reduced cycle life and electrode swelling or shrinkage during charge–discharge cycles. In contrast, EDLCs exhibit superior electrochemical stability and reversibility, making them more suitable for long-term applications. A new class of SCs, known as hybrid capacitors, has been developed to combine the features of both EDLCs and PCs [37], thereby, enhancing SC performance.

The term "hybrid" itself implies that these capacitors combine the advantages of both EDLCs and PCs while addressing their individual limitations. This combination enables them to achieve higher energy and power density without compromising durability or affordability. Carbon-based materials offer high surface area, conductivity, and stability, while redox-active electrodes provide a greater number of charge carriers, ultimately enhancing capacitance. Hybrid capacitors are developed by integrating different materials and can be classified into three main categories: asymmetric SCs, composite electrode-based SCs, and battery-type hybrid SCs [38].

Asymmetric SCs combine both electrostatic and Faradaic charge storage mechanisms across different electrodes. They typically consist of carbon-based and pseudocapacitive electrodes, such as metal oxides or conducting polymers, ensuring high surface area, porosity, electronic conductivity, and cyclic stability. This unique configuration enables an extended potential window, reaching the maximum limit of the electrolyte and significantly enhancing energy and power output [27].

Composite electrode-based capacitors adopt a synergistic approach by combining the benefits of EDLC and pseudocapacitive electrodes. While carbon materials offer a high surface area, porosity, and long lifespan, their capacitance is limited due to surface-concentrated charge storage. Conversely, metal oxides and conductive polymers store charge through Faradaic redox reactions, but they suffer from electrode shrinkage, lower conductivity, and a limited shelf life. To overcome these drawbacks, composite materials are developed by integrating carbonaceous materials with either polymers or metal oxides to form the electrode. Carbon provides mechanical strength, stability, and an efficient electron transport pathway in this setup, while the other component facilitates Faradaic redox reactions. This combination enhances SC performance by reducing electrode resistance and improving durability, allowing for fabrication in both symmetric and asymmetric configurations [39, 40].

The final category of hybrid SCs is the battery-type SC, which adopts a straightforward approach by combining a battery-type electrode with an EDLC-type electrode. The battery-type electrode stores charge via intercalation, enabling high specific capacitance and a higher energy density. Meanwhile, the EDLC-type electrode stores charge through reversible surface adsorption of ions, ensuring excellent cyclic stability and power density. This hybrid design effectively addresses the individual limitations of both electrode materials, resulting in a device with enhanced overall performance [41, 42].

1.4.3 Applications of SCs

As we have discussed the charge storage mechanisms in SCs, it is also important to highlight their diverse applications across various fields. SCs have garnered significant attention due to their high cycle life, rapid charge–discharge rates, and high power densities, making them ideal for backup power sources or high-power pulse applications that require short bursts of energy. They are widely employed in memory backup devices, where power fluctuations could lead to data loss, by providing a temporary power supply. In the realm of portable electronic devices, where conventional power sources require longer charging times, SCs offer a commercial alternative due to their fast-charging capabilities. Their ability to operate across a wide temperature range, maintain stability over millions of cycles, and deliver quick power generation has made them invaluable in military applications. SCs serve as backup sources for military electric vehicles, electronic devices, control systems, helicopter black boxes, tanks, and armored vehicles.

Additionally, they are used in handheld radios, signaling devices for land vehicles, and cold engine starting. While batteries remain the primary power source for many of these applications, their short lifespan, limited mobility, and hazardous nature make electrochemical SCs a promising alternative for the future. Furthermore, the combination of short charging times, extended discharge durations, and long lifespans makes SCs a viable solution for photovoltaic panels, reducing the need for frequent replacements and associated high costs. Beyond these applications, SCs are also utilized in regenerative braking systems, grid energy storage, wearable technology, cameras and flashes, medical implants, hearing aids, smart meters, and traffic signals, among many others [16, 29, 43–46].

1.4.4 Limitations

Although SCs exhibit promising properties, several limitations must be addressed to meet the growing energy demands of modern applications. One of the major drawbacks is their lower energy density compared to batteries; when high energy is required, a large assembly of SCs becomes necessary. Currently, commercially available SCs deliver less than 5 Wh/kg, significantly limiting their widespread adoption. High production and material costs pose another challenge, particularly for large-scale applications. Many high-performance SCs rely on expensive materials such as ruthenium and tantalum-based oxides, making cost-effective alternatives crucial for future development. Another significant limitation is the self-discharge process, which leads to energy loss over time and remains a critical challenge for their long-term usability. Tackling these issues through advancements in materials science, innovative electrode designs, and enhanced electrolytes will be crucial for improving the performance and practicality of SCs in the future [47, 48].

1.5 CONCLUSION AND FUTURE OUTLOOK

In conclusion, this chapter provides a brief overview of the origin of capacitance and the historical development of SCs. It categorizes SC materials based on their charge storage mechanisms and thoroughly examines their various modes of operation. Additionally, it highlights the diverse applications of SCs across multiple fields, emphasizing their advantages. Finally, the chapter outlines key limitations that must be addressed to enhance their performance and ensure their broader adoption. The future of SC technology hinges on overcoming current limitations to realize its full potential as a next-generation energy storage device. Ongoing research in materials science is anticipated to yield advanced, cost-effective electrode materials with higher energy density, bringing the performance of SCs closer to that of batteries. Developing novel nanostructured materials, conductive polymers, and hybrid systems could greatly enhance energy and power density while reducing reliance on costly metals such as ruthenium and tantalum.

Furthermore, electrolyte formulations and cell design innovations are expected to minimize self-discharge and improve overall efficiency. When these technological advances converge, SCs will play a crucial role in applications demanding fast charging, high power, and long life, from electric vehicles and grid storage to wearable electronics and IoT devices. With continued research and industrial investment, SCs could evolve into a mainstream solution for sustainable and high-performance energy storage.

1.6 AUTHOR AI DECLARATION

The authors hereby declare the use of generative AI and AI-assisted technologies in the writing process. While preparing this work, the authors used the ChatGPT tool/service to rephrase several statements in the manuscript. All AI-generated content was reviewed and edited by the authors as needed. The authors take full responsibility for the content of the publication.

ACKNOWLEDGMENT

The editors acknowledge the use of Grammarly as a tool to assist in editing this book chapter. The suggestions made by Grammarly were carefully considered and incorporated where appropriate, but the author is solely responsible for the final content.

REFERENCES

1. Estevão, J., and Lopes, J. D. "SDG7 and renewable energy consumption: The influence of energy sources". *Technological Forecasting and Social Change* 198 (2024): 123004. doi:10.1016/j.techfore.2023.123004.
2. Wang, D., Liu, N., Chen, F., Wang, Y., and Mao, J. "Progress and prospects of energy storage technology research: Based on multidimensional comparison". *Journal of Energy Storage* 75 (2024): 109710. doi:10.1016/j.est.2023.109710.
3. Kumar, R., Sharma, S., Singh, M., Yadav, P. S, Verma, A. K., and Kumar, A. R. "Different energy storage techniques: Recent advancements, applications, limitations, and efficient utilization of sustainable energy". *Journal of Thermal Analysis and Calorimetry* 149 (2024): 12831–12839. doi:10.1007/s10973-023-12831-9.
4. Elalfy, D. A., Gouda, E., Kotb, M. F., Bureš, V., and Sedhom, B. E. "Comprehensive review of energy storage systems technologies, objectives, challenges, and future trends". *Energy Strategy Reviews* 54 (2024): 101482. doi:10.1016/j.esr.2024.101482.
5. Viswanath, N, Yesuraj, J., Ramesh, M., Kim, K., and Biswas, K. "Fundamental aspects of perovskite oxides as an emerging material for supercapacitor applications and its anion intercalation mechanism – Review". *Journal of Energy Storage* 78 (2024): 109968. doi:10.1016/j.est.2023.109968.
6. Banerjee, S., Mordina, B., Sinha, P., and Kar, K. K. "Recent advancement of supercapacitors: A current era of supercapacitor devices through the development of electrical double layer, pseudo and their hybrid supercapacitor electrodes". *Journal of Energy Storage* 108 (2025): 115075. doi:10.1016/j.est.2024.115075.
7. Czagany, M., Hompoth, S., Keshri, A. K., Pandit, N., Galambos, I., Gacsi, Z., and Baumli, P. "Supercapacitors: An efficient way for energy storage application". *Materials* 17 (2024): 702. doi:10.3390/ma17030702.
8. Reenu, Sonia, Phor, L., Kumar, A., and Chahal, S. "Electrode materials for supercapacitors: A comprehensive review of advancements and performance". *Journal of Energy Storage* 84 (2024): 110698. doi:10.1016/j.est.2024.110698.
9. Ho, J., Jow, T. R., and Boggs, S. "Historical introduction to capacitor technology". *IEEE Electrical Insulation Magazine* 26 (2010): 20–25. doi:10.1109/MEI.2010.5383924.
10. Chandra, A. "Supercapacitors: An alternate technology for energy storage". *Proceedings of the National Academy of Sciences, India Section A – Physical Sciences* 82 (2012): 79–90. doi:10.1007/s40010-012-0009-9.
11. Halper, M. S., and Ellenbogen, J. C. "Supercapacitors: A brief overview". MITRE Corporation, McLean, Virginia, USA 1 (2006).
12. Josefowicz, D. G. "Histories of Discovery". *Perspectives on Science* 11 (2003): 346–364. doi:10.1162/106361403773062687.
13. Tan, J., Li, Z., Ye, M., and Shen, J. "Nanoconfined space: Revisiting the charge storage mechanism of electric double layer capacitors". *ACS Applied Materials & Interfaces* 14 (2022): 37259–37269. doi:10.1021/acsami.2c07775.
14. Silva, C. C., and Heering, P. "Re-examining the early history of the Leiden jar: Stabilization and variation in transforming a phenomenon into a fact". *History of Science* 56 (2018): 314–342. doi:10.1177/0073275318768418.
15. Kurra N., and Jiang Q. "18 – Supercapacitors". In Storing Energy Letcher, edited by T. M. Letcher, (2nd ed.). Elsevier, Amsterdam, Netherlands (2022): 383–417. doi:10.1016/B978-0-12-824510-1.00017-9.
16. Poonam, Sharma, K., Arora, A., and Tripathi, S. K. "Review of supercapacitors: Materials and devices". *Journal of Energy Storage* 21 (2019): 801–825. doi:10.1016/j.est.2019.01.010.

17. Shao, Y., et al. "Design and mechanisms of asymmetric supercapacitors". *Chemical Reviews* 118 (2018): 9233–9280. doi:10.1021/acs.chemrev.8b00252.
18. Sharma, P., and Bhatti, T. S. "A review on electrochemical double-layer capacitors". *Energy Conversion and Management* 51 (2010): 2901–2912. doi:10.1016/j.enconman.2010.06.031.
19. Jadhav, V. V., Mane, R. S., and Shinde, P. V. "Bismuth-Ferrite-Based Electrochemical Supercapacitors". SpringerBriefs in Materials (2020): 1–90. Springer International Publishing, Cham. doi:10.1007/978-3-030-16718-9.
20. Volfkovich, Y. M. "Electrochemical supercapacitors (a review)". *Russian Journal of Electrochemistry* 57 (2021): 311–347. doi:10.1134/S1023193521040108.
21. Sharma, S., and Chand, P. "Supercapacitor and electrochemical techniques: A brief review". *Results in Chemistry* 5 (2023): 100885. doi:10.1016/j.rechem.2023.100885.
22. Jalal, N. I., Ibrahim, R. I., and Oudah, M. K. "A review on supercapacitors: Types and components". *Journal of Physics: Conference Series* 1973 (2021): 012015. doi:10.1088/1742-6596/1973/1/012015.
23. Libich, J., Máca, J., Vondrák, J., Čech, O., and Sedlaříková, M. "Supercapacitors: Properties and applications". *Journal of Energy Storage* 17 (2018): 224–227. doi:10.1016/j.est.2018.03.012.
24. Zhang, L. L., and Zhao, X. S. "Carbon-based materials as supercapacitor electrodes". *Chemical Society Reviews* 38 (2009): 2520–2531. doi:10.1039/B813846J.
25. Ojha, K., Arulmozhi, N., Aranzales, D., and Koper, M. T. M. "Double layer at the Pt (111)–aqueous electrolyte interface: Potential of zero charge and anomalous Gouy–Chapman screening". *Angewandte Chemie International Edition* 59 (2020): 711–715. doi:10.1002/anie.201911929.
26. Grahame, D. C. "The electrical double layer and the theory of electrocapillarity". *Chemical Reviews* 41 (1947): 441–501. doi:10.1021/cr60130a002.
27. Wang, G., Zhang, L., and Zhang, J. "A review of electrode materials for electrochemical supercapacitors". *Chemical Society Reviews* 41 (2012): 797–828. doi:10.1039/C1CS15060J.
28. Kuzmina, O., Murali, V. P. K., Sharma, R. K., Ghosh, S., Soni, T. K. A., Kalappattil, M. J. S., Chauhan, A. L., and Kumari, A. P. "Improvements of energy conversion and storage: General discussion". *Faraday Discussions* 190 (2016): 291–306. doi:10.1039/C6FD90029A.
29. Wang, Y., Song, Y., and Xia, Y. "Electrochemical capacitors: Mechanism, materials, systems, characterization and applications". *Chemical Society Reviews* 45 (2016): 5925–5950. doi:10.1039/C5CS00580A.
30. Kim, I. H., and Kim, K. B. "Ruthenium oxide thin film electrodes for supercapacitors". *Electrochemical and Solid-State Letters* 4 (2001): A62–A64. doi:10.1149/1.1359956.
31. Chen, X., Zhu, Y., Zhang, M., Sui, J., Peng, W., Li, Y., Zhang, G., Zhang, F., and Fan, X. "N-butyllithium-treated $Ti_3C_2T_x$ MXene with excellent pseudocapacitor performance". *ACS Nano* 13 (2019): 9449–9456. doi:10.1021/acsnano.9b04301.
32. Yeoh, J. S., Armer, C. F., and Lowe, A. "Transition metal oxalates as energy storage materials: A review". *Materials Today Energy* 9 (2018): 198–222. doi:10.1016/j.mtener.2018.05.010.
33. Tomar, A. K., Joshi, A., Singh, G., and Sharma, R. K. "Perovskite oxides as supercapacitive electrode: Properties, design and recent advances". *Coordination Chemistry Reviews* 431 (2021): 213680. doi:10.1016/j.ccr.2020.213680.
34. Augustyn, V., Simon, P., and Dunn, B. "Pseudocapacitive oxide materials for high-rate electrochemical energy storage". *Energy & Environmental Science* 7 (2014): 1597–1614. doi:10.1039/C3EE44164D.

35. Cherusseri, J., and Kar, K. K. "Hierarchical carbon nanopetal/polypyrrole nanocomposite electrodes with brush-like architecture for supercapacitors". *Physical Chemistry Chemical Physics* 18 (2016): 8587–8597. doi.org/10.1039/C6CP00150E.
36. Eftekhari, A., Li, L., and Yang, Y. "Polyaniline supercapacitors". *Journal of Power Sources* 347 (2017): 86–107. doi:10.1016/j.jpowsour.2017.02.054.
37. Chatterjee, D. P., and Nandi, A. K. "A review on the recent advances in hybrid supercapacitors". *Journal of Materials Chemistry A* 9 (2021): 15880–15918. doi:10.1039/D1TA02505H.
38. Rudra, S., Seo, H. W., Sarker, S., and Kim, D. M. "Supercapatteries as hybrid electrochemical energy storage devices: Current status and future prospects". *Molecules* 29 (2024): 243. doi:10.3390/molecules29010243.
39. Najib, S., and Erdem, E. "Current progress achieved in novel materials for supercapacitor electrodes: Mini review". *Nanoscale Advances* 1 (2019): 2817–2827. doi:10.1039/c9na00345b.
40. Zuo, W., Li, R., Zhou, C., Li, Y., Xia, J., and Liu, J. "Battery-supercapacitor hybrid devices: Recent progress and future prospects". *Advanced Science* 4 (2017): 1–21. doi:10.1002/advs.201600539.
41. Wang, X., and Zheng, J. P. "The optimal energy density of electrochemical capacitors using two different electrodes". *Journal of the Electrochemical Society* 151 (2004): A1683. doi:10.1149/1.1787841.
42. Sun, H., Dai, H., Zhang, G., and Sun, S. "Interface engineering of inorganic solid-state lithium batteries via atomic and molecular layer deposition". *InfoMat* 7 (2025): e12650. doi:10.1002/inf2.12650.
43. Cherusseri, J., Sharma, R., and Kar, K. K. "Nanotechnology Advancements on Carbon Nanotube/Polypyrrole Composite Electrodes for Supercapacitors". In Handbook of Polymer Nanocomposites. Processing, Performance and Application: Volume B: Carbon Nanotube Based Polymer Composites, edited by K. K. Kar, J. K. Pandey, and S. Rana. Berlin, Heidelberg: Springer Berlin Heidelberg, (2015): 479–510. doi:10.1007/978-3-642-45229-1_22.
44. De, B., Nandi, S. P., Saha, P. S., Awasthi, S. K., and Kar, K. K. "Transition Metal Oxide/Graphene/Reduced Graphene Oxide Composites as Electrode Materials for Supercapacitors". In Handbook of Nanocomposite Supercapacitor Materials II: Performance, edited by K. K. Kar. Cham: Springer International Publishing, (2020): 297–331. doi:10.1007/978-3-030-52359-6_12.
45. Nkembi, A. A., Simonazzi, M., Santoro, D., Cova, P., and Delmonte, N. "Comprehensive review of energy storage systems characteristics and models for automotive applications". *Batteries* 10 (2024): 88. doi:10.3390/batteries10030088.
46. Shuja, A., Imran, M., Ali, Z, Asiri, A. M., Khan, M. S., Shaukat, S. D., and Khokhar, R. A. "Supercapacitors for energy storage applications: Materials, devices and future directions: A comprehensive review". *Journal of Alloys and Compounds* 1009 (2024): 176924. doi:10.1016/j.jallcom.2024.176924.
47. Pell, W. G., and Conway, B. E. "Analysis of power limitations at porous supercapacitor electrodes under cyclic voltammetry modulation and DC charge". *Journal of Power Sources* 96 (2001): 57–67. doi:10.1016/S0378-7753(00)00682-0.
48. Khan, H. Rashid, and Latif Ahmad, A. "Supercapacitors: Overcoming current limitations and charting the course for next-generation energy storage". *Journal of Industrial and Engineering Chemistry*. 141 (2025): 46–66. doi:10.1016/j.jiec.2024.07.014.

2 Fundamentals of Supercapacitors

Jaidev Kumbhakar

2.1 INTRODUCTION

Environmental and energy-related concerns have been global issues for many years. Climate change and the depletion of fossil fuel resources have significantly impacted the global economy and ecological balance. As a result, key challenges associated with the energy crisis and global warming include the generation, conversion, storage, and efficient energy use. Electricity, a highly versatile form of energy, plays a crucial role in our daily lives, yet it is not easily stored for immediate use. Therefore, efficient energy storage solutions are essential for reliable energy utilization.

Recently, there has been a notable surge in renewable energy generation, particularly from solar and wind sources, two of the most abundant and sustainable options. However, due to their intermittent nature, there is an urgent need for energy storage technologies that are safe, cost-effective, environmentally friendly, and capable of storing surplus energy for later use during peak demand.

To address this, energy storage systems such as batteries, fuel cells, and supercapacitors (SCs) are becoming increasingly important. SC technology has attracted significant attention from researchers due to its ability to offer much higher capacitance compared to conventional capacitors [1]. Developing efficient energy storage systems is essential to enable large-scale utilization of harvested energy. While batteries are commonly used in industrial electronics due to their high energy density and ability to store substantial power, they typically require long charging times. Similarly, fuel cells offer limited energy density. Although batteries and conventional capacitors have advantages, they fall short in meeting the demands for high energy storage capacity and power output required for effective integration with renewable energy systems [2]. As a result, there is a growing demand for SCs that offer high energy density, long lifespan, compact design, cost efficiency, and environmental sustainability. In SCs, energy storage and release occur through reversible ion adsorption and desorption at the electrolyte and electrode materials interface. Due to the growing interest in enhancing the performance of energy storage systems, SCs have gained attention for their advanced storage capabilities and strong potential. To attain significantly higher capacitance than traditional capacitors, SCs employ electrode materials with large surface areas and ultra-thin dielectric electrolytes [3, 4]. The two key performance indicators of any energy storage device are energy and power density. SCs act as a link between traditional capacitors and fuel cells. In the batteries, the mechanism of charge accumulation occurs in bulk volume and not on surfaces, but the SCs involve the surface for charge accumulation.

 DOI: 10.1201/9781003535836-2

When SCs utilize metal oxides or conducting polymers as electrode materials, charge storage is primarily governed by pseudocapacitance mechanisms [5, 6]. The pseudocapacitor devices employ ruthenium oxide in their electrodes to achieve a notable enhancement in specific capacitance. The performance comparison of different energy storage devices uses a "Ragone plot," where SCs are positioned between traditional capacitors and batteries, highlighting their distinctive ability to deliver high power in short durations [4, 7].

In traditional capacitors, energy is stored by transferring electrons between two electrodes. However, due to their large surface area, energy storage occurs through a mechanism known as the electric double layer in SCs. This concept was introduced in 1853 by the German physicist Hermann von Helmholtz [8]. He found that electric charges could accumulate at the interface between the electrode and electrolyte and on the surface of conductive materials. This discovery laid the foundation for what we now refer to as SCs [3, 9]. The electric double-layer effect produces an electric field across a potential layer, which facilitates the movement of ions into the electrodes' microporous structure. This enhances the ion transfer rate within the thin double layer. In 1957, H.I. Becker showcased this mechanism, introducing a novel charge and discharge process within a cell [10].

2.2 STRUCTURE OF A SUPERCAPACITOR

An SC consists of two electrodes, an electrolyte, and a separator that electrically isolates the electrodes while allowing ionic flow. Among these components, the electrodes are crucial because the overall performance of an SC largely depends on their electrochemical characteristics [11]. These electrodes are responsible for charge storage and significantly influence the device's energy and power density. Techniques such as cyclic voltammetry (CV) and galvanostatic charge–discharge (GCD) tests are commonly employed to evaluate the electrochemical behavior of an SC [12, 13]. The capacitance is typically calculated from the constant current discharge curve and is expressed by the following formula (2.1):

$$C = \frac{I}{\left(\frac{dv}{dt}\right)} \tag{2.1}$$

where I symbolizes the discharge current, and dv/dt is obtained by analyzing the gradient of the discharge curve. Also, the specific capacitance (C_{sp}) in an SC can be measured from the following formula (2.2):

$$Csp = \frac{4C}{m} \tag{2.2}$$

where C represents the capacitance of the two-electrode system, and m denotes the combined mass of the active electrode materials. Depending on the specific application, the mass term may be expressed in volume or surface area.

Equation (2.3) is used to determine the energy storage capacity and power density of an SC:

$$E = \frac{CV^2}{2}, P = \frac{V^2}{4R_s} \tag{2.3}$$

where total capacitance is denoted as C (F/g), cell voltage is indicated by V, and ESR is symbolized as Rs.

2.3 KEY COMPONENTS OF SC

An SC is made up of several essential components, as shown in Figure 2.1, each playing a critical role in its performance:

Electrodes: Typically composed of high surface–area materials like activated carbon (AC), carbon nanotubes (CNTs), or graphene. These materials enable efficient charge storage through electrostatic accumulation or redox reactions.

Electrolyte: This ionic medium allows charge to move between electrodes. Electrolytes can be aqueous, organic, or solid-state, and their conductivity and electrochemical stability directly affect device performance.

Separator: A porous, non-conductive membrane placed between the electrodes to prevent electrical short circuits while still allowing ion flow.

Current Collectors: These components connect the electrodes to the external circuit and are typically made from conductive metals such as aluminum or copper, ensuring efficient charge transport [14].

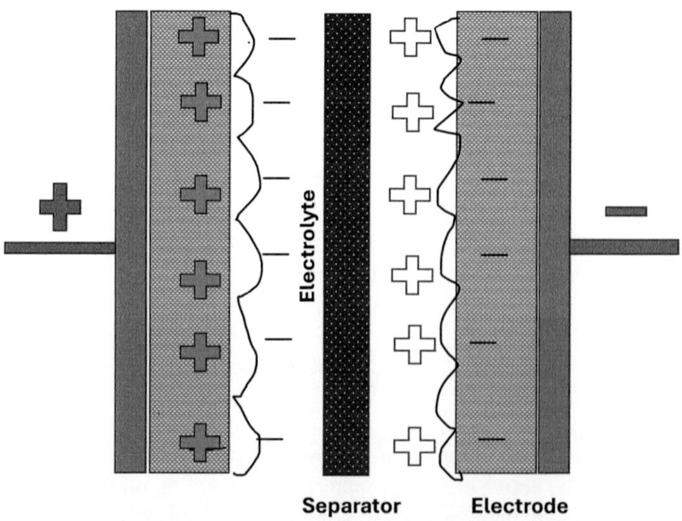

FIGURE 2.1 Structure diagram of a supercapacitor showing key components.

2.4 CAPACITOR TYPES

There are two main categories of capacitors. The first is the conventional capacitor, which consists of two conductive plates separated by an insulating dielectric layer. When a voltage is applied across the capacitor, positive and negative charges build up on the surfaces of the respective electrodes. The dielectric material prevents direct charge flow between the plates, maintaining charge separation and creating an electric field. This setup enables the capacitor to store electrical energy, as Figure 2.2(a)

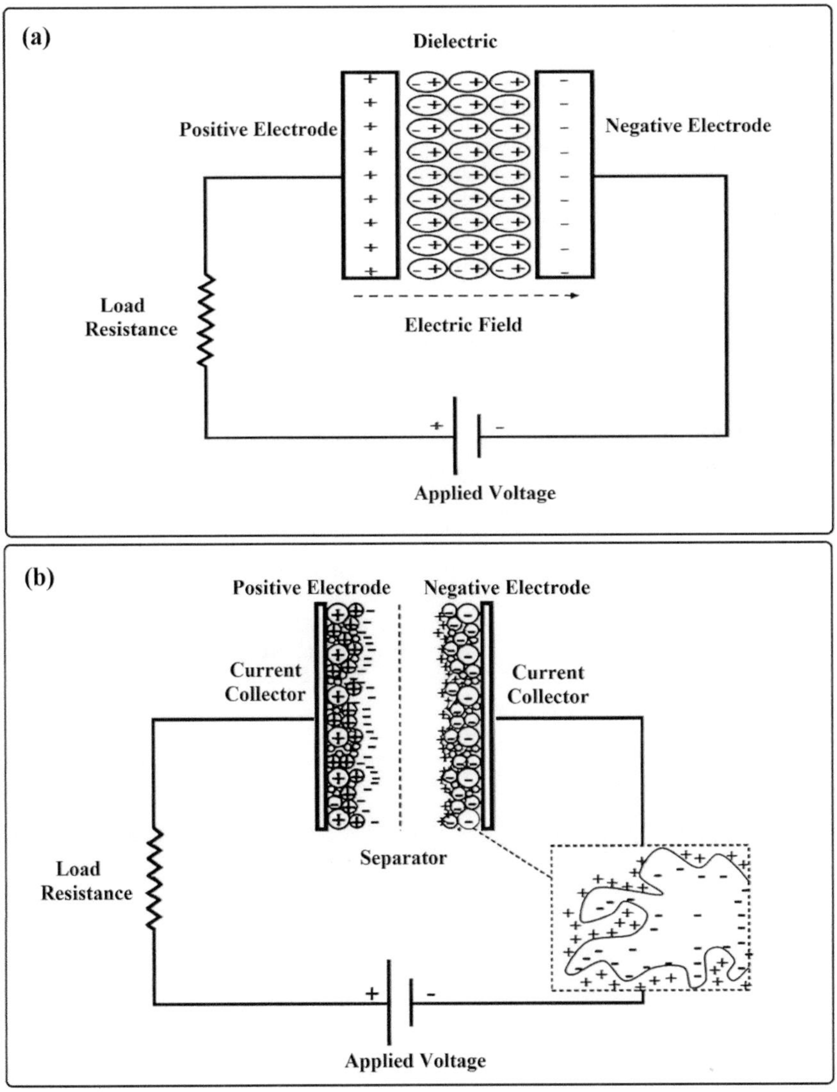

FIGURE 2.2 Structural comparison of capacitor type: (a) conventional capacitor; (b) supercapacitor.

illustrates. Conventional capacitors are known for their relatively high power density. However, they exhibit significantly lower energy density than fuel cells and electrochemical batteries. While batteries can store considerably more energy overall, they are not designed to release that energy quickly, resulting in low power density.

In contrast, capacitors can only store limited energy per unit mass or volume. Still, they can discharge it rapidly, making them ideal for applications requiring high bursts of power [9, 15]. SCs are composed of two electrodes, an electrolyte, and a separator that electrically isolates the electrodes from each other. While they operate on the same fundamental principles as conventional capacitors, SCs differ in that they utilize electrodes with huge surface areas and ultra-thin dielectrics, which minimize the distance between the electrodes. This design allows them to reach similar power densities. Moreover, SCs offer several advantages over fuel cells and traditional batteries, including fast charging, extended cycle life, long shelf life, and high-power output. A schematic representation of an SC is shown in Figure 2.2(b) [9].

2.5 SUPERCAPACITOR TYPES

As illustrated in Figure 2.3, SCs are classified into three categories based on their energy storage mechanisms.

2.5.1 EDLCs

EDLCs consist of two carbon-based electrodes, an electrolyte, and a separator. These devices store energy through electrostatic charge accumulation, relying on a non-faradaic mechanism that does not involve any charge transfer between the electrode and the electrolyte [16, 17]. Energy storage in EDLCs is based on the principle

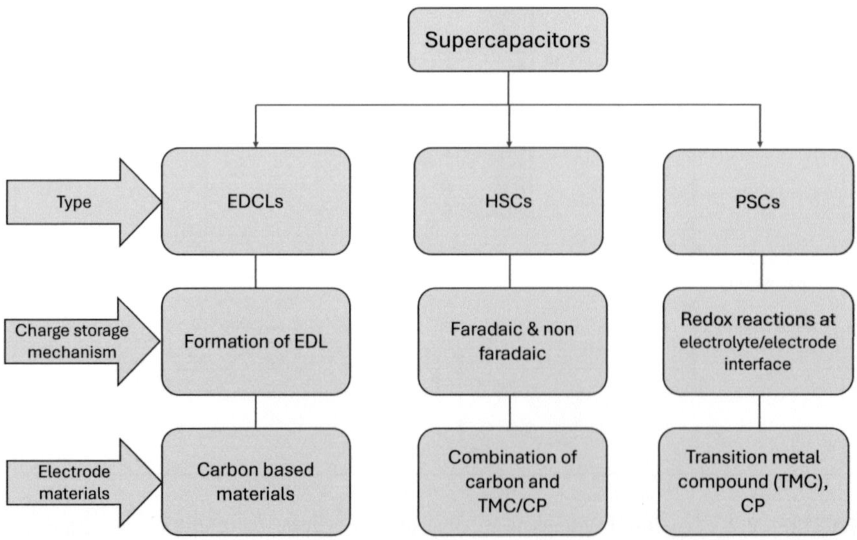

FIGURE 2.3 Conceptual framework of the chapter.

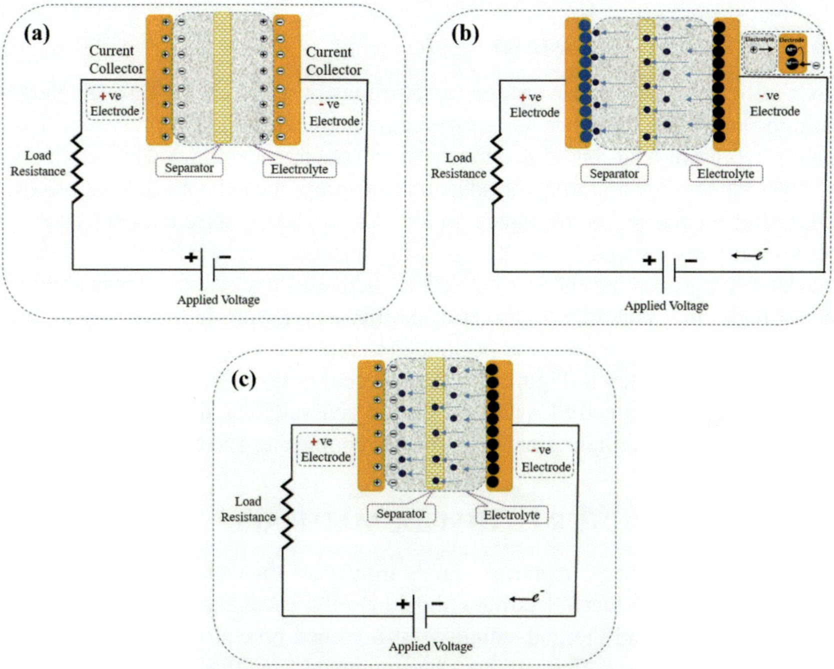

FIGURE 2.4 Schematic diagram of energy storage of SCs: (a) EDLCs, (b) PCs, and (c) HSCs.

of electrochemical double-layer formation. When a voltage is applied, charges do not accumulate directly on the electrode surface. Instead, the potential difference causes opposite charges to be attracted, leading to the migration of electrolyte ions across the separator and into the pores of the oppositely charged electrode, as illustrated in Figure 2.4(a). A double layer of charge forms to prevent the recombination of ions at the electrode surfaces. By increasing the specific surface area and minimizing the distance between the electrodes, this double-layer mechanism allows EDLCs to attain high energy density [9, 18].

2.5.2 Pseudocapacitors

This type of SC utilizes electrodes composed of metal oxides or conductive polymers (CPs), which exhibit high electrochemical pseudocapacitance. Charge storage in these devices occurs through a faradaic process, such as oxidation–reduction (redox) reactions, involving electron transfer between the electrolyte and the electrode, as illustrated in Figure 2.4(b). When a voltage is applied, oxidation and reduction reactions occur on the surface of the electrode, enabling charges to move across the electric double layer. This results in a faradaic current flowing through the SC. Thanks to this mechanism, propylene carbonates (PCs) can achieve significantly higher specific capacitance and energy density compared to EDLCs [18].

2.5.3 Hybrid Supercapacitors

HSCs utilize asymmetric electrodes—one primarily stores charge through electrostatic means, while the other relies on electrochemical (faradaic) processes. These devices represent a fusion of previously unattainable capabilities using traditional SC types alone. By integrating the advantageous properties of both EDLCs and PCs, HSCs offer a balanced performance profile. Although they have not been as extensively researched as EDLCs or PCs, growing interest in enhancing their efficiency and developing accurate models has led to increased exploration. Their ability to deliver high energy density and long cycle life, along with the flexibility to tailor their design and performance, has positioned hybrid SCs as a leading category in the field [19]. As illustrated in Figure 2.4(b), their charge storage mechanism combines those of both EDLCs and PCs [20]. HSCs are generally classified into three types based on electrode configuration: asymmetric, composite, and battery-type.

2.6 SUPERCAPACITOR ELECTRODE MATERIALS

The capacitance of an SC material heavily influences the capacitance of an SC, as shown in Figure 2.5. Ideal electrodes should exhibit excellent electrical conductivity, high chemical and thermal stability, well-defined pore structures, strong resistance to corrosion, ease of processing, environmental friendliness, and affordability. Materials capable of undergoing faradaic charge transfer can significantly enhance capacitance. The pseudocapacitive behavior—achieved through redox reactions, ion intercalation, or electrosorption—depends on the chemical interaction between the electrode material and the adsorbed ions and the structure and size of the electrode pores. The formation of the electric double layer and the amount of charge stored via pseudocapacitance mechanisms are primarily determined by the electrode's available surface area. Therefore, porous and sponge-like materials are often chosen. Optimizing the pore structure to match the size of electrolyte ions can significantly boost electrochemical performance. A well-distributed pore size has a substantial impact on capacitance, energy density, equivalent series resistance, and power density.

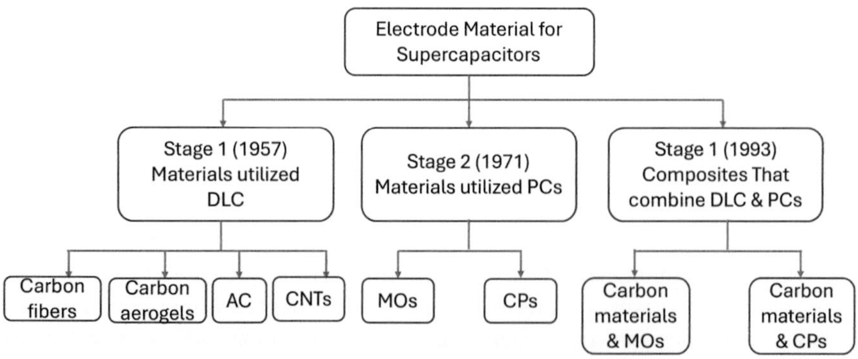

FIGURE 2.5 Classification of SC, along with their electrode materials.

2.6.1 EDLC Electrode Materials

In 1957, General Electric engineers first identified the concept of EDLCs through experiments involving porous carbon electrodes. These studies revealed that energy could be stored within the pores of the carbon material. A few years later, in 1961, the Standard Oil Company of Ohio accidentally rediscovered the principle while working on fuel cell technologies. They built a prototype using activated charcoal as the electrode material, with a thin insulating layer separating the electrodes—a design that still forms the foundation of modern SCs. Although the company did not pursue commercialization, the technology was eventually licensed to NEC, which introduced it to the market in 1978 [21].

2.6.1.1 Carbon

Carbon is considered one of the most cost-effective materials for use in SCs. It is commonly utilized in various forms, including Activated carbons (AC), carbon aerogels, and CNTs, which serve as the primary electrode materials. These carbon-based materials offer a high surface area, and the presence of micropores contributes to pseudocapacitive behavior. Although carbon materials typically have lower electrical conductivity compared to some other options, they are still regarded as highly efficient for SC electrode applications [9, 22].

2.6.1.2 Activated Carbon

AC is a processed form of carbon, produced through relatively simple and low-cost activation methods. It is created by heating carbon-rich materials in the absence of oxygen, followed by physical or chemical treatment to develop a network of fine pores. These pores significantly increase the material's surface area, enhancing its capacity for adsorption and surface reactions. Due to its affordability and large surface area compared to other carbon-based options, AC has become the most widely used electrode material in SCs [15]. It can deliver a specific capacitance ranging from 50 to 150 F/g when used with organic electrolytes and from 100 to 200 F/g with aqueous electrolytes. In particular, AC treated with potassium hydroxide (KOH) can achieve capacitances of approximately 225 and 160 F/g in aqueous systems.

2.6.1.3 Carbon Aerogels

Carbon aerogels feature a continuous framework of conductive carbon nanoparticles, with uniformly distributed mesopores. Because of this cohesive structure and their ability to form chemical bonds with the current collector, there is no need for additional adhesive binders. As binder-free electrodes, carbon aerogels demonstrate a lower ESR than AC. This reduced ESR boosts power output, making it a key focus in SC research involving carbon aerogels [23].

2.6.1.4 Carbon Nanotubes (CNTs)

The discovery of CNTs marked a significant advancement in the science and engineering of carbon-based materials. In SCs, the power density is primarily determined by the overall resistance of the device. Due to their unique pore architecture, exceptional thermal and mechanical strength, and outstanding electrical properties, CNTs have garnered considerable interest as electrode materials for SCs. Unlike other carbon materials, CNTs possess interconnected mesopores, enabling efficient charge distribution across most of the available surface area [24, 25].

CNTs are typically classified into single-walled (SWCNTs) and multi-walled (MWCNTs) types, which have been extensively studied for their potential use in SC electrodes. Their high conductivity and readily accessible surface make them suitable for high-power applications. However, the surface area of CNTs is relatively modest (less than 500 m^2/g), which limits their energy density compared to AC. To enhance their capacitance, CNTs can undergo chemical activation with agents such as KOH, significantly increasing their surface area by two to three times while maintaining their tubular structure [26].

2.6.2 Pseudocapacitor Electrode Materials

Pseudocapacitor electrodes are typically composed of electrically CPs or transition metal oxides (TMOs) due to their ability to undergo redox reactions on the surface of electroactive materials. These redox potentials influence these processes and vary during charge and discharge cycles. This mechanism allows PCs to deliver significantly higher energy density and capacitance than carbon-based EDLCs. However, these advantages are often offset by poor cycling stability. For example, while EDLCs can maintain performance over as many as 500,000 cycles, PCs usually suffer from capacity degradation over time. This degradation is caused by repeated chemical reactions that can alter the morphology of the electrode material, leading to reduced efficiency. To address these issues, carbon-based materials are often incorporated as structural supports to enhance the durability and stability of pseudocapacitive electrodes [27].

2.6.2.1 Metal Oxides

TMOs such as Co_3O_4, NiO, RuO_2, MnO_2, and V_2O_5 are among the most extensively researched materials for pseudocapacitor electrodes. These oxides exhibit increased capacitance due to their ability to undergo redox reactions involving multiple oxidation states at voltage levels. RuO_2, available in crystalline and amorphous forms, is especially significant in theoretical and practical contexts because of its excellent properties, including high electrical conductivity, catalytic efficiency, strong thermal and chemical stability, redox activity, and field emission characteristics. These attributes make RuO_2 suitable for various applications, such as resistive components (thick and thin film), electronic devices, integrated circuits, and ferroelectric film fabrication. Recently, RuO_2 has been adopted as a promising electrode material in SC technology [15]. Among the various metal oxides explored as electrode materials, RuO_2 has proven particularly effective due to its long cycle life, highly reversible redox reactions, broad potential window, and metallic-level electrical conductivity, all contributing to its high specific capacitance. In SC applications, RuO_2 is often synthesized through an electrochemical electrodeposition technique. Electrodes produced using this method have demonstrated excellent cycling stability and delivered a specific capacitance of 498 F/g at a scan rate of 5 F/g [28]. MnO_2 has attracted significant research interest due to its unique chemical and physical properties, making it suitable for various applications such as catalysis, ion exchange, molecular adsorption, biosensing, and energy storage. Its relevance as an electrode material in SCs is especially notable, owing to its low cost, strong capacitive behavior in aqueous electrolytes, and environmentally friendly nature [9, 15]. NiO is a promising electrode material for SCs due to its affordability, ease of synthesis, and ecologically benign characteristics. Electrochemical synthesis methods offer several benefits, including simplicity, cost-effectiveness, reliability, versatility, and precision. This method converts $Ni(OH)_2$ into NiO, producing an ultra-high specific capacitance of 1478 F/g in a 1M KOH aqueous electrolyte solution [15].

2.6.2.2 Conductive Polymer (CPs)

CPs have been extensively explored as materials for PCs, where energy storage occurs through rapid and reversible redox reactions both at the surface and within the bulk of the electrode material. CPs such as polypyrrole (PPy), polyaniline (PANI), and poly(3,4-ethylenedioxythiophene) (PEDOT) can deliver a theoretical specific capacitance of up to 1000 F/g (ranging from 100 to 400 $\mu F/cm^2$), which is nearly double that of typical EDLCs. The high electrical conductivity of CPs is attributed to the delocalization of electrons along their conjugated polymer chains, enabling efficient charge transport in the doped state [29]. The inherent conductivity of CPs can be tailored by altering their chemical structure through cost-effective methods, offering an advantage over TMOs. CPs can be synthesized through chemical polymerization or by electropolymerizing the corresponding monomers. Despite their benefits, such as high electrical conductivity, affordability, and excellent charge storage capacity, CPs often suffer structural degradation, including expansion and contraction during repeated charging and discharging cycles. This structural instability can significantly impact their rate performance and reduce the overall cycling durability of SCs [30, 31]. Therefore, incorporating carbon-based materials such as graphene, porous carbon, CNTs, carbon nanofibers (CNFs), and graphene oxide has been proposed as an effective strategy to enhance the electrochemical stability of CPs [32]. Among the various types of CPs, PANI has emerged as a leading electrode material for SCs due to its ease of synthesis, high electrical conductivity, low cost, and strong energy storage capabilities. However, repeated charge and discharge cycles cause PANI to expand and contract, leading to rapid performance degradation. To address this limitation, integrating carbon-based materials with PANI has proven effective in enhancing its structural stability and overall capacitance [15].

2.6.3 Hybrid Supercapacitor Electrode Materials

Hybrid-type SCs employ composite materials that combine conducting polymers or metal oxides with carbon-based substances. These composites combine the characteristics of both pseudocapacitive and EDLC materials, allowing for a unified electrode that supports both electrochemical and physical charge storage mechanisms [33].

2.7 ELECTROLYTES

An electrolyte comprises a solvent and dissolved compounds that dissociate into positive and negative ions, enabling it to conduct electricity. The level of conductivity is determined by the ion concentration within the solution. The primary role of the electrolyte is to supply the ions necessary for forming the monolayer in the Helmholtz double layer and to support ion transport. The selected electrolyte significantly affects key performance parameters of electrochemical capacitors, including working voltage, specific capacitance, and power density. The ion size and concentration influence the thickness of the electric double layer in the electrolyte. In SCs, energy is stored through electrostatic interactions between electrode surfaces and electrolyte ions, along with redox reactions occurring at the electrode materials. The internal resistance of the SC is closely related to the electrolyte's resistivity and the ease with which ions can move into and out of the porous electrode structures. The maximum voltage an SC can handle is constrained by the electrolyte's breakdown potential, which impacts the achievable energy density. Furthermore, the device's power output is influenced by its ESR, which is primarily determined by the conductivity of the electrolyte [34]. Figure 2.6 illustrates the key characteristics of SCs that are heavily dependent on the properties of the chosen electrolyte [35, 36].

Since redox reactions occur at the interface between the electrode and electrolyte, the electrolyte must possess high ionic conductivity while maintaining minimal or negligible electronic conductivity. The level of ionic conductivity depends on both the concentration of charge carriers and their mobility in an electric field. An ideal electrolyte should be chemically and electrochemically stable, ensuring it does not react with or degrade other components in the capacitor, such as electrodes or current collectors. Additionally, the electrolyte must have low viscosity to penetrate the porous structure of sponge-like electrodes effectively. Other essential characteristics include strong thermal stability, a broad electrochemical

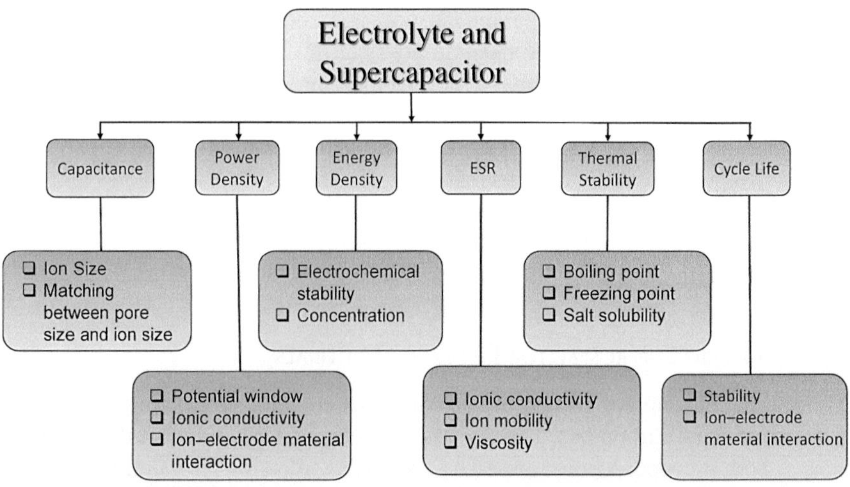

FIGURE 2.6 Supercapacitor performance is reliant on the electrolyte.

voltage window, low flammability, low toxicity, affordability, and environmental safety. Significant research has been dedicated to developing innovative electrolyte formulations to enhance the performance of electrolytes—and by extension, that of SCs. In general, three major categories of electrolytes are employed in electrochemical SCs.

2.7.1 Aqueous Electrolytes

Aqueous electrolytes are solutions where water acts as the solvent. Among these, acidic and alkaline electrolytes are most frequently utilized due to their high ionic conductivity and unique proton transport characteristics. The small size and high mobility of protons in water-based systems allow them to chemisorb onto individual oxide ions. However, using aqueous electrolytes restricts the SC's operating voltage, typically around 1.0 V, because the electrochemical stability range of water is limited to approximately 1.23 V [37]. When the applied voltage surpasses 0.8 V, one of the electrodes may reach a potential that exceeds the thermodynamic stability limit of water, leading to its decomposition [38]. Highly concentrated sulfuric acid is commonly employed as the electrolyte for RuO_2-based electrodes [39]. For MnO_2-based SCs, a mild KCl solution has been utilized as an electrolyte. Additionally, sodium sulfate (Na_2SO_4) and sodium chloride (NaCl) are among the commonly used aqueous electrolytes in such systems [40]. K. Fic et al. achieved an impressive operating voltage of 2.2 V, along with excellent cycling stability, using lithium sulfate (Li_2SO_4) as the electrolyte [41]. While chloride-based electrolytes offer superior ionic conductivity, they tend to corrode current collectors like stainless steel. In contrast, sulfate-based salts provide better long-term stability but suffer from lower solubility. Although materials like gold and platinum resist corrosion effectively, their high cost limits their practical application as current collectors [42]. The researcher typically determines the balance between maximizing ionic conductivity and minimizing corrosion. From an industrial standpoint, neutral-pH aqueous electrolytes offer a more attractive solution. These electrolytes are commonly utilized in SCs, prioritizing high specific power over energy density.

2.7.2 Organic Electrolytes

Organic electrolytes, typically containing solvents such as acetonitrile (ACN), PC, diethyl carbonate, γ-butyrolactone, and tetrahydrofuran, are commonly used in SCs. These electrolytes support cell voltages above 2.0 V, enhancing energy density, as energy output is directly proportional to the square of the voltage. However, organic solvents generally exhibit limited salt solubility and higher internal resistance, which leads to reduced power density. Additionally, because organic molecules tend to be larger, electrodes with wider pores must accommodate them. Organic alternatives are more expensive than aqueous electrolytes and require stringent purification and dehydration during manufacturing. Among various options, tetraethylammonium salts ($(C_2H_5)_4BF_4$) are widely chosen due to their decent solubility and reasonably good conductivity [43].

Brandt et al. reported achieving a maximum cell voltage of up to 3.75 V using adiponitrile as the electrolyte solvent [44]. Electrolytes based on PC typically exhibit lower ionic conductivity, while those utilizing ACN tend to result in reduced power output. Despite offering a broader voltage window, the associated ohmic losses must also be considered. Although organic electrolytes are widely employed in SC systems, their drawbacks—including high cost, limited conductivity, lower power and capacitance, and issues related to toxicity and flammability—must be carefully weighed in practical applications.

2.7.3 Ionic Electrolytes

Ionic liquids (ILs), which are salts that remain liquid at room temperature, consist entirely of cations and anions. They possess several characteristics favorable for use in SCs, such as excellent thermal stability, non-flammability, low toxicity, resistance to corrosion at elevated temperatures, and broad electrochemical stability—typically supporting voltage ranges from 2 to 6 V, often around 4.5 V. Despite these advantages, achieving high purity during synthesis can be challenging. Moreover, their relatively high cost and limited conductivity at room temperature restrict their large-scale adoption. Commonly explored ILs for SC applications include imidazolium, pyrrolidinium, and various asymmetric aliphatic quaternary ammonium salts paired with anions like tetrafluoroborate, trifluoromethanesulfonate, hexafluorophosphate, bis(trifluoromethanesulfonyl)imide, and bis(fluorosulfonyl)imide. The performance of ILs is highly temperature-dependent, often exhibiting improved conductivity and stability at elevated temperatures.

2.8 SEPARATOR

The characteristics of the separator significantly influence the overall performance of an SC. Its primary role is to prevent direct electrical contact between the two electrodes while still allowing the passage of ions to facilitate charge transfer. An ideal separator should exhibit high electrical resistance, excellent ionic conductivity, be extremely thin, and possess a porous structure to minimize ESR. It must also be chemically inert to maintain electrolyte stability and conductivity. Typically, ceramic or glass fiber separators are employed with aqueous electrolytes, whereas polymer-based or paper separators are more commonly used in systems utilizing organic electrolytes [36]. Several key factors must be considered when selecting an appropriate separator for SC applications:

- It must be electrically insulating to prevent electron flow between electrodes.
- It should be chemically stable with both electrode materials and the electrolyte.
- It should allow easy passage of electrolyte ions with minimal ionic resistance.
- It should exhibit good wettability with the chosen electrolyte.
- It must possess sufficient mechanical strength to withstand pressure and volume fluctuations, including swelling.

In early SC designs, natural materials like ceramics, glass, and cellulose paper were used as separators. However, polymer-based separators have become more prominent with advancements in the field due to their cost-effectiveness, flexibility, and high porosity. These polymer separators are generally classified into fibrous structures and monolithic networks with defined pore architectures [19].

2.9 APPLICATIONS

SCs are primarily utilized in applications that demand many charge–discharge cycles, an extended operational lifespan, or rapid bursts of power over short durations. Their high power capability makes them highly valuable in power electronics. They are commonly used in systems requiring uninterruptible power supplies (UPS) and security applications. SCs are particularly useful in scenarios where a large current is needed briefly, such as in the starting systems of heavy diesel engines or hybrid electric vehicles. For instance, compact SCs like Gold Caps from Tokin are frequently employed as maintenance-free power sources for microcontrollers and memory storage in integrated circuits. Due to their exceptional power performance and environmental advantages, SCs are considered ideal for starter systems. Wind turbine pitch control is a backup power source, ensuring that blade pitch adjustment remains possible even if the main power supply fails [45]. SCs effectively stabilize power output in systems with variable loads, such as laptops, solar energy setups, GPS units, portable media devices, and other handheld electronics [46]. In UPS systems, SCs have replaced large arrays of electrolytic capacitors, leading to a reduced cost per cycle, extended battery lifespan, and lower maintenance needs. SCs enhance system reliability and performance by buffering power during brief outages and absorbing high current spikes. They are also used in defibrillators, which can discharge up to 500 joules to restore a normal heart rhythm. Additionally, SCs serve as short-term energy storage in energy harvesting systems, with the stored energy capable of powering application-specific integrated circuits (ASICs) for over 10 hours.

2.10 CONCLUSION

SCs represent a significant advancement in energy storage technology, offering an effective balance between traditional capacitors and batteries. Their ability to provide high power output, rapid charging and discharging, and a long operational life makes them ideal for numerous modern applications. These include uses in backup power systems, hybrid vehicles, renewable energy grids, and compact electronic devices. The energy storage mechanisms in SCs are primarily classified into electric double-layer capacitance and pseudocapacitance. EDLCs rely on carbon-based materials with high surface areas, while PCs utilize TMOs and conducting polymers to store charge through fast and reversible redox reactions. Hybrid SCs, which combine the benefits of EDLCs and PCs, have shown great potential to enhance overall performance.

Electrode material selection is a critical factor in optimizing SC efficiency. Carbon materials such as AC, graphene, and CNTs offer high conductivity and large surface areas. Metal oxides like MnO_2, RuO_2, and NiO and polymers like PANI and PPy contribute to increased capacitance through faradaic processes. Integrating these with carbon supports improves both conductivity and structural stability. Electrolytes also play a vital role, affecting voltage range, energy density, and overall device stability. Aqueous electrolytes are inexpensive and highly conductive but are limited by a narrow voltage window. Organic and IL electrolytes extend the operating voltage but come with trade-offs, including higher costs, lower conductivity, and more complex handling requirements.

Additionally, separators and current collectors are crucial components that affect device reliability, efficiency, and cycle life. Proper design guarantees ion flow while preventing short circuits and minimizing internal resistance. SCs are already employed in various real-world applications, particularly where quick bursts of power and long-term reliability are essential. With ongoing innovations in materials and design, SCs are expected to become even more efficient, cost-effective, and environmentally friendly, solidifying their role in the future of sustainable energy storage technologies.

2.11 AUTHOR AI DECLARATION

The authors hereby declare the use of generative AI and AI-assisted technologies in the writing process. While preparing this work, the authors used the ChatGPT tool/service to rephrase several statements in the manuscript. All AI-generated content was reviewed and edited by the authors as needed. The authors take full responsibility for the content of the publication.

ACKNOWLEDGMENT

The Editors acknowledge the use of Grammarly as a tool to assist in editing this book chapter. The suggestions made by Grammarly were carefully considered and incorporated where appropriate, but the author is solely responsible for the final content.

REFERENCES

1. Li, L., Song, B., Maurer, L., Lin, Z., Lian, G., Tuan, C.-C., Moon, K.-S., & Wong, C.-P. (2016). Molecular engineering of aromatic amine spacers for high-performance graphene-based supercapacitors. Nano Energy, 21, 276–294. https://doi.org/10.1016/j.nanoen.2016.01.028
2. Najib, S., & Erdem, E. (2019). Current progress achieved in novel materials for supercapacitor electrodes: Mini review. Nanoscale Advances, 1(8), 2817–2827. https://doi.org/10.1039/c9na00345b
3. Burke, A. (2000). Ultracapacitors: Why, how, and where is the technology. Journal of Power Sources, 91(1), 37–50. https://doi.org/10.1016/s0378-7753(00)00485-7
4. Kötz, R., & Carlen, M. (2000). Principles and applications of electrochemical capacitors. Electrochimica Acta, 45(15–16), 2483–2498. Elsevier BV. https://doi.org/10.1016/s0013-4686(00)00354-6
5. Armutlulu, A., Kim, J. K., Kim, M., Allen, S. A. B., & Allen, M. G. (2013). Nickel-oxide-based supercapacitors with high aspect ratio concentric cylindrical electrodes. 2013 Transducers & Eurosensors XXVII: The 17th International Conference on Solid-State Sensors, Actuators and Microsystems (TRANSDUCERS & EUROSENSORS XXVII).

6. Lokhande, C. D., Dubal, D. P., & Joo, O.-S. (2011). Metal oxide thin film based supercapacitors. Current Applied Physics, 11(3), 255–270. Elsevier BV. https://doi.org/10.1016/j.cap.2010.12.001
7. Ciocan, I., Farcas, C., Grama, A., & Tulbure, A. (2016). An improved method for the electrical parameters identification of a simplified PSpice supercapacitor model. In 2016 IEEE 22nd International Symposium for Design and Technology in Electronic Packaging (SIITME). 2016 IEEE 22nd International Symposium for Design and Technology in Electronic Packaging (SIITME). IEEE. https://doi.org/10.1109/siitme.2016.7777270
8. Helmholtz, H. (1853). Ueber einige Gesetze der Vertheilung elektrischer Ströme in körperlichen Leitern mit Anwendung auf die thierisch-elektrischen Versuche. Annalen der Physik, 165(6), 211–233. Wiley. https://doi.org/10.1002/andp.18531650603
9. Halper, M.S., & Ellenbogen, J.C. (2013). "Supercapacitors: a brief overview," https://www.mitre.org/publications/technical-papers/supercapacitors-a-brief-overview
10. "Low voltage electrolytic capacitor," US2800616A, 1957 https://patents.google.com/patent/US2800616A/en
11. Pope, M. A., Korkut, S., Punckt, C., & Aksay, I. A. (2013). Supercapacitor electrodes produced through evaporative consolidation of graphene oxide-water-ionic liquid gels. Journal of The Electrochemical Society, 160(10), A1653–A1660. The Electrochemical Society. https://doi.org/10.1149/2.017310jes
12. Winter, M., & Brodd, R. J. (2004). What are batteries, fuel cells, and supercapacitors? Chemical Reviews, 104(10), 4245–4270. American Chemical Society (ACS). https://doi.org/10.1021/cr020730k
13. Simon, P., & Gogotsi, Y. (2008). Materials for electrochemical capacitors. Nature Materials, 7(11), 845–854. Springer Science and Business Media LLC. https://doi.org/10.1038/nmat2297
14. Wasterlain, S., Guven, A., Gualous, H., Fauvarque, J. F., & Garmanage. (2006). Hybrid power source with batteries and supercapacitor for vehicle applications. In L2ES Laboratory UFC-UTBM.
15. Iro, Z. S., Subramani, C., & Dash, S. S. (2016). A brief review on electrode materials for supercapacitor. International Journal of Electrochemical Science, 11(12), 10628–10643. Elsevier BV. https://doi.org/10.20964/2016.12.50
16. Czagany, M., Hompoth, S., Keshri, A. K., Pandit, N., Galambos, I., Gacsi, Z., & Baumli, P. (2024). Supercapacitors: An efficient way for energy storage application. Materials, 17(3), 702. MDPI AG. https://doi.org/10.3390/ma17030702
17. Kiamahalleh, M. V., Zein, S. H. S., Najafpour, G., Sata, S. A., & Buniran, S. (2012). Multiwalled carbon nanotubes based nanocomposites for supercapacitors: A review of electrode materials. Nano, 07(02), 1230002. World Scientific Pub Co Pte Ltd. https://doi.org/10.1142/s1793292012300022
18. Choi, H., & Yoon, H. (2015). Nanostructured electrode materials for electrochemical capacitor applications. Nanomaterials, 5(2), 906–936. MDPI AG. https://doi.org/10.3390/nano5020906
19. Sharma, P., & Kumar, V. (2020). Current technology of supercapacitors: A review. Journal of Electronic Materials, 49(6), 3520–3532. Springer Science and Business Media LLC. https://doi.org/10.1007/s11664-020-07992-4
20. Vangari, M., Pryor, T., & Jiang, L. (2013). Supercapacitors: Review of materials and fabrication methods. Journal of Energy Engineering, 139(2), 72–79. American Society of Civil Engineers (ASCE). https://doi.org/10.1061/(asce)ey.1943-7897.0000102
21. Sevilla, M., & Mokaya, R. (2014). Energy storage applications of activated carbons: Supercapacitors and hydrogen storage. Energy & Environmental Science, 7(4), 1250–1280. Royal Society of Chemistry (RSC). https://doi.org/10.1039/c3ee43525c

22. Adam Marcus, N. A survey of electrochemical supercapacitor technology (PDF) (Technical report). Retrieved 2015-02-21.
23. Aricò, A. S., Bruce, P., Scrosati, B., Tarascon, J.-M., & Van Schalkwijk, W. (2010). Nanostructured materials for advanced energy conversion and storage devices. In Materials for Sustainable Energy (pp. 148–159). Co-Published with Macmillan Publishers Ltd, UK. https://doi.org/10.1142/9789814317665_0022
24. Cheng, Q., Tang, J., Ma, J., Zhang, H., Shinya, N., & Qin, L.-C. (2011). Graphene and carbon nanotube composite electrodes for supercapacitors with ultra-high energy density. Physical Chemistry Chemical Physics, 13(39), 17615. Royal Society of Chemistry (RSC). https://doi.org/10.1039/c1cp21910c
25. Li, J., Cheng, X., Shashurin, A., & Keidar, M. (2012). Review of electrochemical capacitors based on carbon nanotubes and graphene. Graphene, 01(01), 1–13. Scientific Research Publishing, Inc. https://doi.org/10.4236/graphene.2012.11001
26. Pandolfo, A. G., & Hollenkamp, A. F. (2006). Carbon properties and their role in supercapacitors. Journal of Power Sources, 157(1), 11–27. Elsevier BV. https://doi.org/10.1016/j.jpowsour.2006.02.065
27. Yi, C., Zou, J., Yang, H., & Leng, X. (2018). Recent advances in pseudocapacitor electrode materials: Transition metal oxides and nitrides. Transactions of Nonferrous Metals Society of China, 28(10), 1980–2001. Elsevier BV. https://doi.org/10.1016/s1003-6326(18)64843-5
28. Gujar, T. P., Kim, W.-Y., Puspitasari, I., Jung, K.-D., & Joo, O.-S. (2007). Electrochemically deposited nanograin ruthenium oxide as a pseudocapacitive electrode. International Journal of Electrochemical Science, 2(9), 666–673. Elsevier BV. https://doi.org/10.1016/s1452-3981(23)17102-1
29. Syed Zainol Abidin, S. N. J., Mamat, S., Abdul Rasyid, S., Zainal, Z., & Sulaiman, Y. (2017). Fabrication of poly(vinyl alcohol)-graphene quantum dots coated with poly(3,4-ethylenedioxythiophene) for supercapacitor. Journal of Polymer Science Part A: Polymer Chemistry, 56(1), 50–58. Wiley. https://doi.org/10.1002/pola.28859
30. Potphode, D. D., Mishra, S. P., Sivaraman, P., & Patri, M. (2017). Asymmetric supercapacitor devices based on dendritic conducting polymer and activated carbon. Electrochimica Acta, 230, 29–38. Elsevier BV. https://doi.org/10.1016/j.electacta.2017.01.168
31. Asen, P., Shahrokhian, S., & zad, A. I. (2018). Ternary nanostructures of Cr2O3/graphene oxide/conducting polymers for supercapacitor application. Journal of Electroanalytical Chemistry, 823, 505–516. Elsevier BV. https://doi.org/10.1016/j.jelechem.2018.06.048
32. Mohd Abdah, M. A. A., Zubair, N. A., Azman, N. H. N., & Sulaiman, Y. (2017). Fabrication of PEDOT coated PVA-GO nanofiber for supercapacitor. Materials Chemistry and Physics, 192, 161–169. Elsevier BV. https://doi.org/10.1016/j.matchemphys.2017.01.058
33. Gupta, A., Sardana, S., Dalal, J., Lather, S., Maan, A. S., Tripathi, R., Punia, R., Singh, K., & Ohlan, A. (2020). Nanostructured polyaniline/graphene/Fe_2O_3 composites hydrogel as a high-performance flexible supercapacitor electrode material. ACS Applied Energy Materials, 3(7), 6434–6446. American Chemical Society (ACS). https://doi.org/10.1021/acsaem.0c00684
34. Joshi, P.S., Jogade, S.M., Gothe, S.D., & Sutrave, D.S. (2017). Electrochemical performance of sol-gel spin coated RuO_2 electrodes reliant on aqueous electrolytes. Asian Journal of Chemistry, 29(1), 203–208.
35. Zhong, C., Deng, Y., Hu, W., Qiao, J., Zhang, L., Zhang, J., Zhong, C., et al. (2015). A review of electrolyte materials and compositions for electrochemical supercapacitors. Chemical Society Reviews, 44(21), 7484–7539.
36. Sunil Joshi, P. (2019). Supercapacitor: Basics and overview. Journal of Information and Computational Science, 1548–7741. https://www.researchgate.net/publication/337917546_Supercapacitor_Basics_and_Overview

37. Conway, B. E. (1999). Electrochemical Supercapacitors. Springer US. https://doi.org/10.1007/978-1-4757-3058-6
38. Thesis: Optimizing carbon/carbon supercapacitors in aqueous and organic electrolytes, Qiang Gao, 2013.
39. Sugimoto, W., Yokoshima, K., Murakami, Y., & Takasu, Y. (2006). Charge storage mechanism of nanostructured anhydrous and hydrous ruthenium-based oxides. Electrochimica Acta, 52(4), 1742–1748. Elsevier BV. https://doi.org/10.1016/j.electacta.2006.02.054
40. Xie, X., & Gao, L. (2007). Characterization of a manganese dioxide/carbon nanotube composite fabricated using an in situ coating method. Carbon, 45(12), 2365–2373. Elsevier BV. https://doi.org/10.1016/j.carbon.2007.07.014
41. Fic, K., Lota, G., Meller, M., & Frackowiak, E. (2012). Novel insight into neutral medium as electrolyte for high-voltage supercapacitors. Energy & Environmental Science, 5(2), 5842–5850. Royal Society of Chemistry (RSC). https://doi.org/10.1039/c1ee02262h
42. Frackowiak, E., Fic, K., Meller, M., & Lota, G. (2012). Electrochemistry serving people and nature: High-energy ecocapacitors based on redox-active electrolytes. ChemSusChem, 5(7), 1181–1185. Wiley. https://doi.org/10.1002/cssc.201200227
43. Jiang, D., Jin, Z., Henderson, D., & Wu, J. (2012). Solvent effect on the pore-size dependence of an organic electrolyte supercapacitor. The Journal of Physical Chemistry Letters, 3(13), 1727–1731. American Chemical Society (ACS). https://doi.org/10.1021/jz3004624
44. Brandt, A., Isken, P., Lex-Balducci, A., & Balducci, A. (2012). Adiponitrile-based electrochemical double layer capacitor. Journal of Power Sources, 204, 213–219. Elsevier BV. https://doi.org/10.1016/j.jpowsour.2011.12.025
45. Maxwell Technologies Ultracapacitors (ups power supply) Uninterruptible Power Supply Solutions. Maxwell.com. Retrieved 2013-05-29.
46. Graham Pitcher If the cap fits… New Electronics. 26 March 2006.

3 Types of Supercapacitors

Ravi Kumar Sahu, Sanjay Khatri, Manish Kumar, Bhupendra Ghritalahre, Sandip Gangil, Vishal Choudhary, and Parmanand Sahu

3.1 INTRODUCTION

The rapid depletion of fossil fuels, the significant rise in CO_2 levels contributing to global warming, and the surge in global energy consumption driven by population growth, industrial expansion, international economic development, and increased reliance on energy-intensive domestic appliances demand investigation into renewable energy sources and the development of efficient, environmentally sustainable energy storage technologies to tackle social and environmental issues [1]. The World Energy Council predicts global energy consumption will double by 2050 [2]. Consequently, if energy production depends solely on fossil fuel combustion, it will harm human health, the environment, security, global ecology, and energy. The primary solutions to this pressing issue include the extensive use of alternative energy sources, energy conservation, energy storage, and the adoption of efficient energy-driven technologies [3]. Solar and wind power are the primary renewable energy sources; however, their intermittent nature poses a significant challenge to effective utilization. Energy storage devices play a vital role in addressing this challenge, as they can store energy and enhance the power delivered to the electric grid.

Furthermore, these devices can capture the kinetic energy generated by vehicles during deceleration or halting, thus improving their overall efficiency. Batteries are commonly used for low-power applications, while supercapacitors are gaining attention as superior energy storage solutions for enhancing vehicle efficiency. Their advantages include high power density, lightweight design, extended lifespan, and eco-friendly attributes.

Supercapacitors (SCs) are large-capacity capacitors with higher specific capacitance than electrostatic capacitors and can endure more charging–discharging cycles than batteries. They can recharge and discharge faster than batteries but not as quickly as regular capacitors, thereby narrowing the divide between batteries and conventional capacitors by resolving the current power–energy inequalities. Additionally, they require no maintenance, are safe, and do not experience memory effects, making them strong candidates for sophisticated hybrid systems for energy storage, suitable for stationary and onboard applications [4]. SCs exceed traditional SCs in energy density and outperform batteries in power delivery [1]. They also function as an intermediate energy storage device, balancing the rapid power supply of capacitors with the energy density of batteries. Figure 3.1 illustrates the various types of supercapacitors, categorized by their charge storage mechanisms (electrochemical double-layer capacitors (EDLCs), pseudocapacitors, and hybrid supercapacitors), and the electrode configurations, specifically symmetric and asymmetric.

DOI: 10.1201/9781003535836-3

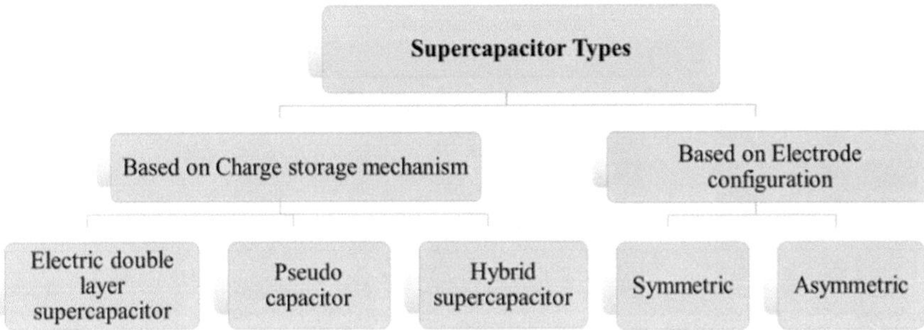

FIGURE 3.1 Classification of the supercapacitor.

3.2 TYPES OF SUPERCAPACITORS

Supercapacitors can be categorized in several ways, including size, form, electrode assembly, and charge storage techniques (Table 3.1).

3.2.1 Based on the Charge Storage Mechanism

As shown in Figure 3.2, supercapacitors are divided into three main types based on their charge storage mechanisms: EDLCs, pseudo, and hybrid capacitors. Each category's unique operating principles and performance characteristics determine their suitability for various energy storage applications.

3.2.1.1 EDLCs

Electric double-layer supercapacitors store charge at the electrode–electrolyte interface through the physical adsorption and desorption of ions [5]. These devices boast extended cyclic stability and can charge within seconds, demonstrating a remarkably high power density and a storage capacity up to an order of magnitude greater than that of ordinary rechargeable batteries [3]. The Helmholtz double-layer model characterizes the electrode/electrolyte interface by a fixed charge distribution, in

TABLE 3.1
Supercapacitors Classified According to Their Method of Charge Storage

Supercapacitor Types	Charge Storage Mechanism	Reaction	Process	Materials
EDLCs	Electrostatic		Non-Faradic	CNTs, activated carbon, rGO, carbon aerogel
Pseudocapacitor	Electrochemical	Underpotential deposition, redox, intercalation	Faradic	Conducting polymer, transition metal oxides
Hybrid supercapacitor	Electrostatic and electrochemical	Battery type, asymmetric, composite based	Faradic and non-Faradic	PbO_2// Graphene, CNT//MnO_2, TiO_2@ rGO// CNT

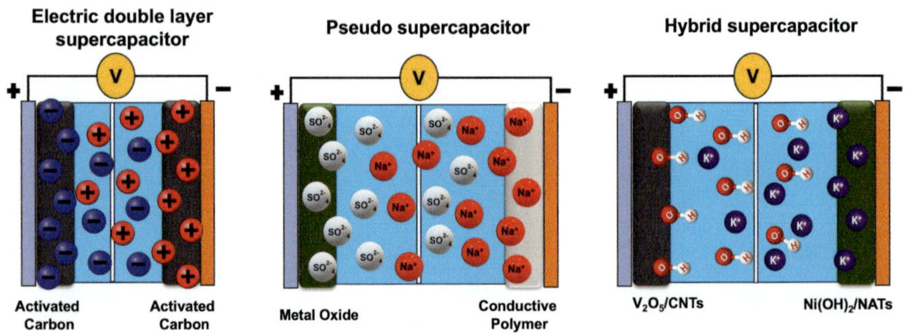

FIGURE 3.2 EDLCs, pseudocapacitors, and hybrid supercapacitors.

which oppositely charged layers are separated by only a few angstroms and remain stationary [6]. EDLCs do not involve a Faradaic redox reaction in their electrostatic charge storage mechanism; instead, it predominantly occurs due to the adsorption of electrolyte ions onto the electrode surfaces, as indicated by the rectangular shape of their CV curves [2]. The electrodes of EDLCs are fabricated using materials with a high specific surface area (SSA) and diverse pore structures, including graphene, activated carbon, and carbon nanofibers [7]. EDLCs are free from swelling, as no Faradaic reactions occur at the electrodes, unlike batteries, which may expand during charge and discharge cycles. Both aqueous and organic electrolytes can promote ionic transport; however, organic electrolytes are generally preferable due to their higher charge storage capacities through ion adsorption [1]. This benefit is partly due to their greater voltage range, often reaching between 2.0 and 3.0 V. When a voltage is applied between the electrodes of EDLCs, a double-layered structure forms, enabling the flow of electrolyte ions over the surface of this layer. The specific capacitance of this capacitor is determined by the interaction of the two charged surfaces. At the same time, the ionic movement creates an electric potential resulting from the compression of the double layers. This phenomenon may be explained by the zeta potential associated with the electrodes. EDLCs have been effectively used in various applications, such as load-lifting machinery, Light-emitting diode (LED) flash devices, computers, and memory backup power systems.

3.2.1.2 Pseudocapacitors

Pseudo (Faradaic) capacitors function through swift redox reactions that involve charge transfer at the electrode–electrolyte boundary, facilitating a Faradic charge exchange between electrodes and electrolytes for storing energy [3]. Pseudocapacitive electrodes, made of nanomaterials that are redox-active, such as conductive polymers and transition metal compounds, store charge through intercalation/de-intercalation and redox reaction processes rather than simple electrostatic adsorption, with common examples, including MnO_2 and RuO_2 [8]. Within the potential range, the capacitance of these capacitors remains consistent. The pseudocapacitors offer high capacitance but lower power density compared to EDLCs [2]. Batteries and EDLCs represent the two extremes, while pseudocapacitance occupies a middle ground in the energy storage spectrum. There are three types of pseudocapacitance: monolayer adhesion, intercalation, and the surface Faradaic reaction [6].

Monolayer adsorption occurs when two metals with distinct redox potentials create a monolayer adsorbent on each other's surfaces, a process referred to as underpotential deposition. Surface Faradaic redox pseudocapacitance in the electrode results from charge-transfer mechanisms that lead to the adsorption of alkali ions on or near the electrode's surface [6]. Pseudocapacitance arises when the intercalation and de-intercalation processes are not limited by the gradual solid-state diffusion, which typically occurs as alkali ions migrate through the lattice structure of the electrode material. The insertion of cations into the bulk lattice of the electrode is known as intercalation and is observed in metal oxides (e.g., V_2O_5, RuO_2, IrO_2) and conducting polymers (e.g., Reduced graphene oxide-functionalized/polyaniline (rGO-f/PANI), Polyaniline (PANI)). Bulk intercalation pseudocapacitance replaces surface redox pseudocapacitance specific to the electrode surface or its immediate vicinity [6]. Pseudocapacitors exhibit considerably high energy density; however, the rapid redox reactions during charging and discharging lead to the accumulation of irreversible components, which subsequently reduce their cyclic stability. They demonstrate extended cycle life, significant electrical conductivity, high power density, high capacitance value, and low current density [3].

3.2.1.3 Hybrid Supercapacitor

The term hybrid indicates that it combines elements of both EDLCs and pseudocapacitive materials, allowing it to mimic the high power density of EDLCs and the superior energy density of pseudocapacitors or batteries [2]. EDLC electrode materials encompass graphene, CNTs, and AC, while pseudocapacitive electrodes are generally formed of metal oxides and conductive polymers [9]. Instead of using transition metal–based compounds, conducting polymers, and carbon separately, researchers have developed hybrid composite materials for hybrid supercapacitors from these active materials. These composites are mechanically flexible, chemically stable, cost-effective, and exhibit high electrical conductivity [3]. A hybrid supercapacitor construction utilizes activated carbon as the negative electrode (cathode) and a conducting polymer as the positive electrode (anode), ensuring balanced kinetics in the charge/discharge process [6]. The anode stores charge through intercalation (a Faradic process), while the cathode stores charge through surface adsorption–desorption (a non-Faradic process), and utilizing both non-Faradic and Faradic processes, these supercapacitors transcend the constraints of EDLCs and pseudocapacitors. EDLCs and pseudocapacitors are connected to manage the energy storage mechanism of hybrid supercapacitors [10]. Since pseudocapacitors do not have the limitations of EDLCs, and vice versa, combining these components conceals their restrictions, leading to increased capacitance. Hybrid supercapacitors may be categorized into these three types depending on the electrode materials employed as composite hybrids, battery-type hybrids, and asymmetric hybrids such as rGO-f/PANI, graphene, or Fe_3O_4 [11]. Additionally, hybrid supercapacitors can be symmetric or asymmetric, depending on the assembly architecture. The electrochemical performance of hybrid supercapacitors improves when they are constructed using separate electrodes derived from different materials, compared to when these electrodes are used to enhance capacitance values individually. Supercapacitors consist of a positive electrode made with redox-active materials and a negative electrode by integrating a carbon material, with a separator in between and electrolyte. They are called asymmetric hybrid supercapacitors. Movement of positive and negative electrolyte ions toward the positive and negative electrodes during charging–discharging creates EDLC behavior through the adsorption–desorption of positive electrolyte ions at the surface of the negative electrode, while pseudocapacitive behavior is driven by rapid and reversible Faradaic reactions involving negative electrolyte ions at the surface of the positive electrode [3]. This asymmetric design represents a substantial breakthrough toward long-lasting, sustainable, and high-performance energy storage devices. Hybrid supercapacitors exhibit higher energy density than symmetric EDLCs, owing to their superior specific capacitance and broader operating voltage range [12]. The performance of these devices is mainly determined by the electrode materials, where a high SSA and large pore volumes promote electrolyte ion accessibility, thereby boosting capacitance. Additionally, careful control of electrode shape plays a key role in maximizing performance. Morphological manipulation influences the properties of the nanomaterials; altering nanostructures reveals unique crystal facets and active spots that may dramatically influence their electrochemical characteristics. Hybrid supercapacitors offer several advantages, including cost-effectiveness, high power and energy density, excellent cyclic stability, mechanical durability, low resistance, and enhanced performance [12]. Their primary application is in vehicles, where they serve as efficient power sources.

3.2.2 Based on the Type of Materials Used for the Electrodes

3.2.2.1 Aqueous Electrolyte–Based Asymmetric Supercapacitors (AASCs)

Aqueous electrolyte–based asymmetric supercapacitors (AASCs) exploit the complementary electrochemical behaviors of dissimilar electrodes—pairing a high surface–area EDLC-type carbon electrode with a pseudocapacitive or battery-type Faradaic electrode—to extend the cell voltage and boost energy density beyond that of symmetric devices [13]. By employing benign aqueous electrolytes such as KOH, Na_2SO_4, or H_2SO_4, these systems achieve low internal resistance and excellent rate performance, albeit within the water-splitting limit of ~1.0–1.8 V [14]. RuO_2 was one of the first pseudocapacitive materials used as the positive electrode in asymmetric configurations; when anchored onto laser-scribed graphene (LSG), hydrous RuO_2 achieves specific capacitances exceeding 1,100 F/g and an operating voltage of 1.8 V, owing to the graphene scaffold's suppression of nanoparticle aggregation and facilitation of rapid electron transport [15, 16]. However, RuO_2's high cost has driven the search for earth-abundant alternatives such as MnO_2, which delivers up to 21 Wh/kg in 0.5 M K_2SO_4 and retains >94% capacitance after 23,000 cycles; performance is maximized with smaller hydrated cations (e.g., K^+) that more readily penetrate the δ-MnO_2 lattice [17].

3.2.2.2 Faradaic Material-Based Aqueous Hybrid Capacitors

Transition metal oxides and hydroxides such as NiO, $Ni(OH)_2$, Co_3O_4, and $Co(OH)_2$ serve as Faradaic electrodes in hybrid aqueous capacitors, quantifying storage in

C/g or mAh/g rather than F/g due to their diffusion-controlled charge mechanisms. NiO nanoflake arrays on reduced graphene hydrogel extend the voltage window to 1.7 V and deliver areal capacities up to 1.2 C/cm^2 at 8 mA/cm^2, with capacity even increasing from 0.28 to 0.76 C/cm^2 over 100,000 cycles as active NiO forms continuously [17]. Likewise, three-dimensional (3D) C/$CoNi_3O_4$ nano-wall arrays exhibit gravimetric capacities of 1,299 C/g and surface areas of 128 m^2/g, capitalizing on their integrated nano-wall architecture for rapid ion diffusion and electron transfer [17]. Battery-type hydroxides like $Ni(OH)_2$, with lamellar MO_6 structures and >7 Å interlayer spacing, facilitate reversible Ni^{2+}/Ni^{3+} redox in alkaline electrolytes; β-$Ni(OH)_2$ on Ni foam paired with activated carbon shows specific capacities up to 790 C/g and stable 1.4 V operation [18]. $Co(OH)_2$ nanosheets on N-doped graphene yield capacities of 1,420 C/g, benefiting from fully exposed electroactive sites and fast ion access[18]. Mixed transition metal systems further elevate performance: spinels such as $NiCo_2O_4$ (theoretical 780 C/g) provide robust 3D diffusion pathways and multi-electron redox, with porous $NiCo_2O_4$ nanowires on Ni foam reaching 740 C/g and enabling 1.6 V hybrid operation with activated carbon [19]. Coordination polymer–derived onion-shaped Ni–Co oxides and hierarchical $NiMoO_4$ nanosheets on conductive substrates achieve working voltages up to 1.7 V and high capacitance through well-defined nanosheet architectures [20]. Sulfide analogues like $NiCo_2S_4$ on N-doped carbon foam double capacity to 614 C/g at 20 A/g, owing to their superior electrical conductivity and rich redox chemistry [17]. Finally, conductive polymer hybrids (e.g., CoO@polypyrrole) combine high polymer conductivity (10–100 S/cm) with oxide pseudocapacitance to achieve capacities >1,500 C/g and stable 1.8 V cycling [16]. Collectively, these advances in Faradaic/carbon composite electrodes underscore the critical role of tailored nanostructures and hybrid architectures in realizing high-power, high-energy aqueous asymmetric supercapacitors.

3.2.2.3 Redox-Active Electrode–Based Nonaqueous Asymmetric Supercapacitors

These represent a significant advancement in energy storage by combining the high-voltage window of nonaqueous electrolytes (e.g., organic solvents, ionic liquids) with asymmetric configurations that integrate EDLC-type and Faradaic electrodes. Unlike aqueous systems, nonaqueous systems offer a wider electrochemical stability window (up to 3.5–4.5 V) and thermal stability (–30 to 70°C), enabling higher energy densities [13]. A typical Li-ion capacitor pairs a Faradaic-type anode (e.g., $Li_4Ti_5O_{12}$, WO_2, graphite, or Li_3VO_4) with an EDLC-type cathode like activated carbon in a $LiPF_6$- or $LiClO_4$-based organic electrolyte [21]. Faradaic charge storage through Li-ion intercalation at the anode is balanced by fast ion adsorption at the cathode, effectively merging the high power density of supercapacitors and the high energy density of batteries. Innovative pre-lithiation strategies, including sacrificial lithium salts such as Li_2DHBN, have resolved lithium-source limitations in graphite-based systems. Notably, gel polymer electrolytes like poly(vinylidene difluoride-co-hexafluoropropylene) [P(VDF-HFP)] have emerged to enhance safety while maintaining rapid charge–discharge capability and long-term cycling, achieving energy densities exceeding 100 Wh/kg [22].

Sodium ion capacitors, inspired by their Li-ion counterparts, utilize low-cost, earth-abundant sodium and similar intercalation chemistry. Despite the challenge of sodium's poor intercalation into graphite, transition metal oxides such as α-V_2O_5 and disordered carbon structures have shown promise as negative electrodes [23]. In these systems, Na^+ ions intercalate into redox-active hosts, while ClO_4^- ions adsorb onto porous carbon cathodes. Enhanced by conductive networks like CNTs or graphene, these hybrid systems offer fast ion transport and pseudocapacitive behavior. The use of gel polymer electrolytes in quasi-solid-state Na-ion capacitors addresses the safety issues associated with organic electrolytes, enabling devices with high-voltage windows (up to 4.2 V), excellent cycling stability, and energy densities approaching 168 Wh/kg [24]. These hybrid systems exemplify how strategic electrode–electrolyte design in nonaqueous asymmetric configurations can significantly advance next-generation energy storage technologies.

3.2.2.4 Redox Electrodes-Based Hybrid Asymmetric Supercapacitors

Redox-electrode hybrid asymmetric supercapacitors (HASCs) combine the rapid electrostatic charge storage characteristic of EDLCs with the enhanced energy density provided by redox-active materials, thus overcoming the energy limitations inherent in symmetric supercapacitor designs [13]. In a typical HASC configuration, a carbon-based EDLC negative electrode—such as activated carbon, reduced graphene oxide, or carbon nanotubes—operates via ion adsorption mechanisms. The positive electrode, in contrast, utilizes transition metal oxides or sulfides that undergo fast, reversible Faradaic reactions. For instance, manganese dioxide (MnO_2), which exploits the Mn^{4+}/Mn^{3+} redox couple, is capable of achieving theoretical capacitances exceeding 1,000 F/g, along with practical energy densities of approximately 21 Wh/kg, while maintaining over 94% capacitance retention after numerous cycles in neutral aqueous electrolytes [14]. NiO nanoflakes supported by graphene hydrogels can expand the operating voltage to 1.7 V and yield areal capacities of up to 1.2 C/cm^2 under high current densities, with capacity enhancements observed over extended cycles due to the ongoing activation of NiO sites [25].

Mixed-metal spinel compounds, such as $NiCo_2O_4$, which can be synthesized into hierarchical nanowires or mesoporous films, exhibit enhanced electrical conductivity and facilitate multi-electron redox reactions, achieving specific capacitances exceeding 1,000 F/g with stable operation up to 1.6 V in alkaline environments. Additionally, sulfide-based materials like $NiCo_2S_4$ on nitrogen-doped carbon foams demonstrate superior redox chemistry and significantly enhanced electronic conductivity, delivering capacities greater than 600 C/g at high discharge rates, effectively outperforming their oxide equivalents [26].

Furthermore, cobalt oxide/carbon nitride oxide asymmetric systems enhanced with redox-active electrolytes achieve energy densities exceeding 40 Wh/kg alongside power densities in the range of tens of kW/kg when operated in aqueous KOH solutions, thereby underscoring the advantages of synergistic electrode–electrolyte designs [27]. The incorporation of conductive polymers, such as polypyrrole coatings on metal oxide structures, further elevates electrode conductivity and results in specific capacities surpassing 1,500 C/g while maintaining stable cycling behavior around 1.8 V. Additionally, the utilization of gel and quasi-solid-state electrolytes formulated from polymer matrices addresses leakage issues and extends cell operating voltages to approximately 2.0 V, facilitating energy densities up to 70 Wh/kg while sustaining performance over extended cycling periods about 1,000 cycles [28]. The deliberate implementation of redox mediator additives, such as iodide/triiodide pairs, enhances overall capacitance by engaging both the electrodes and electrolyte in charge storage mechanisms [29]. Achieving a balance in charge storage capabilities between the positive and negative electrodes through precise mass loading and alignment of potential windows is essential for reducing polarization effects and ensuring longevity in cycle performance about 10,000 cycles [30]. Last, advancements in cell architecture—including 3D carbon foams and flexible fiber and textile electrodes—pave the way for the development of wearable, high-power HASCs, thereby highlighting the versatile potential of redox-electrode HASCs in the future of energy storage technology [31].

3.2.2.5 Other Asymmetric Supercapacitors

Asymmetric supercapacitors are broadly classified into three main types—EDLC-based configurations, redox-functional group–based systems, and surface charge–optimized devices—each leveraging distinct mechanisms to extend the operating voltage and enhance energy density [32]. EDLC-based asymmetric cells achieve higher voltages by balancing the capacitances and potential ranges of two carbon electrodes with unequal mass loadings [33]. Pseudocapacitive systems incorporate redox-active surface groups (e.g., quinones, hydroxyls) to introduce Faradaic charge storage, thereby boosting capacitance beyond purely electrostatic behavior [34]. Surface charge optimization via controlled electrochemical charge injection (ECI) repositions the potential of zero charge (E_0V), enabling stable cell voltages up to 4.5 V without electrolyte decomposition [35]. Collectively, these strategies have yielded asymmetric supercapacitors with energy densities improved by more than 30% and cycle lives exceeding 10,000 cycles [36].

3.2.2.5.1 EDLC-Based Asymmetric Supercapacitors

EDLC-based asymmetric supercapacitors employ two carbon electrodes with deliberately mismatched capacitances (C_p and C_n) and mass loadings to satisfy the charge balance $Q = C_pE_p = C_nE_n$, thereby extending the full-cell voltage window [32]. The operating voltage (E_{device}) can be tailored via the capacitance ratio C_p/C_n, allowing organic-electrolyte cells to function safely beyond 2.5 V [33]. For instance, pairing hierarchical carbon (CS-P) as the positive electrode with activated carbon as the negative electrode yielded a specific energy of 27.95 Wh/kg at 624.96 W/kg, demonstrating the efficacy of capacitance ratio optimization [37]. Moreover, selection of ionic liquids such as N-butyl-N-methylpyrrolidinium ($Pyr14^+$) with bis(trifluoromethylsulfonyl)imide ($TFSI^-$) has extended the stable voltage window up to 3.5 V by minimizing solvent–ion interactions at the carbon interface [32, 33].

3.2.2.5.2 Redox-Functional-Group–Based Asymmetric Supercapacitors

Pseudocapacitive asymmetric supercapacitors integrate carbon electrodes bearing redox-active surface groups (e.g., quinone/hydroquinone, phenolic hydroxyl, carboxyl) to introduce reversible Faradaic reactions in addition to EDLC behavior [38].

The activated carbons with differing oxygen functionalities (designated Aox and Box) produce distinct redox peaks, enabling an aqueous asymmetric cell to operate stably up to 1.6 V and achieve an energy density of ~30 Wh/kg. Mass balancing of these electrodes further optimized the potential windows and capacitance, resulting in >90% retention of initial capacitance after 10,000 cycles in neutral aqueous electrolyte. Beyond carbon, incorporation of conductive polymers (e.g., polyaniline) and transition metal oxides (e.g., MnO_2, RuO_2) as pseudocapacitive components has driven energy densities above 50 Wh/kg in hybrid architectures [39].

3.2.2.5.3 Surface Charge Optimization in Asymmetric Supercapacitors

Surface charge optimization techniques focus on shifting the potential of zero charge (E_0V) of carbon electrodes to maximize the usable voltage window of the full cell [35]. Controlled ECI can selectively modify surface functionalities, relocating E_0V to more favorable positions and enabling organic-electrolyte cells to reach voltages of up to 4.5 V without triggering parasitic reactions [13]. By reducing polarization asymmetry between the two electrodes, ECI enhances Coulombic efficiency and prevents one electrode from exceeding its thermodynamic stability limit during cycling [32]. Similar strategies applied in aqueous systems have achieved voltage windows exceeding 2.2 V and demonstrated superior long-term cycling stability in neutral electrolytes [40].

3.2.3 Based on Electrode Assembly

Supercapacitors are typically classified into symmetric and asymmetric categories based on their electrode configuration.

3.2.3.1 Symmetric Supercapacitor

Symmetrical supercapacitors consist of positive and negative electrodes made from the same material, typically activated carbon [41]. The active materials in a symmetric supercapacitor's positive and negative electrodes are the same, with equal mass loading ($m+ = m-$) [6]. Generally, materials used in EDLC or those contributing to EDLC and Faradaic capacitance are utilized in constructing symmetric supercapacitors. Symmetric supercapacitors that solely rely on Faradaic materials are seldom used because redox reactions are restricted to either positive or negative voltages [42]. However, as indicated in Figure 3.3, symmetric and asymmetric supercapacitors can be combined with different electrode materials.

3.2.3.2 Asymmetric Supercapacitor

On the other hand, asymmetric supercapacitors incorporate two types of active electrode materials within the same electrolyte, one being EDLC and the other pseudocapacitive [39]. In this supercapacitor, both electrode materials have identical active materials but differing material masses ($m+ \neq m-$), requiring careful optimization of the mass ratio between the negative and positive electrodes, even in EDLC-based devices. This occurs because the electrolyte's hydrated cations and anions have different sizes, which alters their accessible surface area [3]. The reduction directly improves the size and kinetics of hydrated ions. Therefore, using the same mass for

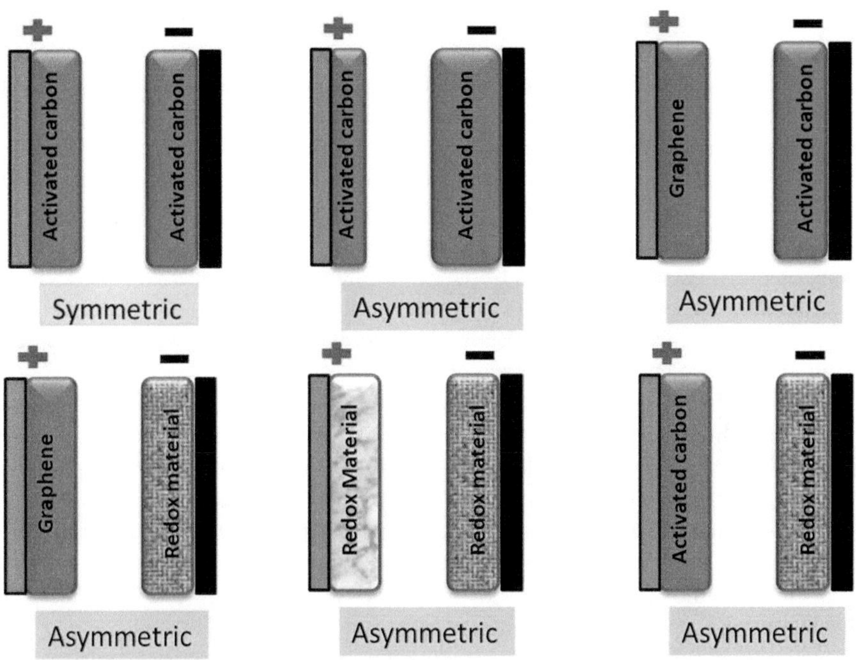

FIGURE 3.3 Symmetric and asymmetric supercapacitor.

both electrode materials could prevent the EDLC-based supercapacitor from achieving its maximum effective capacitance capability [40]. Thus, the mass loading of both negative and positive electrodes must be matched to maximize the device's specific capacitance and energy.

3.3 CONCLUSION

The continuous advancement of supercapacitors highlights their pivotal role in bridging the divide between traditional capacitors and batteries, achieving an admirable balance of high power density, extended cycle life, and rapid charge–discharge capabilities. This chapter systematically examines the various classifications of supercapacitors from three critical perspectives: charge storage mechanisms, material advancements, and electrode assembly architectures. Through an in-depth analysis of EDLCs, pseudocapacitors, and hybrid systems—including symmetric, asymmetric, and redox-active configurations—we articulate the diverse and adaptable frameworks of contemporary energy storage technologies. The spectrum of electrode materials, encompassing high surface–area carbon nanostructures, redox-active transition metal oxides, sulfides, and advanced conductive polymers, has significantly accelerated the development of next-generation energy storage devices. When strategically arranged into optimized assemblies, these materials facilitate remarkable improvements in energy and power densities, voltage stability, rate capabilities, and longevity. Innovations such as hierarchical nanostructures, flexible substrates, and engineered redox-active electrolytes further enhance electrochemical performance and broaden applications to include miniaturized devices, wearables, and grid-scale storage solutions. The adoption of gel-based and quasi-solid-state electrolytes has also contributed to the evolution of supercapacitors into safer, more reliable, and compact systems, paving the way for innovations in flexible electronics and smart textiles. As the energy storage landscape becomes increasingly complex, supercapacitors are recognized not merely as supplementary systems but as integral components of hybrid energy storage frameworks. Their versatile characteristics allow for seamless integration with batteries, fuel cells, and renewable energy sources, making them indispensable for meeting both immediate high-power demands and sustained performance over extended periods.

Looking ahead, the collaboration between nanotechnology, material science, and engineering practices is poised to drive further advancements in this field. Future research aimed at optimizing electrode–electrolyte interactions, improving charge balance, and creating cost-effective, scalable manufacturing techniques will be crucial in realizing the full commercial potential of supercapacitor technology. This chapter serves not only as a reflection of the current technological landscape but also as a strategic guide for researchers and engineers seeking to push the boundaries of energy storage capabilities through innovative supercapacitor solutions.

3.4 AUTHOR AI DECLARATION

The authors hereby declare the use of generative AI and AI-assisted technologies in the writing process. While preparing this work, the authors used the ChatGPT tool/service to rephrase several statements in the manuscript. All AI-generated content was reviewed and edited by the authors as needed. The authors take full responsibility for the content of the publication.

ACKNOWLEDGMENT

The Editors acknowledge the use of Grammarly as a tool to assist in editing this book chapter. The suggestions made by Grammarly were carefully considered and incorporated where appropriate, but the author is solely responsible for the final content.

REFERENCES

1. Sahu, Ravi K., Sandip Gangil, Vinod K. Bhargav, Parmanand Sahu, and Bhupendra Ghritalahre. "Synthesizing Biomass into Nano Carbon for Use in High-Performance Supercapacitors – A Brief Critical Review." *Journal of Energy Storage* 72(2023): 108348. doi: 10.1016/j.est.2023.108348.
2. Banerjee, Soma, Bablu Mordina, Prerna Sinha, and Kamal K. Kar. "Recent Advancement of Supercapacitors: A Current Era of Supercapacitor Devices through the Development of Electrical Double Layer, Pseudo and Their Hybrid Supercapacitor Electrodes." *Journal of Energy Storage* 108(2025): 115075. doi.org/10.1016/j.est.2024.115075.
3. Sharma, Swati, and Prakash Chand. "Supercapacitor and Electrochemical Techniques: A Brief Review." *Results in Chemistry* 5(2023): 100885. doi.org/10.1016/j.rechem.2023.100885.

4. Zhang, Li, and X. S. Zhao. "Carbon-Based Materials as Supercapacitor Electrodes." *Chemical Society Reviews* 38(2009): 2520–2531. doi: 10.1039/b813846j.
5. Yan, Jun, Qian Wang, Tong Wei, and Zhuangjun Fan. "Recent Advances in Design and Fabrication of Electrochemical Supercapacitors with High Energy Densities." *Advanced Energy Materials* 4(2014): 1300816. doi: 10.1002/aenm.201300816.
6. Aziz, Md. Abdul, and Syed Shaheen Shah. *Biomass-Based Supercapacitors*. Wiley, 2023.
7. Rawat, Shivam, Rakesh K. Mishra, and Thallada Bhaskar. "Biomass Derived Functional Carbon Materials for Supercapacitor Applications." *Chemosphere* 286(2022): 131961. doi: 10.1016/j.chemosphere.2021.131961.
8. Saini, Sunaina, Prakash Chand, and Aman Joshi. "Biomass Derived Carbon for Supercapacitor Applications: Review." *Journal of Energy Storage* 39(2021): 102646. doi.org/10.1016/j.est.2021.102646.
9. Poonam, Kriti Sharma, Anmol Arora, and S. K. Tripathi. "Review of Supercapacitors: Materials and Devices." *Journal of Energy Storage* 21(2019): 801–825. doi: 10.1016/j.est.2019.01.010.
10. Lokhande, Prasad Eknath, Umesh S. Chavan, and Abhishek Pandey. *Materials and Fabrication Methods for Electrochemical Supercapacitors: Overview*. Vol. 3. Springer Singapore, 2020.
11. Yusin, S. I., and A. G. Bannov. "Synthesis of Composite Electrodes for Supercapacitors Based on Carbon Materials and the Metal Oxide/Metal Hydroxide System." *Protection of Metals and Physical Chemistry of Surfaces* 53(2017): 475–482. doi: 10.1134/S2070205117030261.
12. Rajagopal, Sivakumar, Rameez Pulapparambil Vallikkattil, M. Mohamed Ibrahim, and Dimiter Georgiev Velev. "Electrode Materials for Supercapacitors in Hybrid Electric Vehicles: Challenges and Current Progress." *Condensed Matter* 7(2022): 6. doi: 10.3390/condmat7010006.
13. Shao, Yuanlong, Maher F. El-Kady, Jingyu Sun, Yaogang Li, Qinghong Zhang, Meifang Zhu, Hongzhi Wang, Bruce Dunn, and Richard B. Kaner. "Design and Mechanisms of Asymmetric Supercapacitors." *Chemical Reviews* 118(2018): 9233–9280.
14. Zhang, Jie, Min Gu, and Xi Chen. "Supercapacitors for Renewable Energy Applications: A Review." *Micro and Nano Engineering* 21(2023): 100229. doi: 10.1016/j.mne.2023.100229.
15. Hwang, Jee Y., Maher F. El-Kady, Yue Wang, Lisa Wang, Yuanlong Shao, Kristofer Marsh, Jang M. Ko, and Richard B. Kaner. "Direct Preparation and Processing of Graphene/RuO_2 Nanocomposite Electrodes for High-Performance Capacitive Energy Storage." *Nano Energy* 18(2015): 57–70. doi: 10.1016/j.nanoen.2015.09.009.
16. Lee, Seung Hwan, Jin Hyeon Kim, and Jung Rag Yoon. "Laser Scribed Graphene Cathode for Next Generation of High-Performance Hybrid Supercapacitors." *Scientific Reports* 8(2018): 8179. doi: 10.1038/s41598-018-26503-4.
17. Xu, Yuanyuan, Sheng Yu, Hannah M. Johnson, Yutong Wu, Xiang Liu, Baizeng Fang, and Yi Zhang. "Recent Progress in Electrode Materials for Micro-Supercapacitors." *IScience* 27(2024).doi: https://doi.org/10.1016/j.isci.2024.108786.
18. Alem, Asab Fetene, Ababay Ketema Worku, Delele Worku Ayele, Tessera Alemneh Wubieneh, Alebel abebaw Teshager, Tadele mihret kndie, Bimrew Tamrat Admasu, Minbale Admas Teshager, Addisu Alemayehu Asege, Mehary Dagnew Ambaw, Misganaw Alemu Zeleke, Alemayehu Kifle Shibesh, and Temesgen Atnafu Yemata. "Ag Doped Co_3O_4 Nanoparticles for High-Performance Supercapacitor Application." *Heliyon* 9(2023). doi: 10.1016/j.heliyon.2023.e13286.
19. Guo, Guanlun, Yilong Mei, Xu Chen, Jun Liu, and Wentao Liu. "Performance Degradation Study of $NiCo_2O_4$-Based Asymmetric Supercapacitors." *RSC Advances* 13(2023): 25018–20528. doi: 10.1039/d3ra05013k.

20. Zhang, Nan, Yifan Li, Junyuan Xu, Junjie Li, Bin Wei, Yu Ding, Isilda Amorim, Rajesh Thomas, Sitaramanjeneya Mouli Thalluri, Yuanyue Liu, Guihua Yu, and Lifeng Liu. "High-Performance Flexible Solid-State Asymmetric Supercapacitors Based on Bimetallic Transition Metal Phosphide Nanocrystals." *ACS Nano* 13(2019): 10612–10621. doi: 10.1021/acsnano.9b04810.
21. Zhao, Lichen, Huifang Zhang, and Boxiang Ma. "Formation of Carbon-Incorporated NiO@Co_3O_4 Nanostructures via a Direct Calcination Method and Their Application as Battery-Type Electrodes for Hybrid Supercapacitors." *ACS Omega* 8(2023): 10503–10511. doi: 10.1021/acsomega.3c00254.
22. Zhang, Jintao, Jianwen Jiang, Hongliang Li, and X. S. Zhao. "A High-Performance Asymmetric Supercapacitor Fabricated with Graphene-Based Electrodes." *Energy and Environmental Science* 4(2011): 4009–4015. doi: 10.1039/c1ee01354h.
23. Wang, Faxing, Chun Wang, Yujuan Zhao, Zaichun Liu, Zheng Chang, Lijun Fu, Yusong Zhu, Yuping Wu, and Dongyuan Zhao. "A Quasi-Solid-State Li-Ion Capacitor Based on Porous TiO_2 Hollow Microspheres Wrapped with Graphene Nanosheets." *Small* 12(2016): 6207–6213. doi: 10.1002/smll.201602331.
24. Chen, Zheng, Veronica Augustyn, Xilai Jia, Qiangfeng Xiao, Bruce Dunn, and Yunfeng Lu. "High-Performance Sodium-Ion Pseudocapacitors Based on Hierarchically Porous Nanowire Composites." *ACS Nano* 6(2012): 4319–4327. doi: 10.1021/nn300920e.
25. Wang, Faxing, Xiaowei Wang, Zheng Chang, Xiongwei Wu, Xiang Liu, Lijun Fu, Yusong Zhu, Yuping Wu, and Wei Huang. "A Quasi-Solid-State Sodium-Ion Capacitor with High Energy Density." *Advanced Materials (Deerfield Beach, Fla.)* 27(2015): 6962–6968. doi: 10.1002/adma.201503097.
26. Wang, Meilong, Linsong Li, Zhentao Liu, Fuzhong Wu, Huixin Jin, Yi Wang, and Siyu Cai. "Multicomponent Co_2O_3@$CoMo_2S_4$ Core-Shell Structures as a Binder-Free Electrode for Cycling Stability Supercapacitors." *ACS Omega* 10(2025): 8901–8910. doi: 10.1021/acsomega.4c05172.
27. Sadavar, Shrikant Vaiju, Seul Yi Lee, and Soo Jin Park. "Advancements in Asymmetric Supercapacitors: From Historical Milestones to Challenges and Future Directions." *Advanced Science* 11(2024): 2403172.
28. Dhakal, Ganesh, Debananda Mohapatra, Tensangmu Lama Tamang, Moonyong Lee, Yong Rok Lee, and Jae Jin Shim. "Redox-Additive Electrolyte–Driven Enhancement of the Electrochemical Energy Storage Performance of Asymmetric Co_3O_4//Carbon Nano-Onions Supercapacitors." *Energy* 218(2021): 119436. doi: 10.1016/j.energy.2020.119436.
29. Chodankar, Nilesh R., Deepak P. Dubal, Abhishek C. Lokhande, Amar M. Patil, Jin H. Kim, and Chandrakant D. Lokhande. "An Innovative Concept of Use of Redox-Active Electrolyte in Asymmetric Capacitor Based on MWCNTs/MnO_2 and Fe_2O_3 Thin Films." *Scientific Reports* 6(2016): 39205. doi: 10.1038/srep39205.
30. Hu, Lintong, Lanlan Xue, Jinzhou Fu, Ping Xiao, Tianyou Zhai, and Huiqiao Li. "A Versatile Capacity Balancer for Asymmetric Supercapacitors." *Advanced Energy Materials* 10(2020): 2001608. doi: 10.1002/aenm.202001608.
31. Alagar, Srinivasan, Rajesh Madhuvilakku, Ramalakshmi Mariappan, Chelladurai Karuppiah, Chun Chen Yang, and Shakkthivel Piraman. "Ultra-Stable $Mn_{1-x}Ni_xCO_3$ Nano/Sub-Microspheres Positive Electrodes for High-Performance Solid-State Asymmetric Supercapacitors." *Scientific Reports* 10(2020): 8871. doi: 10.1038/s41598-020-64867-8.
32. Dubey, Prashant, Vishal Shrivastav, Priyanka H. Maheshwari, and Shashank Sundriyal. "Recent Advances in Biomass Derived Activated Carbon Electrodes for Hybrid Electrochemical Capacitor Applications: Challenges and Opportunities." *Carbon* 170(2020a): 1–29.

33. Singh, Neeraj, Virendra Singh, Neeraj Bisht, Puneet Negi, Archana Dhyani, Rajat Kumar Sharma, and B. S. Tewari. "A Comprehensive Review on Supercapacitors: Basics to Recent Advancements." *Journal of Energy Storage* 121(2025): 116498.
34. Czagany, Mate, Szabolcs Hompoth, Anup Kumar Keshri, Niranjan Pandit, Imre Galambos, Zoltan Gacsi, and Peter Baumli. "Supercapacitors: An Efficient Way for Energy Storage Application." *Materials* 17(2024): 702. doi: 10.3390/ma17030702.
35. Gou, Qianzhi, Shuang Zhao, Jiacheng Wang, Meng Li, and Junmin Xue. "Recent Advances on Boosting the Cell Voltage of Aqueous Supercapacitors." *Nano-Micro Letters* 12(2020): 1–22.
36. Arsha, M. S., and V. Biju. "A 3.5 V Supercapacitor with Ultrahigh Energy and Power Capabilities Using Thermally Deoxygenated Graphite Oxide Electrodes and Water-in-Salt Electrolyte." *Energy and Fuels* 38(2024): 19076–19087. doi: 10.1021/acs.energyfuels.4c02984.
37. Samdhyan, Kajal, Prakash Chand, and Hardeep Anand. "High-Performance Asymmetric Supercapacitor Based on Hierarchical Hydrothermally Grown CuS Nanostructures." *Journal of Electroanalytical Chemistry* 968(2024): 118504. https://doi.org/10.1016/j.jelechem.2024.118504.
38. Lakshmi, K. C. Seetha, and Balaraman Vedhanarayanan. "High-Performance Supercapacitors: A Comprehensive Review on Paradigm Shift of Conventional Energy Storage Devices." *Batteries* 9(2023): 202.
39. Fang, Haiqiu, Dezhi Li, Mingyao Zhao, Yang Zhang, Junwei Yang, and Kai Wang. "Research Progress and Prospect of Hybrid Supercapacitors as Boosting the Performance." *Energy Sources, Part A: Recovery, Utilization and Environmental Effects* 46(2024): 5849–5866.
40. Weng, Zhe, Feng Li, Da Wei Wang, Lei Wen, and Hui Ming Cheng. "Controlled Electrochemical Charge Injection to Maximize the Energy Density of Supercapacitors." *Angewandte Chemie – International Edition* 52(2013): 3722–3725. doi: 10.1002/anie.201209259.
41. Surendran, Vishnu, Raveendran S. Arya, Thazhe Veettil Vineesh, Binson Babu, and Manikoth M. Shaijumon. "Engineered Carbon Electrodes for High Performance Capacitive and Hybrid Energy Storage." *Journal of Energy Storage* 35(2021): 102340. doi: https://doi.org/10.1016/j.est.2021.102340.
42. Ariharan, Arjunan, and Sung Kon Kim. "Three-Dimensional Hierarchical Porous Carbons Derived from Betelnut Shells for Supercapacitor Electrodes." *Materials* 14(2021): 7793. doi: 10.3390/ma14247793.

4 Materials for Supercapacitor Electrodes

Paricha Jebin, Md. Rakib Khan, Nipa Debnath, and A. J. Saleh Ahammad

4.1 INTRODUCTION

Growing concern about rising environmental pollution and the rapid depletion of non-renewable resources has driven researchers to develop sustainable energy solutions. Fossil fuels are the most used energy sources today. However, the widespread use of these non-renewable sources leads to the emission of various air pollutants, primarily carbon dioxide, which is directly involved in the greenhouse effect. The burning of petroleum pollutes the air, water, and soil. The urgency of these concerns highlights the importance of sustainable energy development from renewable resources. Renewable energy, instead of non-renewable sources, is a better way to meet energy and power demands. However, this is not straightforward due to the limitations of renewable resources. For example, solar energy is generated during the day when the sun is shining, while wind energy is not consistently available.

Additionally, energy production from renewable sources fluctuates, making it challenging to utilize the generated energy consistently. Consequently, developing a new renewable energy storage device capable of storing and supplying energy is essential. Among various electrochemical devices, batteries, fuel cells, and supercapacitors (SCs) are eco-friendly and efficient. The comparison between them is shown in Table 4.1. That comparison indicates that batteries and fuel cells have higher energy density (ED) values, but their energy delivery time is slow. In contrast, capacitors can deliver charge quickly, but their storage capacity is low. Furthermore, SCs exhibit high cyclic performance, rapid charging, and greater power values, yet they lack ED. Due to these advantages over batteries and fuel cells, SCs significantly attract researchers' attention. SCs are favored for applications that require high power density (PD) and function well when integrated with batteries and fuel cells. This is why their application has been restricted. Therefore, researchers are concentrating on developing innovative materials that can enhance energy storage and maintain the specific power of SCs.

Among various factors, electrode materials play a crucial role in the charging and discharging. Based on the charging–discharging mechanism, SCs are categorized into three groups: electrical double layer capacitors (EDLCs), pseudocapacitors (PCs), and hybrid supercapacitors (HSCs), which combine both EDLC and PC characteristics. The various types of materials used in SCs are illustrated in Figure 4.1.

 DOI: 10.1201/9781003535836-4

TABLE 4.1
Comparison between Capacitor, Supercapacitor, and Battery

Properties	Capacitor	Supercapacitor	Battery
Electrode material	Two conducting metal plates are used	Carbon materials, CPs, TMOs, and composites are used as electrode materials	Hazardous materials such as Li, Ni–Cd, and nickel metal hybrid lead compounds are used
Charge storage	Stores charge in the form of an electric field between metal plates	Charge is stored as EDL capacitance, pseudocapacitance, or both	Batteries undergo chemical reactions to store energy
Cyclic stability	Moderate	High	Due to electrochemical reactions electrode degradation occurs over time
Capacitance	Low	High	Very low
Energy density	Low	Moderate	High
Power density	High	Very high	Moderate
Costing	Cheaper	Expensive	Cheaper

EDLCs utilize a double layer to store charge at the interface, while PCs involve the faradaic process [1].

Several factors influence the overall performance of SCs, including the selection of materials such as electrodes, separators, current collectors, and electrolytes. This chapter focuses on various electrode materials, including carbon-based, transition metal oxides (TMOs)-based, conducting polymer-based, and composite materials, discussing their advantages and disadvantages. Additionally, modern materials such

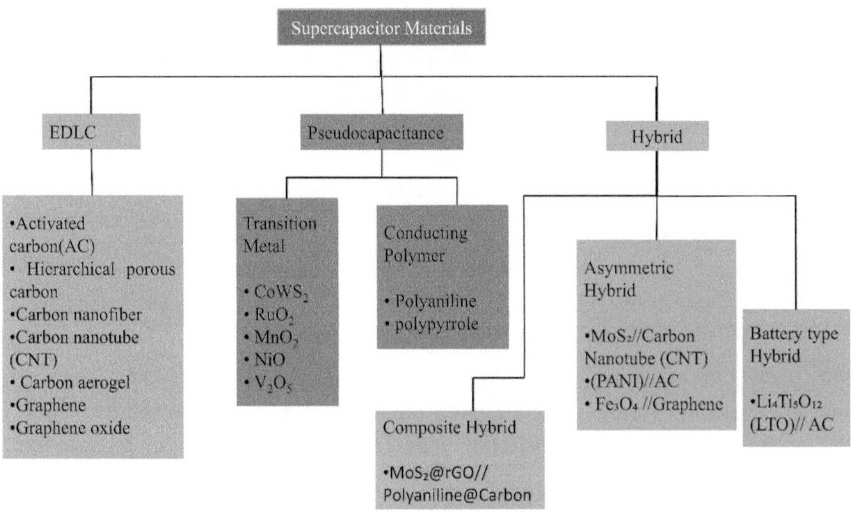

FIGURE 4.1 Classification of electrode material for SCs.

as metal–organic frameworks (MOFs), covalent organic frameworks (COFs), and MXenes are summarized. Alongside the examination of different electrode materials, this chapter also addresses recent challenges and future perspectives aimed at enhancing the properties of SCs.

4.2 BASIC COMPONENTS OF SCs

SCs are composed of four essential components: the current collector, electrodes, electrolyte, and separator. The electrodes are vital for charge storage, whereas electrolytes provide the ions needed for diffusion. Separators help prevent short circuits within SCs, and the current collector aids in electron transfer from the electrodes. Below is a brief overview of these components.

4.2.1 Electrodes

The primary component of the SC is the electrode, which converts energy through an electrochemical reaction. The electrodes are immersed in an electrolyte and separated by a separator. Separators prevent short circuits while allowing ion movement to occur. When a voltage is applied to the cell, mass transfer occurs, and charge is stored efficiently, enabling rapid energy release. SC electrodes are typically made from materials with ideal characteristics, such as high conductivity, a large surface area, and excellent electrochemical stability [1]. AC [2], graphene oxide (GO) [3], carbon nanotubes (CNTs) [4], and TMOs [5] are the choice of electrode material that directly impacts the values of capacitance (C), PD, ED, and cyclic performance of the device. In some cases, composite electrodes are used to enhance the performance of devices composed of multiple materials. The electrode exhibits several key characteristics, including a large surface area, a higher number of reactive sites, chemical and thermal stability, a porous structure, high conductivity, good adsorption–desorption properties, and excellent redox activity.

Materials vary in surface area, with highly porous types offering a larger surface area. Depending on pore size, materials can be categorized into three types: micropores, mesopores, and macropores. Among these, mesoporous structures are the most desirable for ion transport [1]. The porous structure provides extensive space for ion accumulation, thereby increasing charge storage capacity through the adsorption of ions into the pores. Therefore, porosity enhances the electrochemical performance of the SC device by increasing the SSA.

4.2.2 Electrolytes

The electrolyte plays a crucial role in energy storage technology. The electrolyte ion is essential for maintaining charge between electrodes. Therefore, the types of electrolytes significantly influence the interactions at the electrode–electrolyte interfaces, ultimately affecting the overall performance of the devices. Three types of electrolytes are used for SCs: ionic, organic, and aqueous electrolytes.

The properties and performance of various electrolytes have been extensively analyzed in numerous research studies. Among them, aqueous electrolytes operate at a lower operating voltage because the water electrolysis has a ΔV of 1.23 V. The advantages of aqueous electrolytes include cost-effectiveness, higher electrical conductivity, and a large capacitance value, but they exhibit a lower ED. In contrast, the other two types of electrolytes can function at a wider voltage range, albeit with lower conductivity. Handling organic electrolytes is challenging, and they may present toxicity concerns. Additionally, ionic electrolytes cover only a small surface area at the electrode due to their higher cost and the presence of larger ions [6].

4.2.3 Separators

An additional vital component of SCs is the separator, which acts as a barrier between the electrodes while facilitating ion transport. Its main function is to prevent short circuits in SCs. The separator is chemically inert and does not engage in electrochemical reactions. Key characteristics of separators in SCs should include being non-conductive, exhibiting very low ionic resistance, enabling quick ion transfer for charge–discharge cycles, offering mechanical strength to avoid damage, and ensuring electrochemical and thermal stability.

The choice of separator greatly influences the internal resistance, PD, and electrochemical performance of SCs. Hence, selecting the appropriate separator is crucial for enhancing energy storage efficiency and guaranteeing long-term durability. Options include materials like glass, paper, polymeric membranes (such as polypropylene and polyethylene), cellulose-based substances, gel polymer electrolytes, and ceramic-coated materials.

4.2.4 Current Collector

The fourth component of SCs is the current collector, which facilitates the transfer of electrons between the device and the external circuit. It must have high electrical conductivity to minimize resistance for efficient electron transport. Given the wide variety of materials used in SCs, the current collector must also possess excellent corrosion resistance. Furthermore, since it provides the entire structural support to the cell, its mechanical strength must be high. Commonly used materials for current collectors include aluminum, iron, and steel alloys. Recently, materials such as metal meshes, foams, and carbon cloths have been frequently employed as current collectors due to their large surface area, enhancing electrode materials' dispersion and reducing contact resistance.

4.3 ELECTRODE MATERIALS

The electrode materials of SCs are of two types based on their charge storage mechanism, which are (a) EDLC-type material and (b) pseudocapacitive-type electrode material.

EDLC-type electrode materials store charges by the electrical double layers in a non-faradaic manner. The ideal EDLC-type materials must have a large specific surface area with a tunable porous structure [7]. EDLC behavior is exhibited mostly by carbon-based materials, such as CNTs [8], hierarchically porous carbon [9], GO and reduced graphene oxide (rGO) [3], carbon nanofibers [10], and activated carbon (AC) [2]. Nanoscale carbons are primarily used due to their large surface area, highly porous network, lower charge transfer resistance, and high electrical conductivity. Since the maximum carbon materials are obtained from the environment, their allotropes are both eco-friendly and corrosion-resistant [1]. All these materials are discussed briefly in this chapter. On the contrary, pseudocapacitive-type materials store charges by faradaic processes. Here, redox reactions occur at the electrode–electrolyte interface. Since transition metals exhibit variable oxidation states, their oxides, hydroxides, sulfides, or other compounds behave as PC-type materials. Like the metals, the conducting polymer shows pseudocapacitive behavior. Since doping can enhance the active site of the polymer, it enhances the overall performance of the polymer. Combining heteroatoms and polymers results in a high surface area for charge storage, indicating an improvement in both the materials' capacitance and specific energy [11–14]. Among various metal compounds, metal oxides demonstrate the most promising properties for SCs, as previously reported [15]. The classification of electrode materials for SCs is illustrated in Figure 4.1.

While these materials have numerous benefits, they also present some drawbacks. Carbonaceous materials provide extensive active sites and excellent electrical conductivity, yet they exhibit low capacitance values. Transition metal compounds, apart from RuO_2, suffer from poor cyclic stability and limited conductivity, despite showing higher capacitance. Conducting polymers undergo swelling and contraction during charging and discharging, resulting in degradation after several cycles. A promising approach to address these issues is the utilization of composite materials that fuse both EDLC and pseudocapacitive components. Below, each SC material is outlined, highlighting its advantages and disadvantages.

4.3.1 Carbon-Based Electrode Material

Due to their unique properties, such as a highly porous structure, higher conductivity, light weight, and chemical inertness, carbon-based electrode materials are widely used in the field of energy storage. It has been widely used in many applications. Besides being used as electrode materials, it is also used in civil infrastructures, chemical, automobile, food industries, purifying water, etc. [1]. The commercially available carbon is expensive and not eco-friendly because it is obtained from fossil fuel precursors and non-renewable sources. However, natural carbon sources, such as biomass, cooking waste, and microorganisms, attract researchers due to their natural abundance, cost-effectiveness, and highly porous structures. Since pure carbon materials have low capacitance, various heteroatoms such as oxygen, nitrogen, and sulfur are doped to enhance capacitance. Different types of carbon materials, some of which are discussed below, are widely used in SCs. This chapter briefly discusses activated carbon, hierarchical porous carbon, carbon aerogel, Carbon nanotubes (CNT), carbon nanofiber, graphene, and rGO.

4.3.1.1 Activated Carbon

Among different types of carbon materials, the most commonly used material in energy storage devices is AC. Due to its low cost, high abundance, renewability, and eco-friendliness, AC is mainly obtained from biomass precursors. Today, primary sources of carbon production include plant and animal waste, microorganisms, forestry residues, and agricultural waste. Activated carbon materials are produced via carbonization or hydrothermal carbonization, followed by an activation process [1, 16]. Physical, chemical, or self-activation methods do it [1, 17]. The charge is stored electrostatically with the bi-layer formation at the electrode–electrolyte interface. Mostly, the charge storage mechanism depends on the surface area of the materials. The higher the specific surface area, the higher the capacitance of the prepared material. The time of activation is very important. Large activation causes low conductivity due to a large pore size [1]. It also lowers the specific energy and power of the SCs. However, the overall performance of SCs depends on several factors such as the size and shape of the pore, either micro (<2 nm)/meso (2–50 nm)/microporous (2–50 nm), the specific surface area of the electrode, the functional group attached with the surface, etc. The different pore sizes are optimized in different electrode materials to gain maximum electrochemical performance.

T. Islam et al. [7] reported an activated porous coal carbon obtained from the coal of the Boropukuria coal mine in Bangladesh. The activated carbon shows high electrochemical performance for SCs. Since they contain not only active porous structures but also heteroatoms like oxygen and nitrogen, both EDLC and PC have contributed to the overall capacitance of these materials. The values of ED and PD for these materials are 39.04 Wh/kg and 1237.5 W/kg, respectively. The high performance of these materials is attributed to the lower value of charge transfer resistance, facilitated by a porous network that enables the diffusion of ions at the interface. The large surface area, highly porous structure, and the pseudocapacitive behavior of the heteroatoms O and N play a vital role in enhancing the overall performance of the SCs. Here, charge transfer occurs through both the redox process and electron-donating heteroatoms within the ACCN's structure.

A. Adan-Mas et al. [2] investigated the coffee-derived material in 1 M Na_2SO_4 after chemical activation and derived by coffee-ash carbon in high-performance SCs. Activated carbon is produced from a byproduct. This activated carbon features a highly porous and disordered structure with reduced graphitization levels, offering a substantial surface area. This configuration minimizes the distance for ion diffusion at the interface, thus improving the performance of these materials.

4.3.1.2 Hierarchical Porous Carbon

Due to its unique morphology, ease of synthesis, and well-developed porosity, hierarchical porous carbon is widely used as a superior electrode material for SCs [1]. The pore structure is classified into four categories: macropores, mesopores, micropores, and a localized graphitic structure. The macropores serve as a reservoir for ion buffering, mesopores facilitate rapid ion transfer, micropores enhance charge storage, and the graphitic structure reduces charge transfer resistance [1].

Zhimin Chen et al. [9] reported rice husk-based hierarchical porous carbon (RHPC) having macropores (from natural vascular bundles) that act as ion buffers, larger mesopores (3–50 nm) formed during carbonization, and micropores and small mesopores (2–3 nm) developed during activation. Lignin adds to the formation of micropores and mesopores, cellulose boosts conductivity, and hemicellulose reduces resistance, collectively making RHPC an excellent choice for energy storage applications. The presence of micropores offers more adsorption sites for ions from electrolytes, resulting in increased capacitance values of the material. Additionally, mesopores connected with macropores shorten the ion diffusion path and improve the cyclic performance of RHPC.

4.3.1.3 Carbon Aerogel

Among various carbon materials, these are highly porous and lightweight. Carbon aerogel exhibits lower Rct values and boasts a large specific surface area and high conductivity, making it an ideal material for SCs, batteries, and sensors. It is primarily derived from resorcinol and formaldehyde through polycondensation [1]. Mesopores require pyrolysis under N_2. Unique properties of carbon aerogel include ultra-low density, high surface area, good electrical conductivity, and electrochemical stability. Lei E et al. [18] reported a carbon aerogel material that is nitrogen-doped. The biomass precursors are cellulose in that case, and a fine interconnected structure is obtained that provides large active sites. Both combinations of EDLC and pseudocapacitance due to heteroatoms exhibit excellent electrochemical performance. The porous 3D percolation network of the aerogel is crucial in this context.

4.3.1.4 CNTs

CNTs are also a promising material for SCs. CNT is classified as single-walled and multi-walled. G.AM. Ali et al. [4] show that the combination of MWCNT with ferrocene results in impressive performance for SCs. Electrochemical impedance spectroscopy (EIS) data reveal that these composites exhibit low charge transfer resistance, indicating efficient electron transfer. The redox-active ferrocene and MWCNT relationship creates a substantial active surface area. This suggests a strong interaction between the electrode and electrolyte, thereby enhancing the material's conductivity.

Additionally, CNT can be integrated with 3D graphene, leading to enhanced capacitance performance [19]. These composites' unique, interconnected structure is the primary reason for enhancing conductivity compared to pure 3D graphene. The introduction of MWCNT in 3DG fills all the vacancies on the graphene surface, facilitating electron transfer, enhancing effective contact between the electrode and electrolyte, and reducing the diffusion length between ions. CNT also prevents graphene from agglomerating and maintains the interconnected structure, thereby enhancing the overall performance of SCs. The CNT can also form composites with microporous polymer [8]. The tetrabenzonapthalene-based polymer with CNT shows high capacitive performance. The excellent performance of these nanocomposites is attributed to their high porosity, rapid mass transfer, and high conductivity. The combination of Ni–Co–Mn oxides with

CNT also shows high electrochemical activity [20]. The large surface area, conductive network structure, and increased number of active sites make these materials efficient for SCs.

4.3.1.5 Carbon Nanofiber

Carbon nanofiber shows promising behavior toward high-performance SCs. Previous reports have found that carbon nanofibers show high capacitive performance with high energy and specific power. It has been found that carbon nanofibers with a core–shell structure exhibit a high specific capacitance value. The higher surface area is the primary reason for these high capacitance values, providing a longer discharge time, as the galvanostatic charge-discharge (GCD) curve indicates. The optimized core–shell structure provides a lower value of charge transfer resistance. It also facilitates faster electron transfer. The core–shell-structured carbon fiber can be prepared using various processes, such as electrospinning and hydrothermal growth. In that case, the carbon nanofibers were grown in nickel hydroxide and nickel oxide(CHO), and five different metal sulfides were utilized in a hydrothermal growth process. The optimized carbon nanofiber CHO/NiS-3h shows a PD of 800 W/Kg with a specific energy of 37.97 Wh/Kg. Mesoporous carbon nanofibers also show high capacitive performance [21].

K. Wang et al. [22] used mesoporous carbon nanofiber that was prepared by confined self-assembly and shows a high supercapacitive performance. Due to its well-controlled pore size, faster electron transfer, and a small diffusion distance, the outer surface mesoporous opening provides a high specific active surface area. It gives a higher value of both specific power and energy. This also proved that mesoporous carbon nanofibers showed better capacitive performance than mesoporous carbon, which makes the mesoporous carbon nanofibers an efficient material for SCs. The specific capacitance of porous carbon nanofiber materials is also enhanced by using a polymer like polyacrylonitrile [10]. Here, PAN-based porous carbon nanofibers were prepared through a carbonization process. The high PD exhibited by PAN-based carbon nanofibers is attributed to their unique structure, which provides a higher number of active sites and a significant EDLC contribution. This material also shows pseudocapacitance behavior. PAN-based carbon nanofiber shows high ED in both aqueous and organic electrolytes. This is because they have nitrogenated and oxygenated functional groups in the active surface moieties.

Finally, the open mesopores in the active surface enhance ion diffusion and facilitate interfacial charge transfer by increasing electrical conductivity, thereby enhancing the supercapacitive performance. The supercapacitive performance of the polymer-based carbon nanofibers is further enhanced by doping another atom. A previous study showed that PAN/polyvinylpyrrolidone (PVP)-based carbon nanofibers can show higher supercapacitive performance when they contain a higher nitrogen content [23]. The PAN and PVP mixture was used to form the shell, and poly(styrene-co-acrylonitrile) was employed as the core in this structure. The optimized carbon nanofiber exhibits a very high PD of 14,000 W/kg with an ED of 4.12 Wh/Kg. This is due to their hollow structure, which provides a high active surface area, more facile electron transfer, more

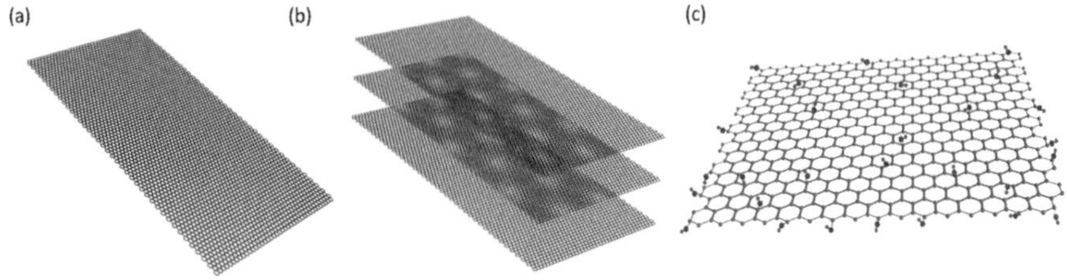

FIGURE 4.2 Schematic representation of (a) single-layer graphene, (b) multiple-layer graphene, and (c) rGO.

crystalline structure, etc. The doping of nitrogen also significantly enhances the pseudocapacitive behavior. The structure provides greater chemical stability to these electrode materials, facilitating the overall electrochemical performance of the SCs.

4.3.1.6 Graphene Oxide and Reduced Graphene Oxide

GO and reduced GO are commonly used carbon-based materials for SC electrodes. The schematic diagrams of GO and rGO are shown in Figure 4.2. These electrode materials' cost-effectiveness and high stability have attracted significant attention for SCs. B. Khademi et al. [3] showed that rGO/GO can be an electrode for binder-free HSCs, where 'water in salt' was used as the electrolyte. The reduced GO provides larger active sites, and the functional groups of GO participate in the charge storage systems. The use of non-conductive adhesives is eliminated by using the GO, which enhances the materials' hydrophilicity. This facilitates the diffusion of ions on the active surface of rGO, which is hydrophobic, improving the energy storage capabilities. Sometimes, the reduced GO composites can provide high-performance electrode materials. When rGO forms composites with sulfur, it provides high supercapacitive performance [24]. The presence of the functional group and sulfur doping provides more active sites. It causes more redox reactions on the electrode surface and enhances the electrical conductivity. The doping of nitrogen also enhances the electrochemical performance of the GO/rGO-based SC [25]. It demonstrates that nitrogen-doped reduced GO exhibits higher electrochemical performance than chemically rGO. This enhancement is attributed to the heteroatoms promoting electron transfer and increasing the active surface area, which boosts the capacitive behavior of the prepared material. Additionally, the GO aerogel shows supercapacitive performance [26]. This indicates that GO is a promising electrode material for SCs.

4.3.2 Transition Metal-Based Electrode

There are so many transition metal-based electrode materials available for SCs. The oxides, hydroxides, and sulfides of nickel, manganese, ruthenium, vanadium, etc., are commonly used for SCs. However, a few other metal compounds are also available for SCs that are not widely used as electrode materials. Some of them are discussed below.

4.3.2.1 Nickel-Based Electrode

The oxides, hydroxides, and sulfides of nickel exhibit promising electrochemical capacitive behavior. Hence, they are used as an electrode material for SCs. It has been found that NiO exhibits promising electrochemical behavior when combined with cobalt. Figure 4.3(a) shows a solid asymmetric SC with a pre-synthesized $NiCo_2O_4$ mesoporous electrode, which was synthesized without binders. A nickel foam, known for its high conductivity, was selected as the current collector, and the $NiCo_2O_4$ provided a large surface area and a unique mesoporous nanostructure. Figure 4.3(b) shows the specific capacitance of $NiCo_2O_4$ with different numbers of electrodeposition cycles at different current densities. $NiCo_2O_4$ exhibits a superior specific capacitance of 1734.9 F/g at a current density of 2 A/g. Figure 4.3(c) shows the X-ray diffraction (XRD) pattern of $NiCo_2O_4$ nanosheets on Ni foam. The peaks at 44.7°, 52.1°, and 76.5°, corresponding to the (111), (200), and (220) planes, marked with asterisks, originate from the nickel foam [27]. This material exhibits high catalytic performance due to the pseudocapacitive behavior, which is facilitated by the presence of NiO and carbon, providing a large surface area. The rectangular shapes of cyclic voltammogram (CV) suggest that mass transfer increases, high-rate capability, and fast electron transfer. These factors affect the overall capacitance of the SCs. The NiO also formed composites with rGO, which shows a higher power

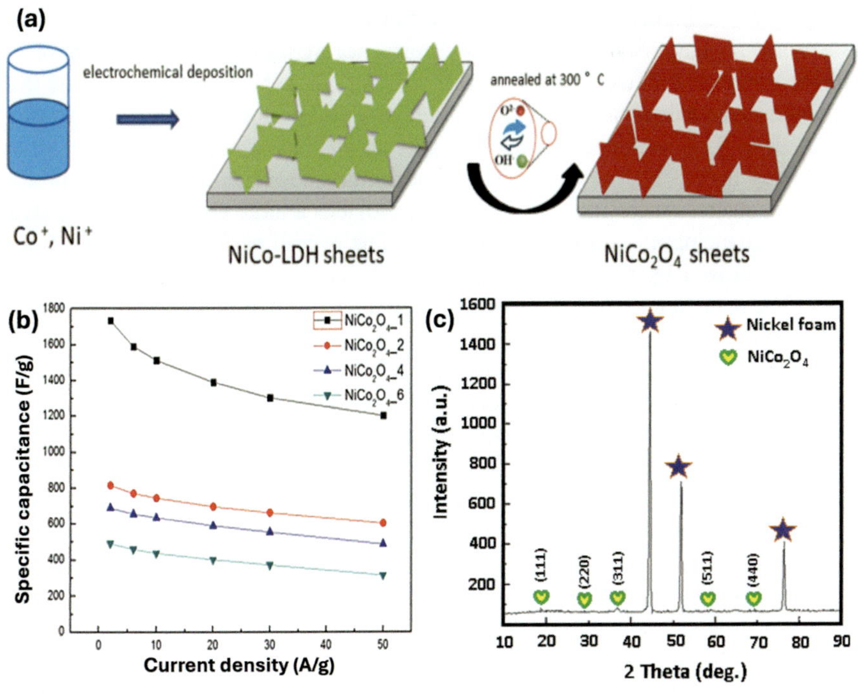

FIGURE 4.3 (a) Synthesis of porous $NiCo_2O_4$ sheets, (b) specific capacitance of $NiCo_2O_4$ with varying numbers of electrodeposition cycles at different CDs, and (c) XRD pattern of $NiCo_2O_4$. Adapted from [27] licensed under CC BY 4.0. (© Nature 2020.)

and ED than individual NiO and rGO [28]. This hybrid SC exhibits values of P and E of 0.43 kW/kg and 5.4 Wh/kg, respectively. The higher capacitance of NiO–rGO nanosheets is due to two reasons: first, the homogenous dispersion of NiO and rGO, and second, the close bonding between them. Reduced graphene shows lower resistance but a higher diffusion rate. On the contrary, the NiO electrode exhibits higher resistance but a lower diffusion rate. When NiO and rGO are combined, the composite exhibits lower resistance than NiO and a higher diffusion rate than rGO, making these electrodes superior to other electrodes of their type. The nickel sulfides also exhibit high catalytic performance.

W. Niu et al. [29] showed that NiS_X exhibits high capacitive behavior when forming composites with carbon materials. Due to its lower diffusion rate and faster charge transfer, the NiS_2-carbon composite electrode exhibits promising electrochemical behavior. The electrochemical performance can be optimized by controlling the preparation temperature and adjusting the sulfur amount. Like nickel oxide and sulfides, nickel hydroxide is a promising material for SCs [30]. The carbon and nickel hydroxide composite structure provides a larger active surface area. It facilitates faster electron transfer and exhibits a lower charge transfer resistance value than the C–Ni composite electrode. The capacitance value of the developed material shows 952 F/g.

4.3.2.2 Ruthenium-Based Electrode

Ruthenium is a highly catalytically active transition metal with compounds acting as a promising electrocatalyst for high-performance SCs. Ruthenium oxide nanoparticles are an excellent electrode material for SCs [15]. B. Nisha et al. [15] show that ruthenium oxide nanoparticles exhibit high catalytic performance for energy storage devices. They synthesize RuO_2 in a greener way using natural plant extract as a reducing agent. The newly prepared ruthenium facilitates electron and proton transfer, enhances the diffusion of ions, and provides a large effective surface area, thereby enhancing its catalytic performance. The overall performance of the RuO_2 can be improved by making a composite with another transition metal or metal oxide. The nanocomposite of tin oxide–ruthenium oxide exhibits higher capacitance performance than the ruthenium oxide nanoparticle [31].

This study utilized carbon cloth as a current collector, leveraging the ionic connection between the composites to improve electronic conductivity at the interface by reducing ionic resistance. Additionally, when combined with RuO_2, tin oxides create a more specific active surface area for intercalation. These factors significantly boost the catalytic activity of these composite materials. Moreover, ruthenium hydroxide is identified as a promising electrode material for SCs. When ruthenium hydroxide aerogel is combined with MWCNT, it demonstrates high capacitive performance [32]. The MWCNT/RuO_2 demonstrates high capacitive performance, primarily due to its structure. The porous configuration, achieved through a 3D architecture, reduces the limiting current via migration. It exhibits a specific capacitance of 420.3 F/g at a current density of 0.5 A/g. The well-ordered 3D structure, decorated with metal oxide, is crucial in achieving this high capacitance. This electrode exhibits elevated specific power and excellent ED, with 8.60 kW/Kg and 36.6 Wh/Kg, respectively. Ruthenium sulfides also exhibit high catalytic performance as an electrode material for SCs, similar to ruthenium oxide and hydroxides. The cubic ruthenium sulfide nanoparticles display strong capacitive performance [33]. The high electroactive surface area and facile electron transfer at the interface make it a highly efficient electrode material for SCs.

4.3.2.3 Manganese-Based Electrode

Like other transition metal oxides, hydroxides, and sulfides, manganese compounds are a standard high-efficiency electrode material for SCs. X. Sun et al. [5] showed that manganese oxide nanoclusters can be a promising electrode material when incorporated with oxidized carbon fiber. The active nanoparticles provide a larger active surface area. It exposed more active sites for ions, which facilitates the electron transfer. The introduction of Mn_3O_4 further enhances the electrical conductivity. Mn_3O_4 nanoflakes provide an active sponge layer that facilitates uniform nucleation and enhances capacitive performance. Manganese oxide nanorods also demonstrate promising potential for high-performance SCs [34]. Its porous nanorod structure significantly contributes to ion diffusion and capacitance. The morphology of this electrode improves ion diffusion from the bulk and facilitates electron transfer, establishing it as a novel material for SCs. Additionally, manganese hydroxides are recognized as promising electrode materials for high-performance SCs. The nickel–manganese-layered double hydroxide is an exceptionally efficient electrode for SCs [35]. The material features a highly porous, honeycomb-like structure with an extensive surface area attributed to the biomass precursors. These cost-effective biomass precursors not only enhance surface area and electrical conductivity but also help prevent the aggregation of NiMn LDH. The NiMn LDH composite with biomass carbon demonstrates high conductivity due to its porous framework, which creates ample active sites that promote rapid electron transfer.

4.3.2.4 Vanadium-Based Electrode

Vanadium compounds are also used as a newly emerging electrode material for SCs. Due to its low cost, variable valency, and lower toxicity, it is considered an efficient electrode material for SCs. A. Jain et al. [36] reported vanadium oxide nanorods as a promising electrode material for SCs. They mainly used polymer gel as an electrolyte. The high capacitive performance of these materials is attributed to the low bulk resistance of the electrolyte, indicating high ionic conductivity, lower values of charge transfer resistance, and rapid electron transfer. All of these effects are due to the unique structure of V_2O_5. The nanorod structure provides a more specific surface area and larger active sites. Here, the high capacitance originates from the contributions of EDLC and PC. The specific power and energy of vanadium oxides can be enhanced for SCs by using an ammonium framework and aqueous storage [37]. The ammonium vanadium oxide framework shows a promising energy storage capacity with an aqueous electrolyte. These aqueous electrolytes offer several advantages, including low cost, user-friendliness, and high safety. The combination of both EDLC and pseudocapacitance shows high capacitive behavior. The small charge transfer resistance indicates facile electron transfer at the interface. Using several electrolytes, they demonstrated that the bulk resistance is significantly smaller in the case of NH_4Cl, resulting in a faster diffusion rate of ions. In this case, the high capacitive performance of vanadium oxide is attributed to the reduction of chemical dissolution, which decreases the diffusion rate of ions and lowers the charge transfer resistance.

4.3.2.5 Other Transition Metal–Based Electrodes

In addition to these commonly used transition metal–based electrode materials, several newly introduced types exist in this field. However, they are not very common. For instance, when cobalt oxide nanoparticles form composites with biomass, the resulting electrode exhibits ultra-capacitive performance [38], which demonstrated an excellent specific capacitance value of 1212.4 F/g. The cobalt oxide NPs, uniformly distributed over the carbon surface, provide a higher specific surface area and enhance conductivity, resulting in high capacitance behavior for this electrode material. Another study reports on defective selenium forming composites with $CoWS_2$ and activated biomass [39]. These materials demonstrate a high energy storage capacity. This structure features a large active site and exhibits a lower Rct value, primarily due to the incorporation of heteroatoms and the introduction of trichalcogenides, which also enhance the surface area and porosity of these materials. Therefore, the high surface area, more abundant active sites, high diffusion rate of ions, and faster electron transfer are crucial for charge storage in these electrode materials. MoS_2 combined with activated carbon is also a highly efficient SC electrode material [40]. The nanoflower-like MoS_2 offers an extensive surface area in the composite, improving the material's electrical conductivity. These composites create additional active sites and provide narrower pathways for ion diffusion, thereby enhancing ion transfer at the interface. Furthermore, it features a reduced Rct value, which boosts the electron transfer rate and elevates supercapacitive performance.

The copper oxide composite with carbon is also employed as a superior supercapacitive electrode material [41]. The composites' micro-mesoporous structure provides many active sites that facilitate electron transfer and enhance their capacitive performance. Copper oxide nanoparticles are homogeneously dispersed on the carbon surface. The EDLC originates from the carbon surface, whereas CuO exhibits faradaic capacitance. Together, they yield a higher capacitance. A schematic diagram of copper oxide is shown in Figure 4.4.

4.3.3 Other Metal-Based Electrodes

Small types of metals other than transition metals are used as electrode materials for SCs. Due to a lack of multiple valencies, low catalytic behavior, and small surface area, most are not promising as electrode materials for SCs. However, few show good electrocatalytic behavior for SCs as well. Biomass-derived calcium oxide with graphene nanosheets shows promising electrochemical behavior for SCs [42]. Due to their porous structure and multiple types of vacancies, the composites exhibit rapid ion transfer, accelerating the interface's diffusion rate. It enhances the electrical conductivity of these materials. S. Kaipannan et al. [43] prepared porous carbon from waste engine oil, which forms a nanocomposite with ZnS. This nanocomposite is an efficient electrode material for SCs due to its highly porous surface, which provides many active sites. The large surface area and many active sites provide high catalytic activity, enhancing the supercapacitive performance.

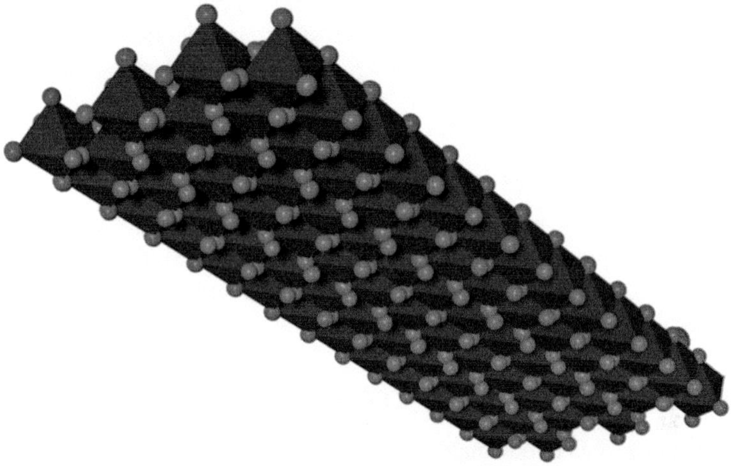

FIGURE 4.4 Schematic representation of copper oxide as an electrode material for a supercapacitor.

4.3.4 Conducting Polymer-Based Electrode

In recent years, conducting polymers such as PPy and PANI have attracted significant attention as electrode materials for SCs due to their unique physical properties. G. Oskueyan et al. [11] prepared a nanocomposite called polyaniline carbon dots/PPy–graphene, providing a wide potential window for SCs. Here, carrot juice is the precursor of carbon dots. Combining conductive polymers with carbon dots and graphene increases electrode conductivity and enhances the capacitance performance of the SCs. However, doping heteroatoms onto the conductive polymer, such as polyaniline, can enhance the energy storage capacity of the electrode [12]. Polyaniline and heteroatoms on carbon surfaces enhance the supercapacitive performance of this electrode material. Both faradaic and electrostatic charge accumulation contribute equally to the high performance of this material. A key factor in its superior performance is that doping heteroatoms into the conductive polymer imparts unique redox-active properties to polyaniline, resulting in higher pseudocapacitive behavior where specific power and energy are retained. Additionally, heteroatoms in polyaniline increase the surface area at the electrode–electrolyte interface and reduce diffusion length, thereby enhancing conductivity. The electrochemical performance of polyaniline heteroatom–based electrodes is improved by doping in carbon nanofibers [13]. Carbon nanofibers offer a large specific surface area, high stability, and renewable properties; this composite demonstrates excellent electrochemical performance. The lower charge transfer resistance, brief diffusion path, and highly porous surface area suggest the high electrical conductivity of the prepared material. Due to heteroatom doping, defect formation occurs at the surface of the electrode material, which also enhances the energy storage system. PANI offers more active sites, and the defects from heteroatom doping together enhance the conductivity of the electrode materials. Q. Bai et al. [14]

used PAN polymer doping with an N atom in the carbon surface. The material's large specific surface area increases active sites, and hierarchical porosity and nitrogen doping improve its conductivity. As a result, it demonstrates excellent electrochemical performance.

4.3.5 Other Electrode Materials

In addition to the above discussion, various electrode materials are available for SCs, including MXene and MOF-based electrode materials [44, 45]. The COF is also used as an electrode material for SCs [46]. Due to less availability and high cost, they are less commonly used for SC electrode material.

4.4 FUTURE PERSPECTIVE AND CHALLENGES

This chapter covers electrode materials for SCs. Carbon-based electrodes are cheap, high in surface area, and porous due to biomass origins. Transition metals provide excellent electrochemical performance thanks to their structure. Doping enhances conductivity but can lower performance if using sulfur or oxygen-containing groups. Researchers face challenges in maintaining carbon material surface area and tuning pore sizes, essential for achieving various pore structures. Transition metal electrodes, although promising with nanostructures, also suffer from low conductivity and scaling issues from lab to commercial use. Layered transition metal compounds improve performance, but controlling interlayer anions remains complex. Efficient energy storage must address waste management, emphasizing the need to remove hazardous waste while producing effective SCs. Selecting cost-effective, high-quality precursors with optimal pore sizes is crucial, while doping strategies must be judiciously managed. Synthesizing nanoscale transition metal-based electrodes enhances their performance through greater surface area and reduced ion diffusion distances. Additionally, combining carbonaceous materials with transition metal compounds, like oxides and sulfides, is essential. Special structures such as porous or core–shell designs can further improve electrical conductivity and reduce charge transfer resistance. Oxygen vacancies in transition metal oxides increase active sites and capacitance, with ternary oxides facilitating faster reactions than single oxides. Focusing on ternary oxide preparation is vital for future advancements in high-performance SCs. Last, recycling SCs in new SC production could help in waste management and reducing carbon emissions.

4.5 CONCLUSION

This chapter describes various types of electrode materials for SCs. In short, several types of electrode materials are available for SCs. Among them, carbon-based and transition metal–based materials are commonly used. Due to their availability, low cost, and high electrochemical performance, they are widely utilized. Most carbon materials are derived from biomass, which makes carbonaceous materials more promising. Various types of carbon-based materials, such as hierarchically porous carbon, carbon aerogel, activated carbon, carbon nanotubes, carbon nanofibers, GO, and rGO, are briefly discussed above as electrode materials for SCs. In addition, various transition metal oxides, hydroxides, and sulfides, such as nickel, ruthenium, manganese, and vanadium, are elaborated upon in this chapter. The effects of doping with different heteroatoms are also discussed above. However, other transition metals and some less commonly used alkali metals are also covered here. Moreover, in addition to these electrode materials, new materials recently utilized as electrode materials for SCs, such as MXene, MOF, and COF, are also discussed here. Despite the widespread advantages of these materials, researchers have faced numerous challenges. Hence, the recent challenges and future perspectives are briefly described above, paving the way for these electrode materials for practical commercial applications in energy storage.

4.6 AUTHOR AI DECLARATION

The authors hereby declare the use of generative AI and AI-assisted technologies in the writing process. While preparing this work, the authors used the ChatGPT tool/service to rephrase several statements in the manuscript. All AI-generated content was reviewed and edited by the authors as needed. The authors take full responsibility for the content of the publication.

ACKNOWLEDGMENT

The Editors acknowledge the use of Grammarly as a tool to assist in editing this book chapter. The suggestions made by Grammarly were carefully considered and incorporated where appropriate, but the author is solely responsible for the final content.

REFERENCES

1. Kar, K. K., Ed., "Handbook of Nanocomposite Supercapacitor Materials II: Performance," Springer International Publishing, Cham, 2020. https://doi.org/10.1007/978-3-030-52359-6
2. Adan-Mas, A., Alcaraz, L., Arévalo-Cid, P., López-Gómez, F. A., and Montemor, F., "Coffee-Derived Activated Carbon from Second Biowaste for Supercapacitor Applications," *Waste Management*, Vol. 120, 2021, pp. 280–289. https://doi.org/10.1016/j.wasman.2020.11.043
3. Khademi, B., Nateghi, M. R., Shayesteh, M. R., and Nasirizadeh, N., "High Voltage Binder Free Hybrid Supercapacitor Based on Reduced Graphene Oxide/Graphene Oxide Electrodes and 'Water in Salt' Electrolyte," *Journal of Energy Storage*, Vol. 43, 2021, p. 103164. https://doi.org/10.1016/j.est.2021.103164
4. Ali, G., Megiel, M., Cieciórski, E., Thalji, P., Romański, M. R., Algarni, J., and Chong, H., "Ferrocene Functionalized Multi-Walled Carbon Nanotubes as Supercapacitor Electrodes," *Journal of Molecular Liquids*, Vol. 318, 2020, p. 114064. https://doi.org/10.1016/j.molliq.2020.114064
5. Sun, X., Wang, J., Chen, B., Dai, G., Situ, Y., and Huang, H., "High-Performance Adjustable Manganese Oxides Hybrid Nanostructure for Supercapacitors," *Electrochimica Acta*, Vol. 381, 2021, p. 138213. https://doi.org/10.1016/j.electacta.2021.138213

6. Ye, W., Wang, H., Ning, J., Zhong, Y., and Hu, Y., "New Types of Hybrid Electrolytes for Supercapacitors," *Journal of Energy Chemistry*, Vol. 57, 2021, pp. 219–232. https://doi.org/10.1016/j.jechem.2020.09.016
7. Islam, T., Hasan, M. M., Shah, S. S., Karim, M. R., Al-Mubaddel, F. S., Zahir, M. H., Dar, M. A., Hossain, M. D., Aziz, M. A., and Ahammad, A. J. S., "High Yield Activated Porous Coal Carbon Nanosheets from Boropukuria Coal Mine as Supercapacitor Material: Investigation of the Charge Storing Mechanism at the Interfacial Region," *Journal of Energy Storage*, Vol. 32, 2020, p. 101908. https://doi.org/10.1016/j.est.2020.101908
8. Samy, M. M., Mohamed, M. G., EL-Mahdy, A. F. M., Mansoure, T. H., Wu, K. C.-W., and Kuo, S.-W., "High-Performance Supercapacitor Electrodes Prepared From Dispersions of Tetrabenzonaphthalene-Based Conjugated Microporous Polymers and Carbon Nanotubes," *ACS Applied Materials & Interfaces*, Vol. 13, No. 44, 2021, pp. 51906–51916. https://doi.org/10.1021/acsami.1c05720
9. Chen, Z., Wang, X., Xue, B., Li, W., Ding, Z., Yang, X., Qiu, J., and Wang, Z., "Rice Husk-Based Hierarchical Porous Carbon for High Performance Supercapacitors: The Structure-Performance Relationship," *Carbon*, Vol. 161, 2020, pp. 432–444. https://doi.org/10.1016/j.carbon.2020.01.088
10. Ra, E. J., Raymundo-Piñero, E., Lee, Y. H., and Béguin, F., "High Power Supercapacitors Using Polyacrylonitrile-Based Carbon Nanofiber Paper," *Carbon*, Vol. 47, No. 13, 2009, pp. 2984–2992. https://doi.org/10.1016/j.carbon.2009.06.051
11. Oskueyan, G., Mansour Lakouraj, M., and Mahyari, M., "Fabrication of Polyaniline–Carrot Derived Carbon Dots/Polypyrrole–Graphene Nanocomposite for Wide Potential Window Supercapacitor," *Carbon Letters*, Vol. 31, No. 2, 2021, pp. 269–276. https://doi.org/10.1007/s42823-020-00162-w
12. Shah, S. S., Alfasane, M. A., Bakare, I. A., Aziz, M. A., and Yamani, Z. H., "Polyaniline and Heteroatoms–Enriched Carbon Derived from Pithophora Polymorpha Composite for High Performance Supercapacitor," *Journal of Energy Storage*, Vol. 30, 2020, p. 101562. https://doi.org/10.1016/j.est.2020.101562
13. Zhu, J., Zhang, Q., Chen, H., Zhang, R., Liu, L., and Yu, J., "*Setaria Viridis* -Inspired Electrode with Polyaniline Decorated on Porous Heteroatom-Doped Carbon Nanofibers for Flexible Supercapacitors," *ACS Applied Materials & Interfaces*, Vol. 12, No. 39, 2020, pp. 43634–43645. https://doi.org/10.1021/acsami.0c10933
14. Bai, Q., Li, H., Zhang, L., Li, C., Shen, Y., and Uyama, H., "Flexible Solid-State Supercapacitors Derived from Biomass Konjac/Polyacrylonitrile-Based Nitrogen-Doped Porous Carbon," *ACS Applied Materials & Interfaces*, Vol. 12, No. 50, 2020, pp. 55913–55925. https://doi.org/10.1021/acsami.0c16752
15. B., Nisha Y., Vidyalakshmi, and Abdul Razack, S., "Enhanced Formation of Ruthenium Oxide Nanoparticles through Green Synthesis for Highly Efficient Supercapacitor Applications," *Advanced Powder Technology*, Vol. 31, No. 3, 2020, pp. 1001–1006. https://doi.org/10.1016/j.apt.2019.12.026
16. Jain, A., Ghosh, M., Krajewski, M., Kurungot, S., and Michalska, M., "Biomass-Derived Activated Carbon Material from Native European Deciduous Trees as an Inexpensive and Sustainable Energy Material for Supercapacitor Application," *Journal of Energy Storage*, Vol. 34, 2021, p. 102178. https://doi.org/10.1016/j.est.2020.102178
17. Hsiao, C., Lee, C., and Tai, N., "Biomass-Derived Three-Dimensional Carbon Framework for a Flexible Fibrous Supercapacitor and Its Application as a Wearable Smart Textile," *RSC Advances*, Vol. 10, No. 12, 2020, pp. 6960–6972. https://doi.org/10.1039/C9RA07441D
18. Sun, L., Gan, J., Wu, W., Xu, Z., Xu, Z., Ma, L., Li, C., and Liu, W., "N-Doped Cellulose-Based Carbon Aerogels with a Honeycomb-Like Structure for High-Performance Supercapacitors," *Journal of Energy Storage*, Vol. 38, 2021, p. 102414. https://doi.org/10.1016/j.est.2021.102414

19. Niu, Z., Zhang, Y., Zhang, Y., Lu, X., and Liu, J., "Enhanced Electrochemical Performance of Three-Dimensional Graphene/Carbon Nanotube Composite for Supercapacitor Application," *Journal of Alloys and Compounds*, Vol. 820, 2020, p. 153114. https://doi.org/10.1016/j.jallcom.2019.153114
20. Shang, Y., Ma, S., Wei, Y., Yang, H., and Xu, Z., "Flower-like Ternary Metal of Ni-Co-Mn Hydroxide Combined with Carbon Nanotube for Supercapacitor," *Ionics*, Vol. 26, No. 7, 2020, pp. 3609–3619. https://doi.org/10.1007/s11581-020-03496-7
21. Fan, P., Wang, J., Ding, W., and Xu, L., "Core–Shell Structured Carbon Nanofiber-Based Electrodes for High-Performance Supercapacitors," *Molecules*, Vol. 28, No. 12, 2023, p. 4571. https://doi.org/10.3390/molecules28124571
22. Wang, K., Wang, Y., Wang, Y., Hosono, E., and Zhou, H., "Mesoporous Carbon Nanofibers for Supercapacitor Application," *The Journal of Physical Chemistry C*, Vol. 113, No. 3, 2009, pp. 1093–1097. https://doi.org/10.1021/jp807463u
23. Kim, J.-G., Kim, H.-C., Kim, N. D., and Khil, M.-S., "N-Doped Hierarchical Porous Hollow Carbon Nanofibers Based on PAN/PVP@SAN Structure for High Performance Supercapacitor," *Composites Part B: Engineering*, Vol. 186, 2020, p. 107825. https://doi.org/10.1016/j.compositesb.2020.107825
24. Tarimo, D. J., Oyedotun, K. O., Mirghni, A. A., and Manyala, N., "Sulphur-Reduced Graphene Oxide Composite with Improved Electrochemical Performance for Supercapacitor Applications," *International Journal of Hydrogen Energy*, Vol. 45, No. 24, 2020, pp. 13189–13201. https://doi.org/10.1016/j.ijhydene.2020.03.059
25. Wang, B., Likodimos, V., Fielding, A. J., and Dryfe, R. A. W., "In Situ Electron Paramagnetic Resonance Spectroelectrochemical Study of Graphene-Based Supercapacitors: Comparison between Chemically Reduced Graphene Oxide and Nitrogen-Doped Reduced Graphene Oxide," *Carbon*, Vol. 160, 2020, pp. 236–246. https://doi.org/10.1016/j.carbon.2019.12.045
26. Lee, S. P., Ali, G. A. M., Hegazy, H. H., Lim, H. N., and Chong, K. F., "Optimizing Reduced Graphene Oxide Aerogel for a Supercapacitor," *Energy & Fuels*, Vol. 35, No. 5, 2021, pp. 4559–4569. https://doi.org/10.1021/acs.energyfuels.0c04126
27. Chen, X., Xie, R., Hui Li, F., Jaber, F., Musharavati, E., Zalnezhad, S., Bae, K. S., Hui, K. N., and Hui. "Supercapacitor Performance of Porous Nickel Cobaltite Nanosheets." *Scientific Reports*, Vol. 10, No. 1, 2020. https://doi.org/10.1038/s41598-020-75946-1
28. Gao, X., Zhang, H., Guo, E., Yao, F., Wang, Z., and Yue, H., "Hybrid Two-Dimensional Nickel Oxide-Reduced Graphene Oxide Nanosheets for Supercapacitor Electrodes," *Microchemical Journal*, Vol. 164, 2021, p. 105979. https://doi.org/10.1016/j.microc.2021.105979
29. Niu, W., Xiao, Z., Wang, S., Zhai, S., Qin, L., Zhao, Z., and An, Q., "Synthesis of Nickel Sulfide-Supported on Porous Carbon from a Natural Seaweed-Derived Polysaccharide for High-Performance Supercapacitors," *Journal of Alloys and Compounds*, Vol. 853, 2021, p. 157123. https://doi.org/10.1016/j.jallcom.2020.157123
30. Cai, J., Zhang, D., Ding, W.-P., Zhu, Z.-Z., Wang, G.-Z., He, J.-R., Wang, H.-B., Fei, P., and Si, T.-L., "Promising Rice-Husk-Derived Carbon/$Ni(OH)_2$ Composite Materials as a High-Performing Supercapacitor Electrode," *ACS Omega*, Vol. 5, No. 46, 2020, pp. 29896–29902. https://doi.org/10.1021/acsomega.0c04117
31. Mevada, C., and Mukhopadhyay, M., "High Mass Loading Tin Oxide-Ruthenium Oxide-Based Nanocomposite Electrode for Supercapacitor Application," *Journal of Energy Storage*, Vol. 31, 2020, p. 101587. https://doi.org/10.1016/j.est.2020.101587
32. Korkmaz, S., Kariper, İ. A., Karaman, C., and Karaman, O., "MWCNT/Ruthenium Hydroxide Aerogel Supercapacitor Production and Investigation of Electrochemical Performances," *Scientific Reports*, Vol. 12, No. 1, 2022, p. 12862. https://doi.org/10.1038/s41598-022-17286-w

33. Krishnamoorthy, K., Pazhamalai, P., and Kim, S. J., "Ruthenium Sulfide Nanoparticles as a New Pseudocapacitive Material for Supercapacitor," *Electrochimica Acta*, Vol. 227, 2017, pp. 85–94. https://doi.org/10.1016/j.electacta.2016.12.171
34. Pandit, B., Goda, E. S., Abu Elella, M. H., Ur Rehman, A., Eun Hong, S., Rondiya, S. R., Barkataki, P., Shaikh, S. F., Al-Enizi, A. M., El-Bahy, S. M., and Ro Yoon, K., "One-Pot Hydrothermal Preparation of Hierarchical Manganese Oxide Nanorods for High-Performance Symmetric Supercapacitors," *Journal of Energy Chemistry*, Vol. 65, 2022, pp. 116–126. https://doi.org/10.1016/j.jechem.2021.05.028
35. Ma, M., Cai, W., Chen, Y., Li, Y., Tan, F., and Zhou, J., "Flower-Like NiMn-Layered Double Hydroxide Microspheres Coated on Biomass-Derived 3D Honeycomb Porous Carbon for High-Energy Hybrid Supercapacitors," *Industrial Crops and Products*, Vol. 166, 2021, p. 113472. https://doi.org/10.1016/j.indcrop.2021.113472
36. Jain, A., Manippady, S. R., Tang, R., Nishihara, H., Sobczak, K., Matejka, V., and Michalska, M., "Vanadium Oxide Nanorods as an Electrode Material for Solid State Supercapacitor," *Scientific Reports*, Vol. 12, No. 1, 2022, p. 21024. https://doi.org/10.1038/s41598-022-25707-z
37. Wang, P., Zhang, Y., Jiang, H., Dong, X., and Meng, C., "Ammonium Vanadium Oxide Framework with Stable NH_4^+ Aqueous Storage for Flexible Quasi-Solid-State Supercapacitor," *Chemical Engineering Journal*, Vol. 427, 2022, p. 131548. https://doi.org/10.1016/j.cej.2021.131548
38. Zou, R., Zhu, L., Yan, L., Shao, B., Cheng, H., and Sun, W., "Co_3O_4 Anchored on Meshy Biomass Carbon Derived from Kelp for High-Performance Ultracapacitor Electrode," *Materials Chemistry and Physics*, Vol. 266, 2021, p. 124556. https://doi.org/10.1016/j.matchemphys.2021.124556
39. Sakthivel, M., Ramki, S., Chen, S.-M., and Ho, K.-C., "Defect Rich Se–$CoWS_2$ as Anode and Banana Flower Skin-Derived Activated Carbon Channels with Interconnected Porous Structure as Cathode Materials for Asymmetric Supercapacitor Application," *Energy*, Vol. 257, 2022, p. 124758. https://doi.org/10.1016/j.energy.2022.124758
40. Wang, F., Ma, J., Zhou, K., and Li, X., "MoS_2/Corncob-Derived Activated Carbon for Supercapacitor Application," *Materials Chemistry and Physics*, Vol. 244, 2020, p. 122215. https://doi.org/10.1016/j.matchemphys.2019.122215
41. Zhan, Y., Bai, J., Guo, F., Zhou, H., Shu, R., Yu, Y., and Qian, L., "Facile Synthesis of Biomass-Derived Porous Carbons Incorporated with CuO Nanoparticles as Promising Electrode Materials for High-Performance Supercapacitor Applications," *Journal of Alloys and Compounds*, Vol. 885, 2021, p. 161014. https://doi.org/10.1016/j.jallcom.2021.161014
42. Liu, H., Chen, W., Zhang, R., Xu, M., Weng, M., Huang, X., Peng, H., Miao, Z., and Huo, C., "Biomass-Derived CaO in Situ Catalyzing Approach towards Hierarchical Porous Graphene Nanosheets for High-Rate Performance Supercapacitors," *Journal of Alloys and Compounds*, Vol. 884, 2021, p. 161127. https://doi.org/10.1016/j.jallcom.2021.161127
43. Kaipannan, S., Ganesh, P. A., Manickavasakam, K., Sundaramoorthy, S., Govindarajan, K., Mayavan, S., and Marappan, S., "Waste Engine Oil Derived Porous Carbon/ZnS Nanocomposite as Bi-Functional Electrocatalyst for Supercapacitor and Oxygen Reduction," *Journal of Energy Storage*, Vol. 32, 2020, p. 101774. https://doi.org/10.1016/j.est.2020.101774
44. Wang, Z., Xu, Z., Huang, H., Chu, X., Xie, Y., Xiong, D., Yan, C., Zhao, H., Zhang, H., and Yang, W., "Unraveling and Regulating Self-Discharge Behavior of $Ti_3C_2T_x$ MXene-Based Supercapacitors," *ACS Nano*, Vol. 14, No. 4, 2020, pp. 4916–4924. https://doi.org/10.1021/acsnano.0c01056

45. Jiao, Y., Pei, J., Chen, D., Yan, C., Hu, Y., Zhang, Q., and Chen, G., "Mixed-Metallic MOF Based Electrode Materials for High Performance Hybrid Supercapacitors," *Journal of Materials Chemistry A*, Vol. 5, No. 3, 2017, pp. 1094–1102. https://doi.org/10.1039/C6TA09805C
46. Kandambeth, S., Jia, J., Wu, H., Kale, V. S., Parvatkar, P. T., Czaban-Jóźwiak, J., Zhou, S., Xu, X., Ameur, Z. O., Abou-Hamad, E., Emwas, A., Shekhah, O., Alshareef, H. N., and Eddaoudi, M., "Covalent Organic Frameworks as Negative Electrodes for High-Performance Asymmetric Supercapacitors," *Advanced Energy Materials*, Vol. 10, No. 38, 2020, p. 2001673. https://doi.org/10.1002/aenm.202001673

5 Electrolytes for Supercapacitors

Anusha Srinivas, Ramesh Chandran K, Sreeja P C, and Sapna Nehra

5.1 INTRODUCTION

Supercapacitors (SCs), also known as electrochemical capacitors (ECs), have gained prominence as advanced energy storage devices due to their high-power density, rapid charge–discharge cycles, and long cycle life compared to conventional batteries [1]. Unlike batteries, which rely on faradaic redox reactions, SCs store energy primarily through electrostatic charge accumulation at the electrode–electrolyte interface [2]. Consequently, the choice of electrolyte plays a critical role in determining the SC's overall performance, efficiency, and durability [3]. Electrolytes serve as ion-conducting media, facilitating charge transfer between electrodes while influencing key performance parameters, including capacitance, ionic conductivity, electrochemical stability, and voltage window [4]. The capacitance of an SC depends on the permittivity of the electrolyte and the distance between charge-separating layers, given by:

$$C = \frac{(\varepsilon_r \varepsilon_0 A)}{d} \tag{5.1}$$

where ε_r is the relative permittivity of the electrolyte, ε_0 is the permittivity of free space, A is the effective surface area of the electrode, and d is the separation distance of the electric double layer (EDL) [5, 6]. Another critical factor is the electrochemical stability window (ESW), which determines the SC's maximum operating voltage (V_{max}). A wider ESW enhances the energy storage capability, as expressed by [7, 8]:

$$E = \frac{1}{2} CV^2 \tag{5.2}$$

However, a trade-off exists between ionic conductivity (σ) and voltage stability, where highly conductive electrolytes, such as aqueous solutions, often exhibit lower voltage windows due to water electrolysis. In contrast, high-voltage electrolytes, such as ionic liquids (ILs), suffer from high viscosity and reduced ion mobility [9, 10]. The ionic conductivity follows the Arrhenius equation:

$$\sigma = \sigma^0 e^{-\frac{Ea}{kT}} \tag{5.3}$$

DOI: 10.1201/9781003535836-5

Ea is the activation energy for ion transport, *k* is Boltzmann's constant, and *T* is the temperature [11, 12].

SC electrolytes can be classified into aqueous, organic, IL, and solid-state (polymer/gel) electrolytes, each offering distinct advantages and limitations. Aqueous electrolytes (e.g., H_2SO_4, KOH, Na_2SO_4) provide high ionic conductivity but are limited by a narrow voltage window (~1.23 V) due to water splitting [13–15]. Organic electrolytes (e.g., ACN, PC with $TEABF_4$) extend the voltage window to ~3.5 V, enabling higher energy densities but pose safety concerns due to flammability [16–18]. ILs (e.g., $EMIMBF_4$, $BMIMPF_6$) offer a wider electrochemical window (~4.5 V) and excellent thermal stability, though at the cost of higher viscosity and lower conductivity [19]. Solid-state and gel electrolytes, such as PVA/H_2SO_4 and PEO/$LiClO_4$, provide flexibility and enhanced safety but generally exhibit lower ionic mobility [20].

Despite significant advancements, several challenges persist in electrolyte development, including stability under high voltages, environmental concerns, limited ionic mobility in high-voltage electrolytes, and performance degradation at extreme temperatures [21, 22]. To overcome these limitations, emerging research focuses on hybrid electrolytes, deep eutectic solvents (DESs), and polymer gel-based systems, offering promising alternatives for next-generation SCs [23, 24].

This chapter provides a comprehensive analysis of electrolyte materials, including their properties, selection criteria, electrochemical performance, and future research directions, as illustrated in Figure 5.1. This discussion aims to guide researchers and engineers in optimizing electrolyte selection for high-performance and sustainable SCs by bridging fundamental principles with practical applications.

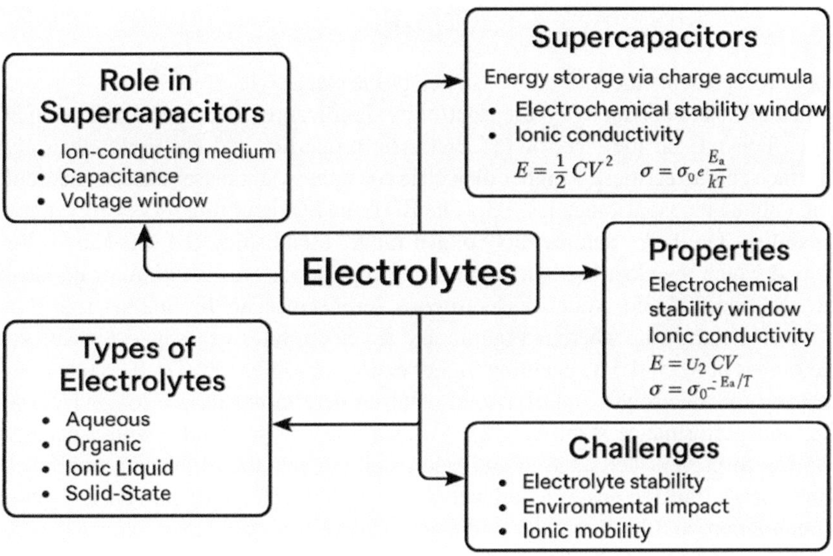

FIGURE 5.1 Overview of electrolytes.

5.2 ROLE OF ELECTROLYTES IN SCs

Electrolytes play a vital role in the performance and efficiency of SCs by facilitating ion transport between the electrodes. They directly influence key parameters such as capacitance, energy density (ED), power density, voltage window, and stability [25]. An optimal electrolyte ensures high ionic conductivity, electrochemical stability, and compatibility with the electrode material, thereby maximizing energy storage and charge–discharge characteristics [26].

The SC's function is based on two primary charge storage mechanisms: EDLC, which relies on electrostatic charge accumulation at the electrode–electrolyte interface, and pseudocapacitance, where fast, reversible faradaic redox reactions occur at the electrode surface [27, 28]. In both cases, the electrolyte provides mobile ions that migrate under an applied electric field, ensuring continuous charge accumulation and rapid energy storage.

The ionic conductivity (σ) of the electrolyte is a key determinant of SC performance, governed by [29, 30]:

$$\sigma = \frac{nz\mu e}{v} \tag{5.4}$$

where n represents ion concentration, z denotes the charge of the ion, μ indicates ion mobility, e signifies the charge of an electron, and v refers to the volume of the electrolyte. Higher ionic conductivity allows for faster charge transport, reducing ESR and improving the device's P, given by [31]:

$$P = \frac{V^2}{4RP} \tag{5.5}$$

where R represents the internal resistance of the electrolyte.

In addition to conductivity, the electrolyte significantly influences the C of an SC, which depends on the electrolyte's dielectric properties and electrode–electrolyte interface characteristics. A higher dielectric constant (ε_r) increases charge accumulation, enhancing capacitance [32, 33]. The ED of an SC determines its energy storage capability. The ESW defines this voltage range, establishing the breakdown limit beyond which the electrolyte undergoes decomposition, gas evolution, or undesired side reactions [34, 35]. Aqueous electrolytes, for instance, are limited to ~1.23 V due to water electrolysis, whereas organic and IL electrolytes can support wider voltage ranges of ~2.5–4.5 V, enabling higher energy densities [36, 37]. The interaction between the electrolyte and electrode material determines device longevity, cycle life, and performance stability [38]. An ideal electrolyte should exhibit chemical stability to prevent degradation and side reactions, maintain high ion mobility for rapid charge transfer, ensure good wettability with the electrode for efficient charge accumulation, and be thermally and electrochemically stable over extended cycles [39–41].

Furthermore, thermal and safety considerations are crucial in selecting an electrolyte. Aqueous electrolytes may freeze at low temperatures, while organic electrolytes are volatile and flammable, posing safety risks [42]. Solid-state and gel polymer electrolytes provide non-flammable, leak-proof alternatives, making them attractive for wearable and high-temperature applications [43].

To address existing limitations, recent research has focused on hybrid electrolytes, DESs, redox-active electrolytes (RAEs), and polymer gel-based systems, which aim to balance ionic conductivity, stability, and environmental safety [44].

Thus, the electrolyte is a fundamental component of SCs, directly affecting energy storage, power output, voltage stability, safety, and lifespan. Advances in novel electrolyte formulations are paving the way for next-generation SCs that offer enhanced ED, flexibility, and environmental sustainability [45, 46].

5.3 TYPES OF ELECTROLYTES FOR SCs

The choice of electrolyte is a critical factor in determining the performance, efficiency, and stability of SCs. Electrolytes influence ionic conductivity, the ESW, the charge storage mechanism, PD, and ED [47, 48]. SC electrolytes can be broadly classified into five main categories: aqueous, organic, IL, solid-state, and gel polymer electrolytes.

Each electrolyte type offers distinct advantages and limitations, depending on application requirements, operating conditions, and safety considerations.

5.3.1 Aqueous Electrolytes

Aqueous electrolytes contain dissolved ionic salts in water, offering high conductivity ($\sigma \approx 1$ S/cm), low cost, and environmental safety [49]. Due to their fast ion transport and low internal resistance, they are commonly used in low-voltage SCs.

Aqueous electrolytes, which are composed of dissolved ionic salts in water, are widely used in SCs because of their high ionic conductivity (~1 S/cm), affordability, and environmental safety [50]. Their rapid ion transport and low internal resistance make them ideal for low-voltage applications. These electrolytes are categorized into acidic, alkaline, and neutral types, each with distinct properties and limitations [51]. Acidic electrolytes, such as H_2SO_4 and HCl, offer high proton conductivity and are commonly used with carbon-based electrodes; however, their electrochemical stability is limited by a narrow voltage window (~1.2 V) due to water electrolysis [52, 53].

Alkaline electrolytes, including KOH and NaOH, are typically used in SCs with nickel and transition metal oxide electrodes; however, they suffer from electrode corrosion and self-discharge issues [54]. Neutral electrolytes, such as Na_2SO_4 and Li_2SO_4, provide a balance between conductivity and electrochemical stability while reducing electrode degradation compared to acidic and alkaline counterparts [55].

Despite their advantages, aqueous electrolytes face limitations, including a restricted electrochemical window (~1.23 V) due to water splitting, the corrosive nature of acidic and alkaline solutions affecting electrode longevity, and their unsuitability for high-energy applications due to low operating voltages [56]. Nevertheless, their safety, non-toxic nature, and compatibility with various electrode materials make them viable for specific energy storage applications.

5.3.2 Organic Electrolytes

Organic electrolytes are dissolved salts in organic solvents such as acetonitrile (ACN) or propylene carbonate (PC). They are widely used in high-voltage SCs (2.5–3.5 V) to enhance ED [57].

Organic electrolytes, primarily based on non-aqueous solvents, offer a significantly wider ESW (~2.7–3.5 V) than aqueous electrolytes, making them ideal for high-energy SC applications [58]. The most commonly used organic electrolyte is tetraethylammonium tetrafluoroborate ($TEABF_4$), dissolved in ACN or PC. $TEABF_4$ in ACN offers high ionic conductivity and a wide operating voltage; however, ACN is flammable, posing safety concerns [59]. In contrast, PC-based electrolytes provide better thermal and chemical stability but have lower conductivity [60]. Lithium salts such as $LiClO_4$, $LiBF_4$, and $LiPF_6$ are employed in hybrid SCs, enhancing capacitance and stability [61]. However, these electrolytes face challenges such as lithium dendrite formation over long-term cycling, affecting performance and safety.

The advantages of organic electrolytes include their high electrochemical stability (~3.5 V), which enables increased ED, and their suitability for high-power applications, such as electric vehicles [62]. They also exhibit good thermal stability under controlled conditions [63, 64]. However, their limitations include lower ionic conductivity (~10^{-3} S/cm) compared to aqueous electrolytes, concerns regarding flammability and toxicity (especially in ACN-based systems), and higher production costs [65]. Despite these drawbacks, organic electrolytes remain a key choice for high-voltage and energy storage efficiency applications.

5.3.3 Ionic Liquid Electrolytes

ILs are room-temperature molten salts made of organic cations and inorganic anions, offering unique advantages for SC electrolytes [66]. Their ultra-wide electrochemical windows (up to 4.5 V) and non-volatility render them highly suitable for long-lasting, high-energy-density SCs [67]. Examples include 1-ethyl-3-methylimidazolium tetrafluoroborate ($EMIMBF_4$), which provides high thermal stability (~300°C) and a wide voltage window (up to 4.0 V), and 1-butyl-3-methylimidazolium hexafluorophosphate ($BMIMPF_6$), which exhibits strong electrochemical stability but suffers from high viscosity, thereby reducing ionic mobility [68]. Pyrrolidinium-based ILs, such as $PYR_{14}TFSI$ and $PYR_{14}FSI$, are gaining attention for their enhanced environmental safety and suitability for high-temperature applications [69].

The advantages of IL-based electrolytes include their non-volatility, non-flammability, high electrochemical stability, and long cycle life. However, their limitations involve high viscosity, which reduces ionic conductivity (typically ~10^{-4} S/cm), increased synthesis costs, and poor wettability with electrode materials, often requiring surface modifications. Recent studies explore optimizing IL properties using mixtures or incorporating additives to enhance conductivity and compatibility with electrode surfaces [70–72].

5.3.4 Solid-State and Gel Polymer Electrolytes

Solid-state and gel polymer electrolytes have emerged as promising alternatives to traditional liquid electrolytes for SCs, owing to their enhanced safety, mechanical stability, and adaptability for flexible and wearable energy storage applications. These electrolytes typically consist of polymeric matrices doped with ionic salts or liquid electrolytes, ensuring ionic conductivity while maintaining structural integrity [73–75]. Among the commonly used gel electrolytes, polyvinyl alcohol (PVA)/H_2SO_4 is widely employed due to its high ionic conductivity and good mechanical properties, making it ideal for flexible SCs [76–78]. Similarly, polyethylene oxide (PEO)/$LiClO_4$ has garnered attention in HSCs due to its compatibility with lithium-based electrodes, balancing electrochemical stability and flexibility [79–81]. Additionally, chitosan-based gel electrolytes provide an eco-friendly alternative with good ionic mobility and biodegradability, making them suitable for sustainable energy storage applications [82–84].

Despite these advantages, solid-state and gel polymer electrolytes face challenges such as lower ionic conductivity compared to liquid electrolytes, reduced performance at low temperatures, and difficulties in large-scale manufacturing. Various strategies, including the incorporation of nanofillers, IL doping, and modifications to polymer blends, have been explored to enhance their electrochemical performance. Hydrogel-polymer electrolytes and their role in enhancing ionic transport and thermal stability in SCs have been well studied by various researchers [85–87]. Moreover, studies emphasize the influence of polymer matrix composition on the electrochemical properties of SC electrolytes [88–90]. These studies collectively indicate that developing advanced solid-state and gel polymer electrolytes will be crucial in improving the safety, flexibility, and long-term stability of next-generation SCs. Table 5.1 depicts the advantages and limitations of various types of electrolytes.

TABLE 5.1
Advantages and Limitations of Various Types of Electrolytes

Electrolyte Type	Voltage Window (V)	Conductivity (S/cm)	Advantages	Limitations
Aqueous	1.0–1.23	>1	High conductivity, safe, low cost	Narrow voltage window, corrosion
Organic	2.5–3.5	10^{-3}	High voltage, good stability	Flammable, expensive
Ionic liquid	3.5–4.5	10^{-4}	Non-volatile, stable	High viscosity, costly
Solid-state	2.0–3.5	10^{-5}	Flexible, safe	Lower conductivity

5.4 SELECTION CRITERIA FOR ELECTROLYTES IN SCs

Selecting a suitable electrolyte is crucial in determining SCs' efficiency, ED, power density, lifespan, and safety. The ideal electrolyte should ensure high ionic conductivity, a wide ESW, excellent compatibility with electrode materials, and thermal/chemical stability (Table 5.3).

5.4.1 Ionic Conductivity and Mobility

Ionic conductivity (σ) directly influences the internal resistance of the SC, affecting its power density and charge/discharge rate [91–93]. A high-conductivity electrolyte enables faster ion transport, reducing ESR and enhancing the capacitor's response time.

Higher ionic mobility ensures rapid charge transport, reducing energy loss. Aqueous electrolytes generally exhibit higher conductivity (~1 S/cm) compared to organic (~10^{-3} S/cm) and IL electrolytes (~10^{-4} S/cm) [94, 95]. The ideal range is $\sigma > 10^{-2}$ S/cm for high-performance SCs.

5.4.2 Electrochemical Stability Window (Voltage Range)

The ESW determines the maximum operating voltage of the SC, directly impacting its ED [35, 96, 97].

Table 5.2 provides an overview of typical ESW electrolytes, including the voltage window and conductivity.

A higher voltage window enables greater energy storage, making organic and IL electrolytes preferable for high-energy applications. The ideal range for high-energy SCs is $V_{max} > 2.5$ V.

5.4.3 Compatibility with Electrode Materials

The electrolyte should ensure efficient ion diffusion and charge storage at the electrode–electrolyte interface, minimizing side reactions, corrosion, and degradation. The compatibility between electrolytes and electrode materials is crucial in maximizing charge accumulation and enhancing the electrochemical performance of SCs [98]. Electrode materials exhibit varying affinities toward specific electrolyte types, influencing charge storage mechanisms and overall efficiency.

Carbon-based electrodes, including AC, graphene, and CNTs, are widely used for their high surface area and excellent electrical conductivity. These materials work well with aqueous, organic, and IL electrolytes, allowing for various operational

TABLE 5.2
Typical ESW Electrolytes

Electrolyte Type	Voltage Window (V)	Conductivity (S/cm)
Aqueous (H_2SO_4, Na_2SO_4)	1.0–1.23	>1
Organic ($TEABF_4$/ACN)	2.5–3.5	10^{-3}
IL ($EMIMBF_4$)	3.5–4.5	10^{-4}
Solid-state (PVA/H_2SO_4)	2.0–3.5	10^{-5}

voltages and energy densities. The choice of electrolyte significantly impacts capacitance and charge storage efficiency, with ILs and organic electrolytes offering a broader ESW, thereby enhancing ED [99].

Metal oxide electrodes, including RuO_2, MnO_2, and $Ni(OH)_2$ exhibit pseudocapacitive behavior, signifying that they store charge through rapid, reversible redox reactions. These materials typically need alkaline electrolytes, such as KOH or NaOH, to promote efficient ion transport and electrochemical activity. Additionally, RAEs can enhance their capacitance by offering extra charge storage through faradaic reactions [100, 101].

Hybrid and lithium-ion SCs operate at higher voltages, requiring improved cycling stability. These devices often use organic electrolytes (e.g., $LiClO_4$ in carbonate solvents) or lithium-based electrolytes, which enhance ionic transport and longevity during repeated charge–discharge cycles [102].

An electrolyte must completely wet the electrode surface for optimal performance to ensure efficient ion transport and charge accumulation. Poor wettability can result in increased internal resistance and decreased capacitance, negatively affecting energy storage efficiency. The development of surface-modified electrodes and functionalized electrolytes remains an active area of research aimed at enhancing electrode–electrolyte interactions and overall SC performance.

5.4.4 Viscosity and Ion Transport Properties

Low viscosity ensures better ion diffusion (DDD) and mobility (μ\muμ), enabling faster charge–discharge cycles [103–105].

$$D = \frac{k_B T}{6\pi\eta r} \tag{5.6}$$

where k_B is the Boltzmann constant, T is the temperature (K), η is the viscosity (Pa·s), and r is the ionic radius (m).

The viscosity of an electrolyte has a significant influence on ion transport, charge storage efficiency, and overall electrochemical performance in SCs. Lower viscosity facilitates faster ion diffusion, reduces internal resistance, and improves the charge–discharge rate, while higher viscosity restricts ion mobility, leading to sluggish electrochemical kinetics.

Aqueous electrolytes have the lowest viscosity (~1 cP), allowing for rapid ion diffusion and minimal resistance. This characteristic makes them ideal for high-power applications that require fast charge storage. However, their limited ESW constrains their ED [106, 107].

Organic electrolytes, exhibiting a viscosity range of ~3–5 cP, balance ion mobility and stability. Although slightly more viscous than aqueous systems, they provide a broader electrochemical window, enhancing energy storage capacity. The choice of solvent (e.g., ACN vs. PC) affects viscosity; ACN-based electrolytes are more conductive but flammable, while PC-based electrolytes are safer yet more viscous.

ILs exhibit high viscosity (~20–50 cP) due to strong ion–ion interactions and large molecular structures, leading to slower ion transport [108, 109]. Although they provide excellent thermal stability and a wide electrochemical window, their high viscosity often requires structural modifications or blending with other solvents to enhance ionic mobility.

The ideal electrolyte viscosity should be $<$10 cP for optimal SC performance, ensuring fast charge storage, stability, and safety. Advanced research explores low-viscosity ILs, hybrid electrolytes, and polymer-based gel systems to enhance ion transport without compromising electrochemical stability [110, 111].

5.4.5 Thermal and Chemical Stability

Electrolytes should remain chemically inert and stable over long cycling periods and under extreme temperatures. Decomposition or evaporation leads to capacitance loss and safety risks [112–114]. The thermal stability of electrolytes is a crucial factor in determining the operating temperature range of SCs, particularly for all-weather applications that require reliable performance in extreme conditions. Different electrolytes exhibit distinct thermal behaviors that impact device efficiency, safety, and longevity. Aqueous electrolytes risk freezing at temperatures below 0°C and evaporating above 100°C, limiting their usability in extreme cold and high-temperature environments [115, 116]. While their high ionic conductivity makes them ideal for room-temperature applications, their thermal constraints require careful thermal management or incorporating antifreeze additives.

Organic electrolytes, widely used in high-voltage SCs, provide a broader ESW [13, 117]. However, they present flammability risks at elevated temperatures ($>$150°C), requiring stringent safety precautions in high-power applications such as electric vehicles and aerospace systems.

IL electrolytes are among the most thermally stable, withstanding temperatures of up to 300°C due to their inherent non-volatility and non-flammability. These characteristics make them the preferred choice for extreme-temperature applications, such as aerospace, military, and industrial environments [118–120].

An ideal electrolyte should provide a wide operating temperature range from 40°C to 100°C, ensuring adaptability to diverse climatic conditions. Research into hybrid electrolytes, DESs, and advanced polymer gel electrolytes aims to expand this range while maintaining high ionic conductivity and safety. Developing thermally stable and non-volatile electrolyte formulations is crucial for advancing next-generation SCs that can withstand extreme environmental conditions [121, 122].

5.4.6 Safety, Toxicity, and Environmental Impact

For SCs, the choice of electrolyte must prioritize safety, cost, and scalability to ensure broad applicability, especially in consumer electronics and wearable devices. Safety considerations require electrolytes to be non-toxic, non-flammable, and environmentally friendly [123, 124]. Among the different electrolyte types, aqueous electrolytes are inherently safe and non-toxic, making them suitable for large-scale applications. In contrast, organic electrolytes, such as those based on ACN and PC, pose flammability and toxicity risks, which limit their use in specific applications. ILs offer a non-flammable alternative, but they are expensive. In contrast, gel polymer electrolytes are leak-proof and flexible, making them ideal for wearable and flexible electronics [125, 126]. The ideal electrolyte should exhibit non-flammability, low volatility, and biodegradability to enhance device safety and environmental sustainability.

From a cost and scalability perspective, the widespread commercialization of SCs necessitates low-cost and easily scalable electrolyte formulations. Aqueous electrolytes,

TABLE 5.3
Summary of Selection Criteria of Electrolytes

Criteria	Ideal Requirement	Best Electrolyte Type
Ionic conductivity (σ)	>10^{-2} S/m	Aqueous and organic
Voltage window (VVV)	>2.5 V	Organic and ILs
Electrode compatibility	Good wettability, minimal side reactions	Aqueous and gel-based
Viscosity (η\etaη)	<10 cP	Aqueous and organic
Thermal stability	>100°C	ILs and PCs
Safety	Non-toxic, non-flammable	Aqueous and polymer gel
Cost	Low cost and scalable	Aqueous and polymer gel

priced at approximately $1/kg, offer an economical solution, making them ideal for large-scale applications [127]. Organic electrolytes, such as ACN and PC-based systems, fall within the moderate cost range of $5–10/kg, balancing performance and affordability [128, 129]. ILs, with costs ranging from $100 to $500/kg, remain limited to niche, high-performance applications due to their expensive synthesis [130, 131]. While moderately priced, gel electrolytes involve complex manufacturing processes, affecting their large-scale feasibility [132, 133]. Achieving an optimal cost-performance balance involves leveraging aqueous electrolytes for cost-sensitive applications and ILs for specialized, high-performance systems. As research advances, developing cost-effective, scalable, and environmentally friendly electrolytes will be crucial in driving the next generation of SCs. Table 5.3 summarizes the selection criteria of the electrolytes.

By considering these factors, one can select the optimal electrolyte based on specific application requirements, such as high ED (organic/ILs), low-cost scalability (aqueous), or flexibility and safety (gel-based electrolytes). Figure 5.2 illustrates the overall impact of different types of electrolytes on various parameters, facilitating the selection of electrolytes.

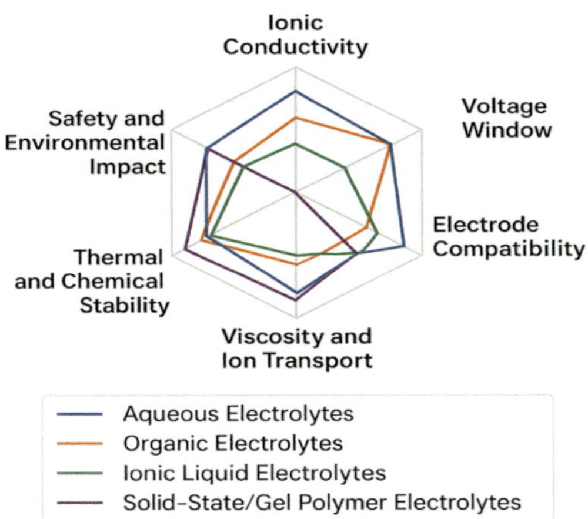

FIGURE 5.2 Impact of various electrolytes on specific parameters.

5.5 ELECTROCHEMICAL PERFORMANCE AND STABILITY OF ELECTROLYTES IN SCs

An electrolyte's electrochemical performance and stability are crucial for determining the energy storage efficiency, power delivery, cycle life, and safety of SCs. The electrolyte's ionic conductivity, ESW, ESR, and long-term stability directly influence the capacitor's efficiency and longevity [134–137]. This section discusses key parameters, including capacitance, ED, power density, cyclic stability, leakage current, and self-discharge rate, supported by governing equations and material comparisons.

5.5.1 Capacitance and Charge Storage Mechanisms

EDLC occurs in both aqueous and organic electrolytes, where ions accumulate at the electrode–electrolyte interface [138, 139]. Pseudocapacitance is common in ILs and RAEs, involving fast redox reactions at the electrode surface, leading to higher capacitance [140, 141]. High-capacitance electrolytes maximize charge storage and ED.

5.5.2 Energy Density and Power Density

ILs and organic electrolytes offer a wider ESW range (~3.5–4.5 V), resulting in higher energy densities (~30–60 Wh/kg), whereas a lower voltage range limits aqueous electrolytes to ~5–10 Wh/kg. Aqueous electrolytes exhibit high power density (>10 kW/kg) due to low ESR, whereas ILs have moderate power densities (~3 kW/kg) owing to higher viscosity [142–145]. Thus, the thumb rule for selection became aqueous electrolytes for high power and ILs for high energy.

5.5.3 ESR and Charge Transport

ESR impacts efficiency and heat generation:

$$\mathrm{ESR} = \frac{V_{drop}}{I} \tag{5.7}$$

where V_{drop} is the voltage drop, and I is the current. Low ESR improves power output and reduces heat loss [146, 147]. Aqueous electrolytes generally have lower ESR (~0.1–1 Ω) due to their superior ion mobility, whereas ILs have higher ESR (>10 Ω). The ideal ESR is less than 1 Ω for high-performance applications (Table 5.4).

TABLE 5.4
Different Electrolytes, Cycle Life, and Retention

Electrolyte	Cycle Life	Retention (%)
Aqueous (H_2SO_4, Na_2SO_4)	$<10^5$ cycles	80–90
Organic ($TEABF_4$/ACN)	10^5–10^6 cycles	90–95
Ionic liquid ($EMIMBF_4$)	$>10^6$ cycles	95–99

5.5.4 Cyclic Stability and Aging

Cycle life is determined by capacity retention (%) after multiple charge–discharge cycles:

$$\% \text{ Retention} = \frac{C_f}{C_i} \times 100 \tag{5.8}$$

C_f is the final capacitance after cycling, and C_i is the initial capacitance. Organic and IL electrolytes exhibit superior cycle life (>10^6 cycles), whereas aqueous electrolytes degrade faster due to corrosion and gas evolution [148–151]. Ideal retention is >90% after 10^5 cycles.

5.5.5 Leakage Current and Self-Discharge

Leakage current (I_L) represents unwanted energy loss when an SC is fully charged:

$$I_L = C \times \frac{dV}{dt} \tag{5.9}$$

Aqueous electrolytes exhibit higher leakage currents due to water electrolysis, whereas ILs display minimal leakage.

Self-discharge occurs due to ion redistribution and faradaic reactions, leading to voltage decay:

$$V(t) = V_{0e^{-\frac{t}{\tau}}} \tag{5.10}$$

where τ is the time constant. IL electrolytes retain charge for a longer period, whereas aqueous electrolytes self-discharge more rapidly [152–155]. Ideal performance is low leakage current (<10 μA) and slow self-discharge.

5.5.6 Thermal and Chemical Stability

A thermally stable electrolyte is crucial for maintaining SC performance across various temperatures. ILs and polymer electrolytes exhibit high thermal stability, often exceeding 300°C, making them suitable for high temperature applications [156, 157]. At the same time, aqueous electrolytes degrade at temperatures below 100°C, limiting their operational stability [158, 159]. Chemical degradation in electrolytes occurs through several mechanisms, including corrosion in aqueous electrolytes, which leads to metal electrode dissolution, solvent evaporation in organic electrolytes that reduces capacitance over time, and electrochemical decomposition that affects all electrolyte types, ultimately leading to performance deterioration and reduced lifespan [160]. An ideal electrolyte should exhibit minimal degradation over 10+ years to ensure long-term stability and efficiency in energy storage applications (Table 5.5).

Each electrolyte presents a distinct trade-off among ED, power performance, stability, and cost. The choice depends on the application—aqueous electrolytes are optimal for high-power uses, organic electrolytes strike a balance between energy and power, and ILs offer superior energy storage coupled with long-term stability.

TABLE 5.5
Summary of Electrochemical Performance in Different Types of Electrolytes

Parameter	Aqueous (H_2SO_4, Na_2SO_4)	Organic ($TEABF_4$/ACN)	IL ($EMIMBF_4$)
Voltage window (V)	1.0–1.2	2.5–3.5	3.5–4.5
Conductivity (S/cm)	>1	10^{-3}	10^{-4}
ED (Wh/kg)	5–10	20–50	30–60
PD (kW/kg)	>10	~5	~3
ESR (Ω)	<1	1–10	>10
Cycle life (cycles)	10^5	10^6	>10^6
Thermal stability (°C)	<100	150–200	>300
Leakage current (μA)	High	Medium	Low

5.6 CHALLENGES AND FUTURE PERSPECTIVES OF ELECTROLYTES IN SCs

Developing electrolytes for SCs presents several critical challenges, including a limited ESW, low ionic conductivity, thermal and chemical instability, high cost, and environmental concerns. The ESW of an electrolyte determines the operating voltage of an SC, directly affecting its ED. Aqueous electrolytes suffer from a narrow ESW (~1.0–1.2 V) due to water decomposition, while organic electrolytes offer a wider range (~2.5–3.5 V) but are flammable and degrade over time. ILs exhibit even higher ESW (~3.5–4.5 V) but have high viscosity and low ionic conductivity, limiting charge transport and power performance. To address this, future research is focusing on solid-state electrolytes, IL modifications, and electrolyte additives to enhance stability and widen ESW.

Another major challenge is low ionic conductivity in non-aqueous electrolytes, which leads to higher ESR and reduced power density. While aqueous electrolytes provide excellent conductivity (>1 S/cm), their low ESW limits energy storage. In contrast, ILs have wider ESW but much lower conductivity (~10^{-4} S/cm). Future advancements involve nanostructured electrolytes, hybrid ILs, and polymer-based gel electrolytes to optimize ion mobility and charge transport. Furthermore, thermal and chemical instability pose significant reliability concerns. Aqueous electrolytes degrade above 80–100°C, while organic solvents suffer from evaporation and flammability. Despite their superior thermal stability (>300°C), ILs face electrochemical degradation at high voltages. Research into DESs, polymeric stabilizers, and electrolyte coatings aims to improve thermal and chemical stability for long-term performance.

Due to complex synthesis and purification processes, the high cost of advanced electrolytes remains a barrier to commercialization, particularly for ILs. Aqueous electrolytes are low-cost but limited in performance, while organic electrolytes offer a balance of performance and cost but suffer from solvent degradation. Future cost-effective strategies include green synthesis methods, electrolyte recycling, and hybrid formulations that optimize cost-to-performance ratios. Environmental and

TABLE 5.6
Challenges and Future Solutions for Electrolytes in SCs

Challenge	Current Limitation	Future Solution
Limited ESW	Aqueous electrolytes decompose at ~1.2 V, while organic and ionic liquids suffer from volatility or viscosity issues	Functionalized ionic liquids, solid-state electrolytes, and electrolyte additives to widen ESW
Low ionic conductivity	High ESR in non-aqueous electrolytes limits charge transport, reducing power density	Hybrid electrolytes, gel-polymer electrolytes, and nanostructured ionic liquids to improve conductivity
Thermal and chemical instability	Aqueous electrolytes degrade above 80°C, and organic electrolytes evaporate or degrade over time	DESs, polymeric stabilizers, and electrolyte coatings to improve longevity
High cost of advanced electrolytes	Ionic liquids and some organic electrolytes are expensive to synthesize and purify	Green synthesis, low-cost ionic liquid formulations, and electrolyte recycling strategies
Environmental and safety concerns	Toxicity, flammability, and leakage risks limit large-scale adoption	Water-in-salt electrolytes, bio-based ionic liquids, and flame-retardant additives
Limited ED	Traditional electrolytes do not contribute to charge storage beyond ion adsorption	Redox-active electrolytes (quinones, vanadium-based), multi-ion charge storage, and hybrid capacitive-redox electrolytes

safety concerns, including toxicity, flammability, and leakage, hinder the large-scale adoption. Water-in-salt (WiS) electrolytes, bio-derived ILs, and flame-retardant additives are being explored as sustainable and safer alternatives. Table 5.6 summarizes the challenges and future solutions for electrolytes in SCs.

To overcome these challenges, researchers are developing advanced hybrid electrolytes that combine the advantages of various electrolyte types, such as IL–organic blends for improved ESW and conductivity. Another promising approach involves the creation of self-healing electrolytes that can regenerate after degradation, thereby enhancing the lifespan of SCs. Moreover, RAEs, which incorporate charge-storing species like quinones or vanadium ions, can significantly increase ED. Machine learning (ML) and artificial intelligence (AI) also revolutionize electrolyte discovery by predicting stability, optimizing formulations, and accelerating the high-throughput screening of novel electrolyte materials.

Looking ahead, the future of SC electrolytes will be defined by high-performance, cost-effective, and environmentally sustainable solutions. Innovations in hybrid electrolytes, AI-driven material discovery, and bio-based alternatives are expected to lead to breakthroughs in energy storage technology. As research progresses, integrating stable, high-energy-density, and scalable electrolytes will enable next-generation SCs to play a crucial role in electric vehicles, grid storage, and wearable electronics, paving the way for a more efficient and sustainable energy future.

5.7 CONCLUSION

The current chapter broadly explores the vital role of electrolytes in SCs regarding their structural design and effectiveness. Different types of electrolytes, including aqueous, organic, ILs, and solid-state or gel-based systems, are thoroughly reviewed for their specific physical and chemical characteristics, which affect the efficacy of the SC. The criteria for selecting the most suitable electrolyte primarily include ionic conductivity, electrochemical stability, compatibility with electrode materials, and environmental safety. This selection is crucial for optimizing the electrochemical interface and ensuring the reliable performance of SCs. Evaluating different types of electrolytes, this chapter also includes key performance metrics such as capacitance, energy and power densities, ESR, and cyclic stability. Limitations, including electrolyte degradation, restricted voltage windows, and safety concerns, need to be addressed for prevalent applications. Future research trends primarily focus on advancements in materials science and nanotechnology. The urgent need is to develop a sustainable electrolyte that offers very high performance, a wider operating window, and greater integration with emerging electrode materials. Developments in the field of electrolytes are crucial in ongoing research aimed at advancing highly efficient next-generation SCs.

5.8 AUTHOR AI DECLARATION

The authors hereby declare the use of generative AI and AI-assisted technologies in the writing process. While preparing this work, the authors used the ChatGPT tool/service to rephrase several statements in the manuscript. All AI-generated content was reviewed and edited by the authors as needed. The authors take full responsibility for the content of the publication.

ACKNOWLEDGMENT

The Editors acknowledge the use of Grammarly as a tool to assist in editing this book chapter. The suggestions made by Grammarly were carefully considered and incorporated where appropriate, but the author is solely responsible for the final content.

REFERENCES

1. Libber, Maitri, Narendra Gariya, and Manoj Kumar. "A Comprehensive Analysis of Supercapacitors with Current Limitations and Emerging Trends in Research." *Journal of Solid State Electrochemistry* 29, 2 (2025): 513–527. https://doi.org/10.1007/s10008-024-06107-x
2. Mehra, Pranathi, Sahaj Saxena, and Suman Bhullar. "A Comprehensive Analysis of Supercapacitors and Their Equivalent Circuits—A Review." *World Electric Vehicle Journal* 15, 8 (2024): 332. https://doi.org/10.3390/wevj15080332
3. Sharma, Priyanka, and Vinod Kumar. "Current Technology of Supercapacitors: A Review." *Journal of Electronic Materials* 49, 6 (2020): 3520–3532. https://doi.org/10.1007/s11664-020-07992-4

4. Abdel Maksoud, Muhammad Ibrahim Ahmed., Ramy Amer Fahim, Ahmed Esmail Shalan, M. Abd Elkodous, S. O. Olojede, Ahmed I. Osman, Charlie Farrell, Ala'a H. Al-Muhtaseb, A. S. Awed, A. H. Ashour, and David W. Rooney. "Advanced Materials and Technologies for Supercapacitors Used in Energy Conversion and Storage: A Review." *Environmental Chemistry Letters* 19 (2021): 375–439. https://doi.org/10.1007/s10311-020-01075-w
5. Conte, Mario. "Supercapacitors Technical Requirements for New Applications." *Fuel Cells* 10, 5 (2010): 806–818. https://doi.org/10.1002/fuce.201000087
6. Olabi, Abdul Ghani, Qasim Abbas, Ahmad Al Makky, and Mohammad A. Abdelkareem. "Supercapacitors as Next Generation Energy Storage Devices: Properties and Applications." *Energy* 248 (2022): 123617. https://doi.org/10.1016/j.energy.2022.123617
7. Ruschhaupt, Peter, Sebastian Pohlmann, Alberto Varzi, and Stefano Passerini. "Determining Realistic Electrochemical Stability Windows of Electrolytes for Electrical Double-Layer Capacitors." *Batteries & Supercaps* 3, 8 (2020): 698–707. https://doi.org/10.1002/batt.202000009
8. Pan, Zhi, Jie Yang, Qian Zhang, Ming Liu, Yifan Hu, Zhen Kou, and Yuhong Zhang. "All-Solid-State Fiber Supercapacitors with Ultrahigh Volumetric Energy Density and Outstanding Flexibility." *Advanced Energy Materials* 9, 9 (2019): 1802753. https://doi.org/10.1002/aenm.201802753
9. Tan, Jing, and Jun Liu. "Electrolyte Engineering toward High-Voltage Aqueous Energy Storage Devices." *Energy & Environmental Materials* 4, 3 (2021): 302–306. https://doi.org/10.1002/eem2.12163
10. MacFarlane, Douglas R., Naoki Tachikawa, Maria Forsyth, Jennifer M. Pringle, Patrick C. Howlett, Gloria D. Elliott, James H. Davis Jr., Masayoshi Watanabe, Patrice Simon, and C. Austen Angell. "Energy Applications of Ionic Liquids." *Energy & Environmental Science* 7, 1 (2014): 232–250. https://doi.org/10.1039/C3EE42099J
11. Chua, Iek-Heng, Christopher S. Kompella, Han Nguyen, Zhuoying Zhu, Sunny Hy, Zhi Deng, and Shyue Ping Ong. "Room-Temperature All-Solid-State Rechargeable Sodium-Ion Batteries With a Cl-Doped Na_3PS_4 Superionic Conductor." *Scientific Reports* 6 (2016): 33733. https://doi.org/10.1038/srep33733
12. Aniya, Masaru, Haruhito Sadakuni, and Eita Hirano. "Ionic Conductors: Effect of Temperature on Conductivity and Mechanical Properties and Their Interrelations." *Crystals* 11, 8 (2021): 1008. https://doi.org/10.3390/cryst11081008
13. Wan, Fang, Jiacai Zhu, Shuo Huang, and Zhiqiang Niu. "High-Voltage Electrolytes for Aqueous Energy Storage Devices." *Batteries & Supercaps* 3, 4 (2020): 323–330. https://doi.org/10.1002/batt.201900229
14. Shinagawa, Tatsuya, and Kazuhiro Takanabe. "Towards Versatile and Sustainable Hydrogen Production through Electrocatalytic Water Splitting: Electrolyte Engineering." *ChemSusChem* 10, 7 (2017): 1318–1336. https://doi.org/10.1002/cssc.201601583
15. Ranganatha, Sudhakar "Aqueous Redox-Active Electrolytes." *Electroceramics for High Performance Supercapacitors* (2023): 233–253. https://doi.org/10.1002/9781394167166.ch10
16. Wang, Ting, Cheng Fu, Xia Liu, Wei Gu, Dong He, Xiaoming Liu, and Hui Pang. "Development of Aromatic Organic Materials for High-Performance Lithium-Ion Batteries: Strategies, Advances and Future Perspectives." *ChemSusChem* (2024): e202402231. https://doi.org/10.1002/cssc.202402231
17. Muench, Simon, Andreas Wild, Christian Friebe, Bernhard Haupler, Tobias Janoschka, and Ulrich S. Schubert. "Polymer-Based Organic Batteries." *Chemical Reviews* 116, 16 (2016): 9438–9484. https://doi.org/10.1021/acs.chemrev.6b00070

18. Tian, Xiaolu, Yikun Yi, Binren Fang, Ping Yang, Ting Wang, Peng Liu, and Shuo Zhang. "Design Strategies of Safe Electrolytes for Preventing Thermal Runaway in Lithium Ion Batteries." *Chemistry of Materials* 32, 23 (2020): 9821–9848. https://doi.org/10.1021/acs.chemmater.0c02428
19. Pereira, José, Rafael Souza, and António Moita. "A Review of Ionic Liquids and Their Composites with Nanoparticles for Electrochemical Applications." *Inorganics* 12, 7 (2024): 186. https://doi.org/10.3390/inorganics12070186
20. Bohnke, Odile, Christophe Rousselot, Pierre A. Gillet, and Christian Truche. "Gel Electrolyte for Solid-State Electrochromic Cell." *Journal of the Electrochemical Society* 139, 7 (1992): 1862–1865. https://doi.org/10.1149/1.2069512
21. Chen, Shimou, Kaihua Wen, Juntian Fan, Yoshio Bando, and Dmitri Golberg. "Progress and Future Prospects of High-Voltage and High-Safety Electrolytes in Advanced Lithium Batteries: From Liquid to Solid Electrolytes." *Journal of Materials Chemistry A* 6 (2018): 11631–11663. https://doi.org/10.1039/C8TA03358G
22. Lin, Xia, Majid Salari, Laxminarayan M. R. Arava, Pulickel M. Ajayan, and Mark W. Grinstaff. "High Temperature Electrical Energy Storage: Advances, Challenges, and Frontiers." *Chemical Society Reviews* 45, 21 (2016): 5848–5887. https://doi.org/10.1039/C6CS00012F
23. Bhat, Tejasvinee Shrikrishna, Pramod Shankarrao Patil, and Raghavan Baby Rakhi. "Recent Trends in Electrolytes for Supercapacitors." *Journal of Energy Storage* 50 (2022): 104222. https://doi.org/10.1016/j.est.2022.104222
24. Samantaray, Somya, Debabrata Mohanty, I-Ming Hung, Md Moniruzzaman, and Santosh Kumar Satpathy. "Unleashing Recent Electrolyte Materials for Next-Generation Supercapacitor Applications: A Comprehensive Review." *Journal of Energy Storage* 72 (2023): 108352. https://doi.org/10.1016/j.est.2023.108352
25. Pathak, Mayank, Diksha Bhatt, Rajesh Chandra Bhatt, Bhashkar Singh Bohra, Gaurav Tatrari, Sravendra Rana, Mahesh Chandra Arya, and Nanda Gopal Sahoo. "High Energy Density Supercapacitors: An Overview of Efficient Electrode Materials, Electrolytes, Design, and Fabrication." *The Chemical Record* 24, 1 (2024): e202300236. https://doi.org/10.1002/tcr.202300236
26. Ates, Murat, Achref Chebil, Ozan Yoruk, Chérif Dridi, and Murat Turkyilmaz. "Reliability of Electrode Materials for Supercapacitors and Batteries in Energy Storage Applications: A Review." *Ionics* 28, 1 (2022): 27–52. https://doi.org/10.1007/s11581-021-04296-3
27. Bhojane, Prateek. "Recent Advances and Fundamentals of Pseudocapacitors: Materials, Mechanism, and Its Understanding." *Journal of Energy Storage* 45 (2022): 103654. https://doi.org/10.1016/j.est.2021.103654
28. Sinha, Prerna, and Kamal K. Kar. "Introduction to Supercapacitors." *Handbook of Nanocomposite Supercapacitor Materials II: Performance* (2020): 1–28. https://doi.org/10.1007/978-3-030-52359-6_1
29. Yue, Xiaoyang, Juntian Fan, and Qiang Xiang. "Internal Electric Field on Steering Charge Migration: Modulations, Determinations and Energy-Related Applications." *Advanced Functional Materials* 32, 12 (2022): 2110258. https://doi.org/10.1002/adfm.202110258
30. Klein, Robert J., Shihai Zhang, Shichen Dou, Brad H. Jones, Ralph H. Colby, and James Runt. "Modeling Electrode Polarization in Dielectric Spectroscopy: Ion Mobility and Mobile Ion Concentration of Single-Ion Polymer Electrolytes." *The Journal of Chemical Physics* 124, 14 (2006). https://doi.org/10.1063/1.2186638
31. Yang, Hui, and Nianqiang Wu. "Ionic Conductivity and Ion Transport Mechanisms of Solid-State Lithium-Ion Battery Electrolytes: A Review." *Energy Science & Engineering* 10, 5 (2022): 1643–1671. https://doi.org/10.1002/ese3.1163

32. Shah, Syed Shaheen, and Md. Abdul Aziz. "Properties of Electrode Materials and Electrolytes in Supercapacitor Technology." *Journal of Chemistry and Environment* 3, 1 (2024): 1–45. https://doi.org/10.56946/jce.v3i1.309
33. Wang, Haifeng, and Liang Yang. "Dielectric Constant, Dielectric Loss, Conductivity, Capacitance and Model Analysis of Electronic Electroactive Polymers." *Polymer Testing* 120 (2023): 107965. https://doi.org/10.1016/j.polymertesting.2023.107965
34. Rizwan, Muhammad., Asma Ayub Ambreen, and Fatima Aleena. "Key Factors for Optimizing Energy Density in High-Performance Supercapacitors." *Electroceramics for High Performance Supercapacitors* (2023): 135–158. https://doi.org/10.1002/9781394167166.ch6
35. Kasnatscheew, Jens, Bastian Streipert, Sebastian Röser, Ralf Wagner, Ivana Cekic Laskovic, and Martin Winter. "Determining Oxidative Stability of Battery Electrolytes: Validity of Common Electrochemical Stability Window (ESW) Data and Alternative Strategies." *Physical Chemistry Chemical Physics* 19, 24 (2017): 16078–16086. https://doi.org/10.1039/C7CP03072J
36. Sui, Yiming, and Xiulei Ji. "Anticatalytic Strategies to Suppress Water Electrolysis in Aqueous Batteries." *Chemical Reviews* 121, 11 (2021): 6654–6695. https://doi.org/10.1021/acs.chemrev.1c00191
37. Yuan, Xinhai, Fuxiang Ma, Linqing Zuo, Jing Wang, Nengfei Yu, Yuhui Chen, Yusong Zhu, Qinghong Huang, Rudolf Holze, and Yuping Wu. "Latest Advances in High-Voltage and High-Energy-Density Aqueous Rechargeable Batteries." *Electrochemical Energy Reviews* 4 (2021): 1–34. https://doi.org/10.1007/s41918-020-00075-2
38. Yu, Xiao, and Arumugam Manthiram. "Electrode–Electrolyte Interfaces in Lithium-Based Batteries." *Energy & Environmental Science* 11, 3 (2018): 527–543. https://doi.org/10.1039/C7EE02555F
39. Cabana, Jordi, Benjamin J. Kwon, and Liangbing Hu. "Mechanisms of Degradation and Strategies for the Stabilization of Cathode–Electrolyte Interfaces in Li-Ion Batteries." *Accounts of Chemical Research* 51, 2 (2018): 299–308. https://doi.org/10.1021/acs.accounts.7b00482
40. Wang, Qingsong, Lihua Jiang, Yan Yu, and Jinhua Sun. "Progress of Enhancing the Safety of Lithium Ion Battery from the Electrolyte Aspect." *Nano Energy* 55 (2019): 93–114. https://doi.org/10.1016/j.nanoen.2018.10.035
41. Du, Hao, Yadong Wang, Yuqiong Kang, Yun Zhao, Yao Tian, Xianshu Wang, Yihong Tan, et al. "Side Reactions/Changes in Lithium-Ion Batteries: Mechanisms and Strategies for Creating Safer and Better Batteries." *Advanced Materials* 36, 29 (2024): 2401482. https://doi.org/10.1002/adma.202401482
42. Li, Matthew, Chunsheng Wang, Zhongwei Chen, Kang Xu, and Jun Lu. "New Concepts in Electrolytes." *Chemical Reviews* 120, 14 (2020): 6783–6819. https://doi.org/10.1021/acs.chemrev.9b00531
43. Al-zubaidi, Ayar, Xixi Ji, and Jie Yu. "Thermal Charging of Supercapacitors: A Perspective." *Sustainable Energy & Fuels* 1, 7 (2017): 1457–1474. https://doi.org/10.1039/C7SE00239D
44. Bai, Yang, Rong Liu, Yang Liu, Yuanming Wang, Xue Wang, Huanhao Xiao, and Guohui Yuan. "Concentrated Hydrogel Electrolyte for Integrated Supercapacitor with High Capacitance at Subzero Temperature." *Science China Chemistry* 64 (2021): 852–860. https://doi.org/10.1007/s11426-020-9950-8
45. Zhang, Kun, Congwei Wang, Xinglin Luo, Yuede Pan, Gang Li, and Kaiying Wang. "Progress on Engineering Low-Temperature Electrolyte for High-Performance Supercapacitors." *ACS Sustainable Chemistry & Engineering* 12, 16 (2024): 6059–6081. https://doi.org/10.1021/acssuschemeng.4c00249

46. Barik, Rasmita, Vaishali Tanwar, and Pravin P. Ingole. "Supercapacitors: Future direction and challenges." In *Nanostructured Materials for Supercapacitors*, pp. 619–644. Cham: Springer International Publishing (2022). https://doi.org/10.1007/978-3-030-99302-3_27
47. Gao, Han, Alvin Virya, and Keryn Lian. "Monovalent Silicotungstate Salts as Electrolytes for Electrochemical Supercapacitors." *Electrochimica Acta* 138 (2014): 240–246. https://doi.org/10.1016/j.electacta.2014.06.127
48. Méry, Adrien, Steeve Rousselot, David Lepage, and Mickaël Dollé. "A Critical Review for an Accurate Electrochemical Stability Window Measurement of Solid Polymer and Composite Electrolytes." *Materials* 14, 14 (2021): 3840. https://doi.org/10.3390/ma14143840
49. Gou, Qianzhi, Shuang Zhao, Jiacheng Wang, Meng Li, and Junmin Xue. "Recent Advances on Boosting the Cell Voltage of Aqueous Supercapacitors." *Nano-Micro Letters* 12 (2020): 1–22. https://doi.org/10.1007/s40820-020-00430-4
50. Verma, Kapil Dev, Alka Jangid, Prerna Sinha, and Kamal K. Kar. "Electrolyte Material Selection for Supercapacitors." *Handbook of Nanocomposite Supercapacitor Materials III: Selection* (2021): 233–270. https://doi.org/10.1007/978-3-030-68364-1_7
51. Zhao, Cuimei, and Weitao Zheng. "A Review for Aqueous Electrochemical Supercapacitors." *Frontiers in Energy Research* 3 (2015): 23. https://doi.org/10.3389/fenrg.2015.00023
52. Fic, Krzysztof, Grzegorz Lota, and Elzbieta Frackowiak. "Electrochemical Properties of Supercapacitors Operating in Aqueous Electrolyte with Surfactants." *Electrochimica Acta* 55, 25 (2010): 7484–7488. https://doi.org/10.1016/j.electacta.2010.04.019
53. Reber, David, Ruben-Simon Kühnel, and Corsin Battaglia. "High-Voltage Aqueous Supercapacitors Based on NaTFSI." *Sustainable Energy & Fuels* 1, 10 (2017): 2155–2161. https://doi.org/10.1039/C7SE00423K
54. Ramachandran, Rajendran, and Fei Wang. "Electrochemical capacitor performance: Influence of aqueous electrolytes." In *Supercapacitors—Theoretical and Practical Solutions*. IntechOpen (2018): 51–68. https://doi.org/10.5772/intechopen.70694
55. Qu, Qunting, Peng Zhang, Bo Wang, Yuhui Chen, Shuang Tian, Yujie Wu, and Rudolf Holze. "Electrochemical Performance of MnO_2 Nanorods in Neutral Aqueous Electrolytes as a Cathode for Asymmetric Supercapacitors." *The Journal of Physical Chemistry C* 113, 31 (2009): 14020–14027. https://doi.org/10.1021/jp8113094
56. Kurzweil, Peter, and Maciej Chwistek. "Electrochemical Stability of Organic Electrolytes in Supercapacitors: Spectroscopy and Gas Analysis of Decomposition Products." *Journal of Power Sources* 176, 2 (2008): 555–567. https://doi.org/10.1016/j.jpowsour.2007.11.078
57. Górska, Barbara, Elzbieta Frackowiak, and François Béguin. "Redox Active Electrolytes in Carbon/Carbon Electrochemical Capacitors." *Current Opinion in Electrochemistry* 9 (2018): 95–105. https://doi.org/10.1016/j.coelec.2018.05.006
58. Mansha, Muhammad, Aqsa Anam, Safyan Akram Khan, Atif Saeed Alzahrani, Majad Khan, Aziz Ahmad, Muhammad Arshad, and Shahid Ali. "Recent Developments on Electroactive Organic Electrolytes for Non-Aqueous Redox Flow Batteries: Current Status, Challenges, and Prospects." *The Chemical Record* 24, 1 (2024): e202300233. https://doi.org/10.1002/tcr.202300233
59. Kurig, Heisi, Tõnis Romann, Aare Jänes, and Enn Lust. "Electrochemical Characteristics of Titanium Carbide Derived Carbon|1-Ethyl-3-Methylimidazolium Tetrafluoroborate Electrical Double Layer Capacitors." *ECS Transactions* 25, 23 (2010): 15. https://doi.org/10.1149/1.3274208
60. Kurig, Heisi, Aare Jänes, and Enn Lust. "Electrochemical Characteristics of Carbide-Derived Carbon|1-Ethyl-3-Methylimidazolium Tetrafluoroborate Supercapacitor Cells." *Journal of the Electrochemical Society* 157, 3 (2010): A272. https://doi.org/10.1149/1.3274208

61. Sillars, Fiona B., Stephen I. Fletcher, Mohammad Mirzaeian, and Peter J. Hall. "Variation of Electrochemical Capacitor Performance with Room Temperature Ionic Liquid Electrolyte Viscosity and Ion Size." *Physical Chemistry Chemical Physics* 14, 17 (2012): 6094–6100. https://doi.org/10.1039/C2CP40089H
62. Lust, Enn, Lauri Siinor, Heisi Kurig, Tõnis Romann, Vladimir Ivaništšev, Carl Siimenson, and Aare Jänes. "Characteristics of Capacitors Based on Ionic Liquids: From Dielectric Polymers to Redox-Active Adsorbed Species." *ECS Transactions* 75, 15 (2016): 161. https://doi.org/10.1149/07515.0161ecst
63. Huang, Pei-Li, Xian-Feng Luo, Yu-Ying Peng, Nai-Wen Pu, Ming-De Ger, Ching-Hsiu Yang, and Jeng-Kang Chang. "Ionic Liquid Electrolytes with Various Constituent Ions for Graphene-Based Supercapacitors." *Electrochimica Acta* 161 (2015): 371–377. https://doi.org/10.1016/j.electacta.2015.01.063
64. Yang, Ching-Hsiu, Pei-Li Huang, Xian-Feng Luo, Cheng-Hung Wang, Chia-Chin Li, Yi-Hsien Wu, and Jeng-Kang Chang. "Holey Graphene Nanosheets with Surface Functional Groups as High-Performance Supercapacitors in Ionic-Liquid Electrolyte." *ChemSusChem* 8, 10 (2015): 1779–1786. https://doi.org/10.1002/cssc.201500030
65. Li, Mengjie, Robert Paul Hicks, Zifeng Chen, Chao Luo, Juchen Guo, Chunsheng Wang, and Yunhua Xu. "Electrolytes in Organic Batteries." *Chemical Reviews* 123, 4 (2023): 1712–1773. https://doi.org/10.1021/acs.chemrev.2c00374
66. Lin, Jia-Hao. "The Anionic Surfactant/Ionic Liquids Intercalated Reduced Graphene Oxide for High-Performance Supercapacitors." *Nanoscale Research Letters* 13 (2018): 1–9. https://doi.org/10.1186/s11671-018-2636-9
67. Brandt, Anja, Stefan Pohlmann, Alessandro Varzi, Andrea Balducci, and Stefano Passerini. "Ionic Liquids in Supercapacitors." *MRS Bulletin* 38, 7 (2013): 554–559. https://doi.org/10.1557/mrs.2013.133
68. Zhu, Yucheng, Hao Shen, Hao Ji, Min Hua, Xuhai Pan, Madeleine Bussemaker, and Lian Liu. "Application of 1-Butyl-3-Menthylimidazolium-Hexafluorophosphate as Flame Retardant in Electrolyte of Lithium Ion Battery." *Journal of Loss Prevention in the Process Industries* 79 (2022): 104837. https://doi.org/10.1016/j.jlp.2022.104837
69. MacFarlane, Douglas R., Jennifer M. Pringle, Patrick C. Howlett, and Maria Forsyth. "Ionic Liquids and Reactions at the Electrochemical Interface." *Physical Chemistry Chemical Physics* 12, 8 (2010): 1659–1669. https://doi.org/10.1039/B923053J
70. Sakaebe, Hikari, and Hajime Matsumoto. "Application of Ionic Liquids to Li Batteries." *Electrochemical Aspects of Ionic Liquids* (2005): 171–186. https://doi.org/10.1002/0471762512.ch14
71. Sultana, Sharmin., Kamran Ahmed, Pramod Kumar Jiwanti, Budi Yuli Wardhana, and Md. Nazrul Islam Shiblee. "Ionic Liquid-Based Gels for Applications in Electrochemical Energy Storage and Conversion Devices: A Review of Recent Progress and Future Prospects." *Gels* 8, 1 (2021): 2. https://doi.org/10.3390/gels8010002
72. Villanueva, Maria., Antonio Coronas, Javier Garcia, and José Salgado. "Thermal Stability of Ionic Liquids for Their Application as New Absorbents." *Industrial & Engineering Chemistry Research* 52, 45 (2013): 15718–15727. https://doi.org/10.1021/ie401656e
73. Wang, Yujia, Ke Xue, Chao Yan, Yaqiong Li, Xin Zhang, Ke Su, and Jun Lang. "Tuning of Ionic Liquid–Solvent Electrolytes for High-Voltage Electrochemical Double Layer Capacitors: A Review." *Batteries* 10, 2 (2024): 54. https://doi.org/10.3390/batteries10020054
74. Ngai, Koh Sing, Subramaniam Ramesh, Kasi Ramesh, and Joon Ching Juan. "A Review of Polymer Electrolytes: Fundamentals, Approaches, and Applications." *Ionics* 22 (2016): 1259–1279. https://doi.org/10.1007/s11581-016-1756-4

75. Aruchamy, Kanakaraj, Subramaniyan Ramasundaram, Sivasubramani Divya, Murugesan Chandran, Kyusik Yun, and Tae Hwan Oh. "Gel Polymer Electrolytes: Advancing Solid-State Batteries for High-Performance Applications." *Gels* 9, 7 (2023): 585. https://doi.org/10.3390/gels9070585
76. Chen, Shuang, Yichen Chen, Xuelin Mu, Peng Wang, Liang Miao, Shigeki Tanemura, and Hongbo Cai. "Strategies for Enhancing Ionic Conductivity and Energy Density of Gel Polymer Electrolytes for Next-Generation Flexible Energy Storage Devices." *Sustainable Materials and Technologies* 36 (2023): e00635. https://doi.org/10.1016/j.susmat.2023.e00635
77. Fard, Hamed Nazarpour, Ghobad Behzadi Pour, Mehdi Nasiri Sarvi, and Parisa Esmaili. "PVA-Based Supercapacitors." *Ionics* 25 (2019): 2951–2963. https://doi.org/10.1007/s11581-019-03059-5
78. Muzaffar, Adnan, Mohammad B. Ahamed, and Kunal Deshmukh. "Conducting Polymer Electrolytes for Flexible Supercapacitors." *Flexible Supercapacitor Nanoarchitectonics*, (2021): 233–262. https://doi.org/10.1002/9781119711469.ch9
79. Keum, Kayeon, Jung Wook Kim, Soo Yeong Hong, Jeong Gon Son, Sang-Soo Lee, and Jeong Sook Ha. "Flexible/Stretchable Supercapacitors with Novel Functionality for Wearable Electronics." *Advanced Materials* 32, 51 (2020): 2002180. https://doi.org/10.1002/adma.202002180
80. Yu, Guihua, Xing Xie, Lijia Pan, Zhenan Bao, and Yi Cui. "Hybrid Nanostructured Materials for High-Performance Electrochemical Capacitors." *Nano Energy* 2, 2 (2013): 213–234. https://doi.org/10.1016/j.nanoen.2012.10.006
81. Chatterjee, Dhruba P., and Arun K. Nandi. "A Review on the Recent Advances in Hybrid Supercapacitors." *Journal of Materials Chemistry A* 9, 29 (2021): 15880–15918. https://doi.org/10.1039/D1TA02505H
82. Alipoori, Saeideh, Saeedeh Mazinani, Seyed Hamed Aboutalebi, and Farhad Sharif. "Review of PVA-based gel polymer electrolytes in flexible solid-state supercapacitors: Opportunities and challenges." *Journal of Energy Storage* 27 (2020): 101072. https://doi.org/10.1016/j.est.2019.101072
83. Rostami, Mohammad Saeid, and Mohammad Mehdi Khodaei. "Recent Advances of Chitosan-Based Nanocomposites for Supercapacitor Applications: Key Challenges and Future Research Directions." *Journal of Energy Storage* 72 (2023): 108344. https://doi.org/10.1016/j.est.2023.108344
84. Roy, Bijoy Krishna, Ishmam Tahmid, and Taslim Ur Rashid. "Chitosan-Based Materials for Supercapacitor Applications: A Review." *Journal of Materials Chemistry A* 9, 33 (2021): 17592–17642. https://doi.org/10.1039/D1TA02997E
85. Mazuki, Norfatihah, Mohd Saiful Asmal Rani, Nuraziliana Muhd Ghazali, Nurhasniza Mamajan Khan, Muhammad Nor Arifin Yaakob, and Ahmad Salihin Samsudin. "Structural and performance of chitosan-based polymer composites for electrical applications." In *Polymer Composites Derived from Animal Sources* (2024): 321–351. https://doi.org/10.1016/B978-0-443-22414-0.00017-X
86. Choudhury, Nurul Alam, Subash Sampath, and Ashok K. Shukla. "Hydrogel-Polymer Electrolytes for Electrochemical Capacitors: An Overview." *Energy & Environmental Science* 2, 1 (2009): 55–67. https://doi.org/10.1039/B811217G
87. Wang, Haoran, Zhaoxia Hou, Yue Wang, Haibo Long, Depeng Zhang, Zhongchao Fu, Nan Wu, Zihui Zhai, and Bo Wang. "3D Hierarchical Porous Hydrogel Polymer Electrolytes for Flexible Quasi-Solid-State Supercapacitors." *Chemical Engineering Journal* 510 (2025): 161766. https://doi.org/10.1016/j.cej.2025.161766
88. Ding, Jiansen, Yang Yang, Jade Poisson, Yuan He, Hua Zhang, Ying Zhang, Yulan Bao, Shuiliang Chen, Yong Mei Chen, and Kai Zhang. "Recent Advances in Biopolymer-Based Hydrogel Electrolytes for Flexible Supercapacitors." *ACS Energy Letters* 9, 4 (2024): 1803–1825. https://doi.org/10.1039/D0TA09500A

89. Suriyakumar, Shruti, Preetam Bhardwaj, Andrews Nirmala Grace, and A. Manuel Stephan. "Role of Polymers in Enhancing the Performance of Electrochemical Supercapacitors: A Review." *Batteries & Supercaps* 4, 4 (2021): 571–584. https://doi.org/10.1002/batt.202000272
90. Cole, Daniel P., Arava Leela Mohana Reddy, Myung Gwan Hahm, Ryan McCotter, Amelia HC Hart, Robert Vajtai, Pulickel M. Ajayan, Shashi P. Karna, and Mark L. Bundy. "Electromechanical Properties of Polymer Electrolyte-Based Stretchable Supercapacitors." *Advanced Energy Materials* 4, 3 (2014): 1300844. https://doi.org/10.1002/aenm.201300844
91. Wee, Gyeong, Olof Larsson, Muthuraja Srinivasan, Magnus Berggren, Xavier Crispin, and Subodh Mhaisalkar. "Effect of the Ionic Conductivity on the Performance of Polyelectrolyte-Based Supercapacitors." *Advanced Functional Materials* 20, 24 (2010): 4344–4350. https://doi.org/10.1002/adfm.201001096
92. Jia, Yanju, Dong Yang, Lisha Zhang, Yaxin Shi, Jun Xie, Fang Lei, and Lijun Fan. "Ionic Conductivity Enhancement Achieved by Binder in Electrodes and Its Influence in Supercapacitor." *ChemElectroChem* 9, 15 (2022): e202200662. https://doi.org/10.1002/celc.202200662
93. Liu, Xiaojun, Xiong Dai, Guo Wei, Yulin Xi, Min Pang, Vadim Izotov, and Wenlong Han. "Experimental and Theoretical Studies of Nonlinear Dependence of the Internal Resistance and Electrode Thickness for High-Performance Supercapacitor." *Scientific Reports* 7, 1 (2017): 45934. https://doi.org/10.1038/srep45934
94. Saha, Manju, Anil Kumar, R. Kanaoujiya, Kishore Behera, and Sushil Trivedi. "A Comprehensive Review of Novel Emerging Electrolytes for Supercapacitors: Aqueous and Organic Electrolytes versus Ionic Liquid-Based Electrolytes." *Energy & Fuels* 38, 10 (2024): 8528–8552. https://doi.org/10.1021/acs.energyfuels.4c00685
95. Bayley, Paul M., George H. Lane, Nathalie M. Rocher, Bronya R. Clare, Adam S. Best, Douglas R. MacFarlane, and MariaForsyth. "Transport Properties of Ionic Liquid Electrolytes with Organic Diluents." *Physical Chemistry Chemical Physics* 11, 33 (2009): 7202–7208. https://doi.org/10.1039/B902200G
96. Sun, Xinqiang, Xiaojing Zhang, Kaixuan Wang, Yumei An, Xiaoyan Zhang, Chao Li, and Yuhui Ma. "Determination Strategy of Stable Electrochemical Operating Voltage Window for Practical Lithium-Ion Capacitors." *Electrochimica Acta* 428 (2022): 140972. https://doi.org/10.1016/j.electacta.2022.140972
97. Chen, Lihua, Shruti Venkatram, Chiho Kim, Rohit Batra, Anand Chandrasekaran, and Rampi Ramprasad. "Electrochemical Stability Window of Polymeric Electrolytes." *Chemistry of Materials* 31, 12 (2019): 4598–4604. https://doi.org/10.1021/acs.chemmater.9b01456
98. Zhong, Cheng, Yuling Deng, Weitao Hu, Jingqiu Qiao, Lijuan Zhang, and Jianguo Zhang. "A Review of Electrolyte Materials and Compositions for Electrochemical Supercapacitors." *Chemical Society Reviews* 44, 21 (2015): 7484–7539. https://doi.org/10.1039/C5CS00303B
99. Mousavi, Maral P. S., Benjamin E. Wilson, Sadra Kashefolgheta, Evan L. Anderson, Siyao He, Philippe Bühlmann, and Andreas Stein. "Ionic Liquids as Electrolytes for Electrochemical Double-Layer Capacitors: Structures That Optimize Specific Energy." *ACS Applied Materials & Interfaces* 8, 5 (2016): 3396–3406. https://doi.org/10.1021/acsami.5b11353
100. Frackowiak, Elzbieta, Małgorzata Meller, Jens Menzel, Dominik Gastol, and Krzysztof Fic. "Redox-Active Electrolyte for Supercapacitor Application." *Faraday Discussions* 172 (2014): 179–198. https://doi.org/10.1039/C4FD00052H
101. Wang, Yujie, Zhiwei Chang, Ming Qian, Zhen Zhang, Jinhui Lin, and Fang Huang. "Enhanced Specific Capacitance by a New Dual Redox-Active Electrolyte in Activated Carbon-Based Supercapacitors." *Carbon* 143 (2019): 300–308. https://doi.org/10.1016/j.carbon.2018.11.054

102. Xu, Hong, and Min Shen. "The Control of Lithium-Ion Batteries and Supercapacitors in Hybrid Energy Storage Systems for Electric Vehicles: A Review." *International Journal of Energy Research* 45, 15 (2021): 20524–20544. https://doi.org/10.1002/er.6609
103. Li, Guang. "Regulating Mass Transport Behavior for High-Performance Lithium Metal Batteries and Fast-Charging Lithium-Ion Batteries." *Advanced Energy Materials* 11, 7 (2021): 2002891. https://doi.org/10.1002/aenm.202002891
104. Sun, Yunjiao, Bo Liu, Lei Liu, and Xudong Yan. "Ions Transport in Electrochemical Energy Storage Devices at Low Temperatures." *Advanced Functional Materials* 32, 15 (2022): 2109568. https://doi.org/10.1002/adfm.202109568
105. Pal, Bhupender, Shengyuan Yang, Subramaniam Ramesh, Venkataraman Thangadurai, and Rajan Jose. "Electrolyte Selection for Supercapacitive Devices: A Critical Review." *Nanoscale Advances* 1, 10 (2019): 3807–3835. https://doi.org/10.1039/C9NA00374F
106. Zheng, Jun, Guang Tan, Ping Shan, Tao Liu, Jianhua Hu, Yanfeng Feng, and Fang Pan. "Understanding Thermodynamic and Kinetic Contributions in Expanding the Stability Window of Aqueous Electrolytes." *Chem* 4, 12 (2018): 2872–2882. https://doi.org/10.1016/j.chempr.2018.09.004
107. Zhang, Hui, Xiaodong Liu, Haifeng Li, Igor Hasa, and Stefano Passerini. "Challenges and Strategies for High-Energy Aqueous Electrolyte Rechargeable Batteries." *Angewandte Chemie International Edition* 60, 2 (2021): 598–616. https://doi.org/10.1002/anie.202004433
108. Dong, Kai, Xiaodong Liu, Hong Dong, Xiaoyan Zhang, and Shuang Zhang. "Multiscale Studies on Ionic Liquids." *Chemical Reviews* 117, 10 (2017): 6636–6695. https://doi.org/10.1021/acs.chemrev.6b00776
109. Fedorov, Mikhail V., and Alexander A. Kornyshev. "Ionic Liquids at Electrified Interfaces." *Chemical Reviews* 114, 5 (2014): 2978–3036. https://doi.org/10.1021/cr400374x
110. Kaur, Harpreet, Abhinay Thakur, Ramesh Chand Thakur, and Ashish Kumar. "A Review on Multifaceted Role of Ionic Liquids in Modern Energy Storage Systems: From Electrochemical Performance to Environmental Sustainability." *Energy & Fuels* 39, 8 (2025): 3703–3734. https://doi.org/10.1021/acs.energyfuels.4c05274
111. Singh, Abhishek, Rakesh Bhardwaj, Rajendra K. Mishra, Arun K. Sundramoorthy, Vikas Gupta, and Sandeep Arya. "Potential of Functional Gel Polymers as Electrolytes for Supercapacitors." *Ionics* 29, 10 (2023): 3831–3851. https://doi.org/10.1021/cr400374x
112. Roth, Eric P., and Charles J. Orendorff. "How Electrolytes Influence Battery Safety." *The Electrochemical Society Interface* 21, 2 (2012): 45. https://doi.org/10.1149/2.F02122if
113. Mazloomian, Katrina, Thomas R. Dore, Mark Buckwell, Liam Bird, Paul R. Shearing, and Thomas S. Miller. "Supercapacitor Safety: Temperature Driven Instability and Failure of Electrochemical Double Layer Capacitors." *Energy Storage Materials* (2025): 104115. https://doi.org/10.1016/j.ensm.2025.104115
114. Shrivastava, Ankur, Mohammad H. Azarian, Carlos Morillo, Bhanu Sood, and Michael Pecht. "Detection and Reliability Risks of Counterfeit Electrolytic Capacitors." *IEEE Transactions on Reliability* 63, 2 (2014): 468–479. https://doi.org/10.1109/TR.2014.2315914
115. Wang, Haizhou, Zhen Chen, Zhiqiang Ji, Pei Wang, Jihua Wang, Weiping Ling, and Yiguang Huang. "Temperature Adaptability Issue of Aqueous Rechargeable Batteries." *Materials Today Energy* 19 (2021): 100577. https://doi.org/10.1016/j.mtener.2020.100577
116. Jiang, Xiaohua, Jia Jia, Yu Zhu, Jing Li, Hong Jia, Chao Liu, and Guojun Zhu. "Beyond Conventional Limits: Advancements and Insights in Broadening Operating Temperature Ranges of Supercapacitors." *Energy Storage Materials* (2024): 103462. https://doi.org/10.1016/j.ensm.2024.103462.

117. Tan, Songhua, Yujia Ji, Zhiqiang Zhang, and Yi Yang. "Recent Progress in Research on High-Voltage Electrolytes for Lithium-Ion Batteries." *ChemPhysChem* 15, 10 (2014): 1956–1969. https://doi.org/10.1002/cphc.201402175
118. Zhou, Tian, Cheng Gui, Lihong Sun, Yulong Hu, Haibo Lyu, Zhiwen Wang, and Guihua Yu. "Energy Applications of Ionic Liquids: Recent Developments and Future Prospects." *Chemical Reviews* 123, 21 (2023): 12170–12253. https://doi.org/10.1021/acs.chemrev.3c00391
119. Zhang, Qi, and John N. M. Shreeve. "Energetic Ionic Liquids as Explosives and Propellant Fuels: A New Journey of Ionic Liquid Chemistry." *Chemical Reviews* 114, 20 (2014): 10527–10574. https://doi.org/10.1021/cr500364t
120. Vishweswariah, Karthik, Anil Kumar Madikere Raghunatha Reddy, and Karim Zaghib. "Beyond Organic Electrolytes: An Analysis of Ionic Liquids for Advanced Lithium Rechargeable Batteries." *Batteries* 10, 12 (2024): 436. https://doi.org/10.3390/batteries10120436
121. Volfkovich, Yury M. "Supercapacitors for Extreme Temperatures: A Review." *Russian Journal of Electrochemistry* 60, 2 (2024): 93–115. https://doi.org/10.1134/S1023193524020022
122. Ike, Ikechukwu S., Ioannis J. Sigalas, Simeon E. Iyuke, and Emmanuel E. Kalu. "The Contributions of Electrolytes in Achieving the Performance Index of Next-Generation Electrochemical Capacitors (ECs)." *Electrochemical Devices for Energy Storage Applications* (2019): 215–248. https://doi.org/10.1134/S1023193524020022
123. Ma, Xinxin, Dongming Zhang, Juewen Wen, Ling Fan, Arun M. Rao, and Bing Lu. "Sustainable Electrolytes: Design Principles and Recent Advances." *Chemistry–A European Journal* 30, 36 (2024): e202400332. https://doi.org/10.1002/chem.202400332
124. Jaumaux, Pierre, Jun Wu, Dheepak Shanmukaraj, Yujie Wang, Dawei Zhou, Ben Sun, and Guangzhi Wang. "Non-Flammable Liquid and Quasi-Solid Electrolytes toward Highly-Safe Alkali Metal-Based Batteries." *Advanced Functional Materials* 31, 10 (2021): 2008644. https://doi.org/10.1002/adfm.202008644
125. Guo, Wei, Cheng Yu, Shuang Li, and Jun Qiu. "Toward Commercial-Level Mass-Loading Electrodes for Supercapacitors: Opportunities, Challenges, and Perspectives." *Energy & Environmental Science* 14, 2 (2021): 576–601. https://doi.org/10.1039/D0EE02649B
126. Xie, Yijun, Hui Zhang, Huan Hu, and Zhen He. "Large-Scale Production and Integrated Application of Micro-Supercapacitors." *Chemistry–A European Journal* 30, 21 (2024): e202304160. https://doi.org/10.1002/chem.202304160
127. Darling, Robert M., Kevin G. Gallagher, Jeffrey A. Kowalski, Seungbum Ha, and Fikile R. Brushett. "Pathways to Low-Cost Electrochemical Energy Storage: A Comparison of Aqueous and Nonaqueous Flow Batteries." *Energy & Environmental Science* 7, 11 (2014): 3459–3477. https://doi.org/10.1149/MA2015-01/3/660
128. Innocenti, Andrea, Henry Adenusi, and Stefano Passerini. "Assessing n-Type Organic Materials for Lithium Batteries: A Techno-Economic Review." *InfoMat* 5, 11 (2023): e12480. https://doi.org/10.1002/inf2.12480
129. Gaines, Linda, and Roy Cuenca. "Life-Cycle Costs of Lithium-Ion Vehicle Batteries." *SAE Transactions* (2000): 1920–1931. https://doi.org/10.4271/2000-01-2503
130. Eshetu, Gebrehiwot G., Michel Armand, Hiroshi Ohno, Bruno Scrosati, and Stefano Passerini. "Ionic Liquids as Tailored Media for the Synthesis and Processing of Energy Conversion Materials." *Energy & Environmental Science* 9, 1 (2016): 49–61. DOI: 10.1039/C5EE02284C
131. Eshetu, Gebrehiwot G., Michel Armand, Bruno Scrosati, and Stefano Passerini. "Energy Storage Materials Synthesized from Ionic Liquids." *Angewandte Chemie International Edition* 53, 49 (2014): 13342–13359. https://doi.org/10.1039/C5EE02284C

132. Fang, Zhaoyang, Peng Li, and Guihua Yu. "Gel Electrocatalysts: An Emerging Material Platform for Electrochemical Energy Conversion." *Advanced Materials* 32, 39 (2020): 2003191. https://doi.org/10.1002/adma.202003191
133. Gür, Turgut M. "Review of Electrical Energy Storage Technologies, Materials and Systems: Challenges and Prospects for Large-Scale Grid Storage." *Energy & Environmental Science* 11, 10 (2018): 2696–2767. https://doi.org/10.1039/C8EE01419A
134. Pandolfo, Tony, Vanessa Ruiz, Seepalakottai Sivakkumar, and Jawahr Nerkar. "General Properties of Electrochemical Capacitors." *Supercapacitors: Materials, Systems, and Applications* (2013): 69–109. https://doi.org/10.1002/9783527646661
135. Chen, Min, Jiamin Zhang, Xiang Ji, Jun Fu, and Guoying Feng. "Progress on Predicting the Electrochemical Stability Window of Electrolytes." *Current Opinion in Electrochemistry* 34 (2022): 101030. https://doi.org/10.1016/j.coelec.2022.101030
136. Li, Kai Bing, Dawei Shi, Zhenyu Cai, Guoliang Zhang, Qiang A. Huang, Dawei Liu, and Chunping Yang. "Studies on the Equivalent Serial Resistance of Carbon Supercapacitor." *Electrochimica Acta* 174 (2015): 596–600. https://doi.org/10.1016/j.electacta.2015.06.143
137. Azmi, Siti, Andrzej Klimek, and Elzbieta Frackowiak. "Why Electrochemical Capacitor Electrolytes Should Not Be Ignored?" *Electrochimica Acta* 452 (2023): 142347. https://doi.org/10.1016/j.electacta.2023.142347
138. Lewandowski, Andrzej, Angelika Olejniczak, Maciej Galinski, and Izabela Stepniak. "Performance of Carbon–Carbon Supercapacitors Based on Organic, Aqueous and Ionic Liquid Electrolytes." *Journal of Power Sources* 195, 17 (2010): 5814–5819. https://doi.org/10.1016/j.jpowsour.2010.03.082
139. Khademi, Mahmoud, and Dominik P. J. Barz. "Structure of the Electrical Double Layer Revisited: Electrode Capacitance in Aqueous Solutions." *Langmuir* 36, 16 (2020): 4250–4260. https://doi.org/10.1021/acs.langmuir.0c00024
140. Sun, Le, Kai Zhuo, Ying Chen, Qiang Du, Shuang Zhang, and Jun Wang. "Ionic Liquid-Based Redox Active Electrolytes for Supercapacitors." *Advanced Functional Materials* 32, 27 (2022): 2203611. https://doi.org/10.1002/adfm.202203611
141. Navalpotro, Paula, Jesús Palma, Marc Anderson, and Rebeca Marcilla. "High Performance Hybrid Supercapacitors by Using Para-Benzoquinone Ionic Liquid Redox Electrolyte." *Journal of Power Sources* 306 (2016): 711–717. https://doi.org/10.1016/j.jpowsour.2015.12.103
142. Hsieh, Yi-Yan, and Hsing-Yu Tuan. "Emerging Trends and Prospects in Aqueous Electrolyte Design: Elevating Energy Density and Power Density of Multivalent Metal-Ion Batteries." *Energy Storage Materials* (2024): 103361. https://doi.org/10.1016/j.ensm.2024.103361
143. Chen, Rong. "Toward High-Voltage, Energy-Dense, and Durable Aqueous Organic Redox Flow Batteries: Role of the Supporting Electrolytes." *ChemElectroChem* 6, 3 (2019): 603–612. https://doi.org/10.1002/celc.201801505
144. Kim, Eunhwan, Juyeon Han, Seokgyu Ryu, Youngkyu Choi, and Jeeyoung Yoo. "Ionic Liquid Electrolytes for Electrochemical Energy Storage Devices." *Materials* 14, 14 (2021): 4000. https://doi.org/10.3390/ma14144000
145. Shahzad, Shazia, Asim Shah, Ehsan Kowsari, Faiza Jan Iftikhar, Ammar Nawab, Bruno Piro, and Yu Zou. "Ionic Liquids as Environmentally Benign Electrolytes for High-Performance Supercapacitors." *Global Challenges* 3, 1 (2019): 1800023. https://doi.org/10.1002/gch2.201800023
146. Satpathy, Sandeep, Sujit Das, and Bidyut Kumar Bhattacharyya. "How and Where to Use Super-Capacitors Effectively: An Integration of Review of Past and New Characterization Works on Super-Capacitors." *Journal of Energy Storage* 27 (2020): 101044. https://doi.org/10.1016/j.est.2019.101044

147. Braham, Anis, Ali Lahyani, Pierre Venet, and Naceur Rejeb. "Recent Developments in Fault Detection and Power Loss Estimation of Electrolytic Capacitors." *IEEE Transactions on Power Electronics* 25, 1 (2009): 33–43. https://doi.org/10.1109/TPEL.2009.2026749
148. Basile, Anthony, Mark Hilder, Fariborz Makhlooghiazad, Carlos Pozo-Gonzalo, David R. MacFarlane, Peter C. Howlett, and Michael Forsyth. "Ionic Liquids and Organic Ionic Plastic Crystals: Advanced Electrolytes for Safer High-Performance Sodium Energy Storage Technologies." *Advanced Energy Materials* 8, 17 (2018): 1703491. https://doi.org/10.1002/aenm.201703491
149. Forsyth, Maria, Patrick C. Howlett, Anthony E. Somers, Douglas R. MacFarlane, and Andrew Basile. "Interphase Engineering of Reactive Metal Surfaces Using Ionic Liquids and Deep Eutectic Solvents—From Corrosion Control to Next-Generation Batteries." *npj Materials Degradation* 1, 1 (2017): 18. https://doi.org/10.1038/s41529-017-0016-z
150. Chen, Shengnan, Min Zhang, Pei Zou, Bin Sun, and Shixin Tao. "Historical Development and Novel Concepts on Electrolytes for Aqueous Rechargeable Batteries." *Energy & Environmental Science* 15, 5 (2022): 1805–1839. https://doi.org/10.1039/D2EE00004K
151. Liao, Yuan, Hui Zhang, Yushi Peng, Yipeng Hu, Jialin Liang, Zhen Gong, and Yicheng Yang. "Electrolyte Degradation during Aging Process of Lithium-Ion Batteries: Mechanisms, Characterization, and Quantitative Analysis." *Advanced Energy Materials* 14, 18 (2024): 2304295. https://doi.org/10.1002/aenm.202304295
152. Galiński, Mariusz, Andrzej Lewandowski, and Iwona Stępniak. "Ionic Liquids as Electrolytes." *Electrochimica Acta* 51, 26 (2006): 5567–5580. https://doi.org/10.1016/j.electacta.2006.03.016
153. Zhou, Wei, Min Zhang, Xue Kong, Wei Huang, and Qi Zhang. "Recent Advance in Ionic-Liquid-Based Electrolytes for Rechargeable Metal-Ion Batteries." *Advanced Science* 8, 13 (2021): 2004490. https://doi.org/10.1002/advs.202004490
154. Huang, Jing, Xian Dong, Zhiqiang Guo, and Yi Wang. "Progress of Organic Electrodes in Aqueous Electrolyte for Energy Storage and Conversion." *Angewandte Chemie* 132, 42 (2020): 18478–18489. https://doi.org/10.1002/ange.202004563
155. Beck, Fritz, and Peter Rüetschi. "Rechargeable Batteries with Aqueous Electrolytes." *Electrochimica Acta* 45, 15–16 (2000): 2467–2482. https://doi.org/10.1016/S0013-4686(00)00344-3
156. Ye, Yun-Sheng, John Rick, and Bing-Joe Hwang. "Ionic Liquid Polymer Electrolytes." *Journal of Materials Chemistry A* 1, 8 (2013): 2719–2743. https://doi.org/10.1039/C2TA00126H
157. Xu, Chao, and Zhijun Cheng. "Thermal Stability of Ionic Liquids: Current Status and Prospects for Future Development." *Processes* 9, 2 (2021): 337. https://doi.org/10.3390/pr9020337
158. Nian, Qi, Tianhao Sun, Shuai Liu, Hui Du, Xinyu Ren, and Zhiqiang Tao. "Issues and Opportunities on Low-Temperature Aqueous Batteries." *Chemical Engineering Journal* 423 (2021): 130253. https://doi.org/10.1016/j.cej.2021.130253
159. Liang, Yan, and Ying Yao. "Designing Modern Aqueous Batteries." *Nature Reviews Materials* 8, 2 (2023): 109–122. https://doi.org/10.1038/s41578-022-00511-3
160. Nowak, Sascha, and Martin Winter. "Chemical Analysis for a Better Understanding of Aging and Degradation Mechanisms of non-Aqueous Electrolytes for Lithium Ion Batteries: Method Development, Application and Lessons Learned." *Journal of the Electrochemical Society* 162, 14 (2015): A2500. https://doi.org/10.1149/2.0121514jes

6 Fabrication Techniques of Supercapacitors

Mohamed Ahmed Belal, Ahmed M. Bayoumy, Mohamed Magdy Mohamed Abdo, and Ahmed Abd El-Moneim

6.1 INTRODUCTION

Contemporary energy storage devices, such as supercapacitors (SCs), have attracted considerable attention due to their remarkable power and energy densities, fast charging and discharging capabilities, long cycling stability, environmental sustainability, and operational safety. In contrast to conventional capacitors, which store charges solely through the electrostatic physical adsorption of charges on the electrode and electrolyte interface, SCs utilize both physical and electrochemical mechanisms, significantly enhancing their power density and cyclic stability. Therefore, SCs are extensively used in various electronic applications in wearable electronics to power transient microscale devices such as strain, bio, gas, and force sensors [1–5]. Their characteristics make them highly attractive for portable electronics, electric vehicles, renewable energy systems, and industrial power backups.

It is important to note that SC fabrication techniques such as screen, 3D, inkjet, gravure, flexographic printing, sputtering, spin and spray coating, lithography, laser engraving, and conventional methods significantly impact both device performance and commercial viability [2–10]. In addition to electrode fabrication methods, several other critical factors influence performance, such as optimizing the active materials, electrolyte formulation, selection of appropriate flexible substrates, and the advancement of packaging techniques, particularly for cylindrical supercapacitor (SC) configurations [11–25].

This chapter provides a comprehensive examination of various SCs fabrication techniques, including traditional approaches and innovative printing technologies. It also discusses the earlier-mentioned factors, including electrode synthesis, electrolyte systems, substrate selection, advanced assembly and packaging processes, as shown in Figure 6.1.

6.2 ELECTRODE FABRICATION TECHNIQUES

The development of SCs has progressed significantly due to the emergence of sophisticated printing and deposition techniques, which enable scalable and cost-effective manufacturing. Several techniques, including inkjet printing, spray and spin coating, and vacuum filtration, are widely used for depositing graphene and

DOI: 10.1201/9781003535836-6

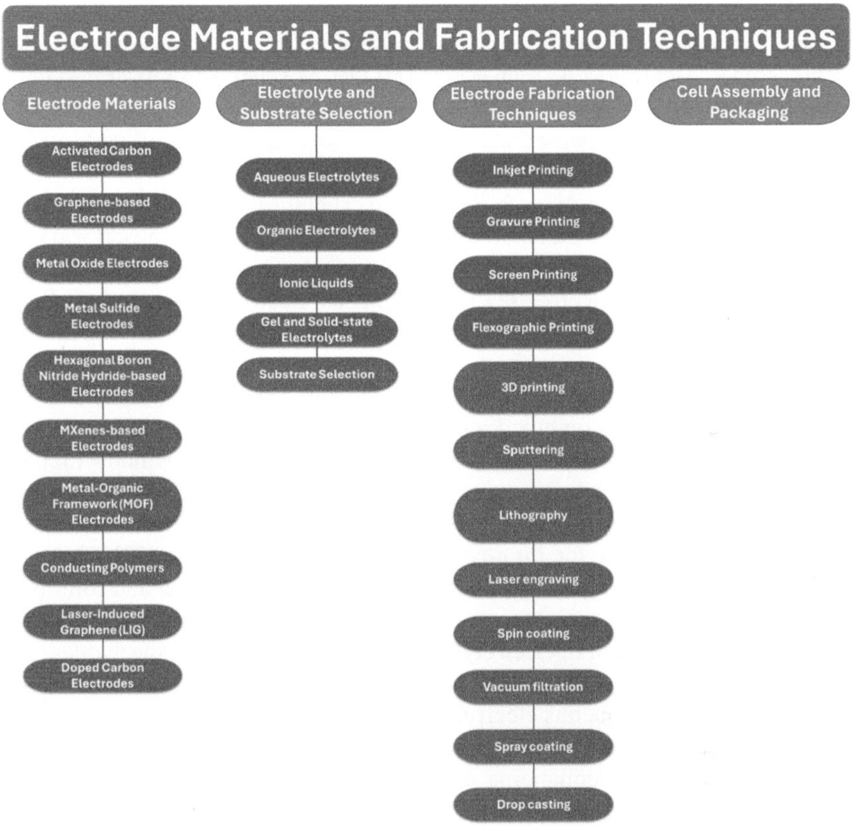

FIGURE 6.1 Schematic diagram of the electrode materials and fabrication techniques of SCs.

carbon-based materials onto flexible substrates, making them particularly suitable for wearable applications, and electrode fabrication processes [5, 23]. Inkjet printing is a prevalent technique for the precise deposition of functional inks onto flexible substrates. For large-scale production, gravure and screen-printing techniques are commonly employed, enabling the transfer of conductive pastes onto electrodes through engraved rollers or mesh screens with optimized rheology. Moreover, flexographic printing, a roll-to-roll (R2R) process, facilitates the continuous application of electrode materials [2]. Additional advanced techniques include 3D printing, which enables the fabrication of customized electrode structures with complex geometries [26]. Sputtering and lithography techniques are employed to deposit thin films and pattern electrodes with nanoscale precision [27]. Furthermore, laser engraving has proven to be an effective method for fabricating carbon-based electrodes and enhancing their electrochemical characteristics [28].

The implementation of advanced printing techniques is crucial for enhancing the performance and scalability of SC manufacturing. Electrode active materials are the primary components of SCs, significantly influencing their capacitance, conductivity, energy and power densities. A wide variety of materials are commonly utilized during the fabrication, including carbon-based materials, metal oxides/hydroxides, metal sulfides, Metal-organic frameworks (MOFs), MXenes, boron hydrides, laser-induced graphene (LIG), and conductive polymers [3–5, 29]. All of these material exhibits distinct electrochemical characteristics that affect the overall performance and efficiency of the SC device.

The selection of electrolytes in SCs is also crucial for determining the charge storage mechanism, ion transport efficiency, and overall device performance. Various electrolyte categories are employed, such as aqueous, organic, ionic liquid, and gel-based electrolytes [30]. Aqueous electrolytes, including KOH, H_2SO_4, and Na_2SO_4, exhibit high ionic conductivity and are extensively utilized in SCs [31; however, their limited potential window constrains the achievable energy density. Organic electrolytes, which include acetonitrile (ACN) and propylene carbonate (PC) combined with dissolved salts like tetraethylammonium tetrafluoroborate ($TEABF_4$), offer a broader operating potential window [32], but require careful handling due to their flammability. Ionic liquids, which are room-temperature molten salts, offer exceptional thermal and electrochemical stability, making them ideal for high-performance SC applications [33]. However, their high cost limits their use in many commercial contexts. Gel and solid-state electrolytes have become increasingly favored for flexible and wearable SCs, addressing issues related to electrolyte leakage while enhancing mechanical stability [23].

To explain more, aqueous electrolytes, including KOH, H_2SO_4, and Na_2SO_4, offer excellent ionic conductivity at a low cost, facilitating rapid ion diffusion and consequently enhancing power density. However, their limited electrochemical potential window, typically around 1.2 V, restricts the achievable energy density in SCs employing these electrolytes [21]. On the other hand, organic electrolytes, made up of solvents such as $TEABF_4$ dissolved in ACN or PC, provide a broader voltage window (~2.5–3 V) in comparison to aqueous electrolytes [34]. The increasing in voltage results in an augmented energy density. However, organic electrolytes introduce challenges of meticulous handling, their flammability, and toxicity which are considered as drawbacks in electrolyte formulation [35].

Ionic liquids consist only of ions and are characterized by their non-volatility and thermal stability as electrolytes. Their electrochemical potential window is remarkably broad, approximately 3.0–6.0 V, which renders them ideal for high-energy SCs [36, 37]. However, their inherently high viscosity can hinder ion transport, thereby affecting power density [34, 37]. Advances in gel and solid-state electrolytes have enhanced both safety and mechanical flexibility. Gel electrolytes are composed of polymer matrices such as polyvinyl alcohol (PVA), facilitating the development of flexible and wearable all-solid-state SCs [3]. Moreover, substrate selection plays a critical role in device performance and flexibility, ranging from rigid conductive metals and steel meshes to flexible, lightweight carbon cloth/paper and polymer films such as polyethylene terephthalate (PET) and polyimide (PI), suitable for wearable technologies due to the high mechanical stability [4].

6.2.1 Inkjet Printing

Inkjet printing is a highly precise, template-free, and non-invasive method for depositing functional inks onto both flexible and rigid surfaces in a controlled manner [38]. This method offers significant advantages for fabricating patterned electrodes with well-defined geometries. Inkjet printing utilizes low-viscosity liquid dispersions containing a wide variety of active materials, including graphene, carbon nanotubes, metal oxides, or conducting polymers, as illustrated in Figure 6.2 [3]. The process is digitally controlled, enabling drop-on-demand (DOD) deposition without the need for physical masks or templates. A key benefit of inkjet printing is its ability to deposit ultrathin, highly uniform layers, thereby minimizing material consumption [4]. Furthermore, it facilitates the fabrication of multi-layered structures, promoting the development of complex electrode architectures with tailored properties [6]. However, challenges such as controlling ink viscosity, preventing nozzle blockages,

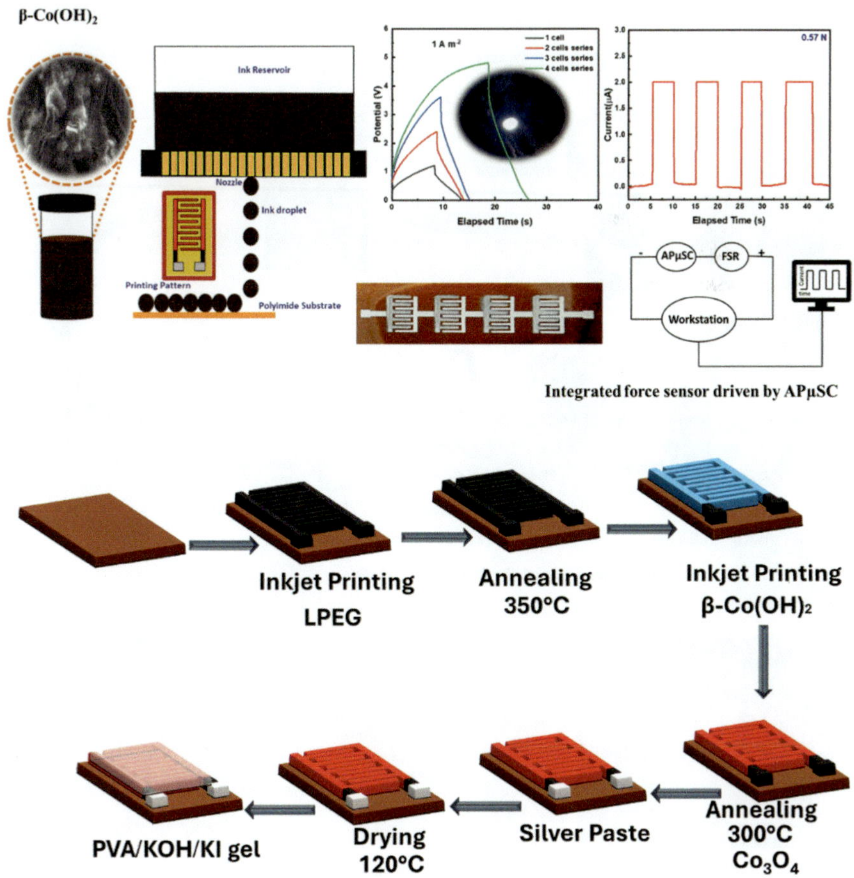

FIGURE 6.2 Schematic diagram of inkjet printing of a supercapacitor. (Reprinted with permission from [3]. © Elsevier 2024.)

and the need for compatible substrates have to be addressed to fabricate high resolution, flexible, lightweight, and high-performance SCs.

Moreover, Belal et al. [3] developed a cost-effective method for synthesizing liquid phase exfoliated graphene (LPEG) using a kitchen blender, which was subsequently formulated into functional inks for inkjet printing of SC electrodes to be integrated with other materials such as metal oxides or sulfides, as shown in Figure 6.3. Reduced graphene oxide (rGO), known for its superior conductivity, is produced from graphene oxide (GO) via chemical, thermal, or electrochemical reduction methods. Moreover, the combination of graphene with metal oxides, metal sulfides, conductive polymers, or carbon nanotubes (CNTs) synergistically enhances electrochemical performance by integrating pseudocapacitive mechanisms with electric double-layer capacitance (EDLC). The exceptional characteristics of graphene such as electrical conductivity, mechanical strength, and large theoretical surface area have made it a focal material for SC applications. A variety of fabrication techniques are utilized to create graphene-based electrodes.

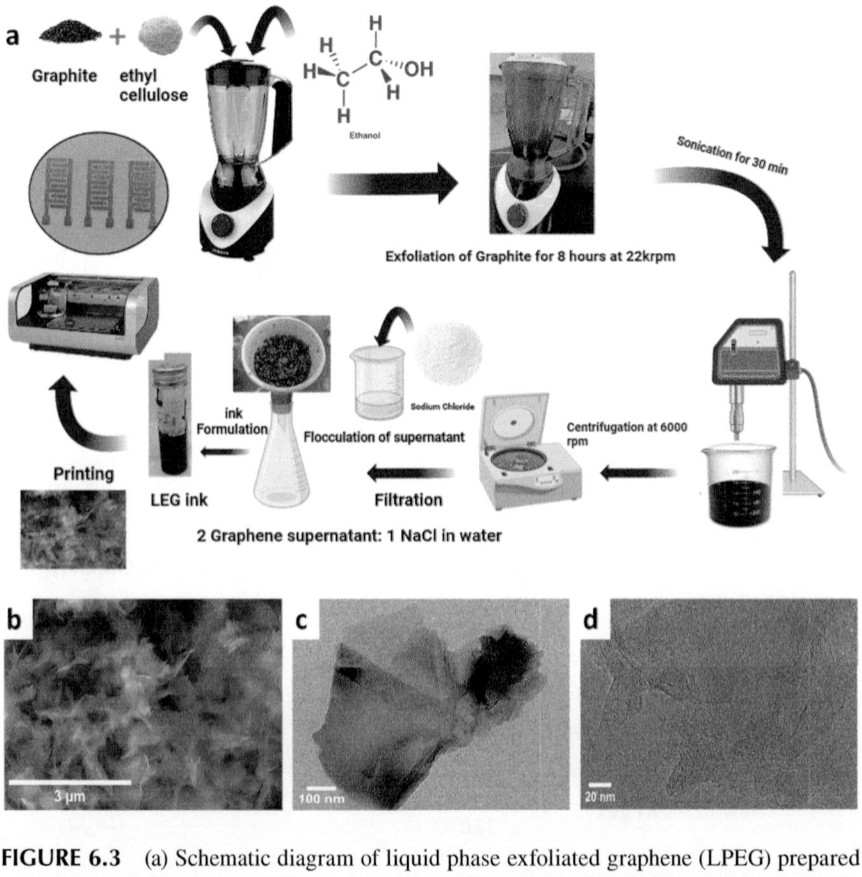

FIGURE 6.3 (a) Schematic diagram of liquid phase exfoliated graphene (LPEG) prepared using a kitchen blender, (b) SEM, and (c and d) TEM. (Reprinted with permission from [3]. © Elsevier 2024.)

Chemical vapor deposition (CVD) is a highly promising technique for producing high-quality graphene films on metal substrates like copper or nickel at high temperature [39]. Additionally, plasma-enhanced CVD has been used to grow vertical graphene (VG), which has attracted significant interest for SC applications [40]. Alternatively, other methods focused on solutions entail the exfoliation of GO, succeeded by chemical, laser, or thermal reduction to yield rGO [41–44], which serves as viable electrode materials. Moreover, blending graphene with GO has been reported to be an efficient way to fabricate high-performance SC devices. For example, Bayoumy et al. [6] employed the MICRODROP Inkjet printing technique to fabricate interdigitated SC electrodes (IDE-SCs) of G/GO blends, achieving high electrochemical performance alongside notable mechanical flexibility.

Conductive polymers, including polyaniline (PANI), polypyrrole (PPy), and Poly(3,4-ethylenedioxythiophene) (PEDOT), demonstrate remarkable capacitance and flexibility, making them highly suitable for wearable and flexible SCs applications [21–23]. Typically, conducting polymer electrodes are synthesized via electropolymerization, wherein monomers are electrochemically polymerized onto conductive substrates, forming uniform films [45]. Alternatively, inkjet printing can be employed, allowing the deposition of polymer inks onto flexible substrates, which are then dried to form conductive layers [21]. These fabrication techniques enable the effective integration of conductive polymers into advanced energy storage devices.

Metal oxides are extensively utilized in SC electrode fabrication due to their ability to store charges through reversible redox reactions, a mechanism known as pseudocapacitance. MOs include several materials, such as manganese dioxide (MnO_2), ruthenium dioxide (RuO_2), nickel oxide (NiO), and cobalt oxide (Co_3O_4), which have been widely studied for their high theoretical capacitance [3, 4, 46–48]. Electrodeposition is a commonly employed method for fabricating metal oxide electrodes, exemplified by PEDOT-coated manganese dioxide nanospheres used in flexible asymmetric planar SCs [49]. This method involves dissolving a metal salt precursor in an electrolyte solution, followed by applying an electric current to deposit a uniform metal oxide film onto a conductive substrate. Furthermore, Xia et al. [50] demonstrated the electrodeposition of mesoporous Fe_2O_3 and MnO_2 nanoflakes on inkjet-printed graphene to create an asymmetric SC, with Fe_2O_3 as the negative electrode, and MnO_2 as the positive electrode. This technique allows precise control over film thickness and morphology, resulting in enhanced electrochemical performance.

Sol–gel processing involves the hydrolysis and condensation of metal precursors to form metal oxide nanoparticles. The resulting sol–gel mixture is subsequently applied onto a substrate and subjected to heat treatment to produce a stable electrode material with well-defined porosity [51]. Additionally, hydrothermal synthesis is commonly employed to produce nanostructured metal oxides [52]. This approach entails the reaction of metal precursors within a sealed autoclave under high temperatures and pressures, resulting in the formation of crystalline nanostructures with high surface area and enhanced ion transport properties [53].

6.2.2 Gravure Printing

Gravure printing is a high-speed, R2R method widely employed for the mass production of printed electronics, such as SC electrodes. This process uses an engraved roller to continuously deposit conductive pastes or inks onto a substrate. The design and intricacies of the engraved cells on the roller play a crucial role in influencing both the thickness and uniformity of the deposited layers. Gravure printing is highly effective in producing thin, uniform films of active materials with high reproducibility and compatibility with various materials, such as carbon-based inks, metal oxides, and polymeric conductors. The main advantage of gravure printing are its high throughput, cost efficiency, and ability to handle large-area substrates. However, achieving optimal deposition requires careful control of ink viscosity and substrate adhesion [54].

Secor et al. [55] developed a graphene-ethyl cellulose (EC) ink and optimized a gravure printing process to fabricate high-resolution, conductive graphene patterns. The ink was prepared by exfoliating graphite in ethanol with EC serving as a stabilizer, resulting in a high dispersion yield. Excess EC was subsequently removed through flocculation, enabling isolating the graphene-EC powder to be formulated as ink. This approach yielded predominantly few-layer graphene flakes, around 2 nm thick and approximately 50 nm × 50 nm in size, which are ideal for achieving the sub-micron precision required for high-resolution gravure printing.

6.2.3 Screen Printing

Screen printing is a well-established fabrication technique employed for the deposition of thick films as SC electrodes [5]. This process involves applying conductive paste or ink through a fine mesh screen onto a substrate, with a stencil outlining the desired pattern [7]. Screen printing offers remarkable versatility, allowing the use of diverse materials, such as carbon-based inks, conducting polymers, and metal oxides [2]. Screen printing's advantages lies in the ability to deposit thick and uniform layers, which improves the electrodes' mechanical stability and charge storage capacity. Additionally, this method is scalable and cost effective, positioning it as an ideal choice for producing commercial SCs, as shown in Figure 6.4. To achieve optimal performance, precise control over parameters such as screen mesh size, paste viscosity, and drying conditions is essential to fabricate high resolution films.

Hexagonal boron nitride (hBN) hydrides, attributed to their distinctive electron-rich structures, exhibit significant potential for achieving high theoretical capacitance and energy density, as demonstrated by Chaney et al. [9]. The fabrication of hBN electrodes involves in a systematic approach beginning with chemical synthesis, followed by electrochemical deposition onto conductive substrates. The properties of these materials indicate significant potential for redox reactions, making them promising candidate for high-performance SCs [7].

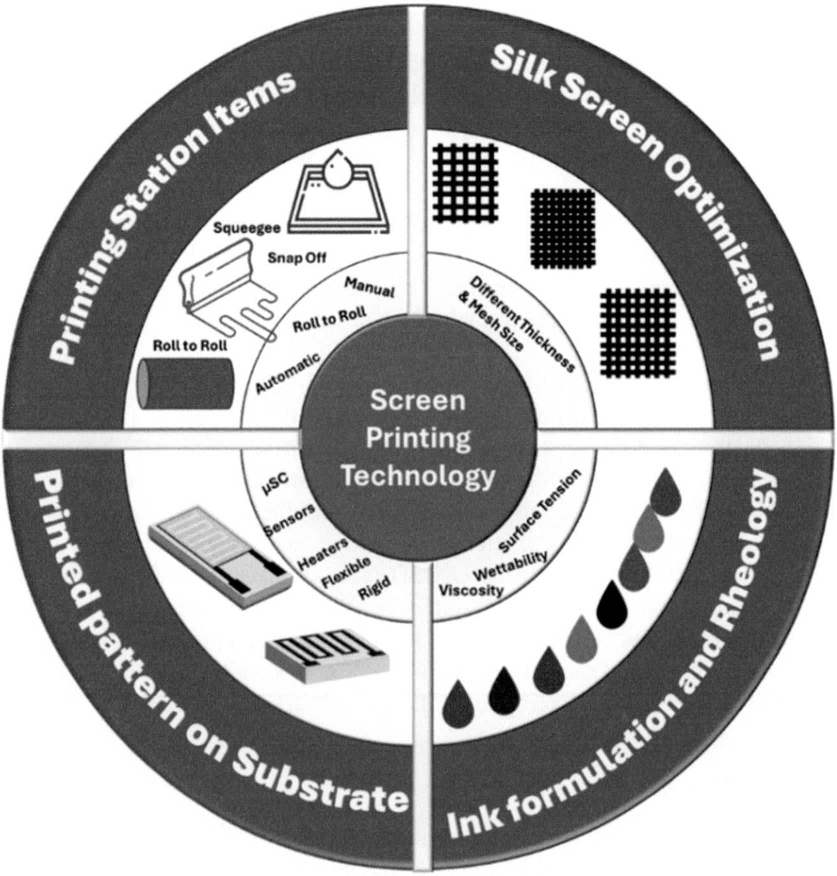

FIGURE 6.4 Schematic diagram of screen-printing technique process. (Reprinted with permission from [2]. © RSC 2025.)

6.2.4 Flexographic Printing

Flexographic printing represents a R2R method that enables the continuous deposition of electrode materials onto flexible substrates. In contrast to gravure printing, which employs engraved rollers, flexographic printing utilizes flexible relief plates to transfer ink onto a flexible substrate. This approach is especially effective for large scale production, due to its high speed and adaptability to different materials. Flexographic printing demonstrates compatibility with a wide range of conductive inks, such as carbon nanotubes (CNTs), graphene, and polymer composites. its ability to print on flexible and lightweight substrates positions it as an excellent choice for fabricating SCs for wearable and portable electronic devices. However, precise control of ink formulation and substrate uniformity is essential to consistent electrical performance [56].

6.2.5 3D Printing

The advent of 3D printing, or additive manufacturing, has revolutionized the fabrication of SC electrodes, allowing for the development of intricate and tailored geometries [26]. In contrast to conventional 2D deposition methods, 3D printing facilitates the layer by layer assembly of three-dimensional electrode structures, enabling customization of porosity and architectural design. This capability permits the fabrication of network-like structures that enhance the diffusion of the electrolyte ions throughout the electrode material, resulting improving electrochemical performance. Moreover, this technique offers a significant advantage by introducing hierarchical porosities within the electrode structure, which enhance charge storage capacity and accelerate ion transport. Materials such as graphene aerogels, metal oxide composites, and conductive polymers are widely employed in 3D printing for the development of high performance SCs. Despite its remarkable design flexibility, challenges remain concerning material compatibility, printing resolution, and scalability for commercial applications. Graphene-based materials have garnered considerable interest in SC research due to their exceptional electrical conductivity, mechanical strength, and high specific surface area. Additive manufacturing techniques, particularly 3D printing, have recently emerged as a leading research focus for the cost-effective and efficient production of flexible and miniaturized electrical components [6]. Graphene, composed of a single layer of carbon atoms arranged in a hexagonal lattice, is especially attractive as an electrode material owing to its exceptional theoretical surface area (2630 $m^2\ g^{-1}$) [57].

6.2.6 Sputtering

Sputtering is a thin film deposition technique in which a target material is bombarded with high-energy ions, resulting in the ejection of atoms that subsequently deposited onto a substrate. This technique is widely employed for the deposition of metal oxide layers, current collectors, and conductive films in SC electrodes. The main advantage of sputtering lies in its ability to produce highly uniform and conformal coatings with precise thickness control. Additionally, sputtering enables the deposition of materials that are difficult to process using solution-based methods. However, the high cost associated with vacuum equipment and the relatively slow deposition rates present notable challenges that need to be considered [27].

6.2.7 Lithography

Lithography is patterning technique that enables the production of microscale and nanoscale features within SC electrodes. This technique utilizes a light-sensitive photoresist exposed to UV light via a mask, resulting in precisely defined patterns that direct the deposition of active materials. Lithography offers exceptional control over the morphology of electrodes, resulting in enhanced electrochemical performance. This method is particularly advantageous for the fabrication of micro-SCs, which require high resolution and precise patterning. However, lithography involves a complex and costly procedure, making it less suitable for large scale production [58].

6.2.8 Laser Engraving

Laser engraving is an advanced technique used to modify electrode surfaces and enhance their electrochemical properties. By employing a high-energy laser beam allows for the selective ablation of a conductive material's surface, creating hierarchical porosity and enhancing the active surface area. A notable application of laser engraving in the fabrication of SCs is laser induced graphene (LIG), where a polymer precursor, such as polyimide (PI), is converted into porous graphene through a template-free and chemicals-free laser irradiation process. This method is rapid, cost effective, and environmentally friendly, making it particularly attractive for flexible and wearable SCs [59].

LIG is synthesized through laser irradiation of polymer substrates such as PI, resulting in porous graphene structures directly integrated onto flexible substrates [59]. This technique offers advantages including rapid, mask-free patterning, excellent electrical conductivity, and controllable porosity. LIG presents an innovative approach that converts polymer surfaces into porous graphene via laser processing [60, 61]. This approach provides an efficient and economical solution to produce high-performance SC electrodes [62, 63]. An innovative approach involves LIG, which utilizes direct laser writing on polymer substrates such as PI or polymer nanofiber to form porous graphene structures exhibiting outstanding electrochemical properties [61, 64]. All the early mentioned fabrication techniques reveal facilitated approaches for the creation of flexible and high-performance electrodes for SC applications.

The incorporation of heteroatoms such as nitrogen, sulfur, or phosphorus into graphene or other carbon-based materials significantly enhances their electrochemical characteristics [65]. Common methods for incorporating these heteroatoms into graphene structures include chemical doping and plasma treatment heteroatoms [28, 66, 67]. Among the various approaches to improve the performance of the carbon-based materials is doping and decoration with metal oxide nanoparticles. For instance, Z. Peng et al. achieved 165.0 F m^{-2} for boron-doped LIG [68], whereas M. Khandelwal et al. reported 1636.0 F m^{-2} for nitrogen and phosphorous co-doped LIG [69] on PI substrates. In their approach, the precursors of the heteroatoms were dispersed in polyamic acid (PAA) and coated on the PI substrate prior to the laser processing, which may represent a challenge for scalable fabrication. In addition, several metal oxides such as MnO_2, SnO_2, TiO_2, and NiO were embedded into the LIG and evaluated for IDE-SC devices. For instance, J. Zhao et al. reported MnO_2/LIG-based IDE-SC that has 150.4 F m^{-2} areal specific capacitance, which is much higher than that of bare LIG [70], while C. Zhu et al. reported a slightly higher areal capacitance of 185.8 F m^{-2} for SnO_2/LIG-based IDE-Sc [71]. Enhancements in the metal oxides/LIG IDE-SC performances were attributed to improved hydrophilicity and electrolyte wettability of LIG-based electrodes, in addition to pseudocapacitive contribution of metal oxide nanoparticles.

6.2.9 Spin Coating

The spin coating is a widely used technique for depositing uniform thin films, particularly for electrode materials and electrolytes. This technique involves applying a dispersed liquid to a rotating substrate, where the centrifugal force ensures even distribution of the material across the surface. The film thickness can be controlled by adjusting the spin speed and the viscosity of the solution. Spin coating is especially effective for fabricating gel and solid-state electrolyte layers in SCs. This technique offers several advantages, such as its simplicity, cost effectiveness, and ability to create highly uniform coatings. However, it is limited to planar substrates and may not be suitable for depositing thick film [72].

Metal sulfides, such as molybdenum disulfide (MoS_2) and nickel cobalt sulfide ($NiCo_2S_4$), have attracted significant interest as SC electrode materials due to their excellent electrical conductivity and outstanding electrochemical properties [43, 73]. These materials demonstrate significant pseudocapacitive characteristics, making them suitable high-energy storage applications [74, 75]. Hydrothermal and solvothermal synthesis techniques are frequently utilized to produce metal sulfide electrodes [76, 77]. These methods involve the reaction of metal precursors with sulfur-containing compounds under controlled temperature and pressure conditions, yielding nanostructured sulfides. The synthesized materials are subsequently incorporated into SC electrodes using techniques such as drop and spray coating, or inkjet printing. Alternatively, CVD can be employed, where sulfur vapors react with metal substrates to form thin metal sulfide films. This approach enables the fabrication of electrodes with exceptional uniformity with tunable thickness and morphology.

6.2.10 Vacuum Filtration

Vacuum filtration is a widely employed technique for fabricating nanostructured electrode films by filtering a suspension of active materials through porous membranes. This method is commonly utilized for the assembly of carbon nanotube films, graphene membranes, and MXenes electrodes. The primary advantage of vacuum filtration is its ability to produce films that are highly compact and uniform, with precise control over their thickness [78].

MXenes, a class of two-dimensional transition metal carbides and nitrides, exhibit outstanding electrical conductivity, hydrophilicity, and chemical stability, making them promising candidates for SC electrode materials [79, 80]. The fabrication of MXenes-based electrodes typically includes selective etching, wherein MAX phases (ternary carbides and nitrides) undergo treatment with hydrofluoric acid (HF) or alternative etching agents to eliminate undesired layers, yielding exfoliated MXenes sheets [81]. Moreover, Vacuum filtration is subsequently used to assemble MXenes films with tailored thickness and porosity. These electrodes demonstrate outstanding electrochemical performance, particularly in aqueous electrolyte systems [82].

6.2.11 Spray Coating

Spray coating is an efficient and scalable technique that involves the atomization of a liquid dispersion, which is then applied onto a substrate [83]. This approach is particularly effective for applying nanomaterial-based inks to large electrode surfaces. Spray coating enables precise control over thickness and ensures high uniformity, making it well-suited for flexible and wearable SCs. However, optimization of the spray parameters and ink formulation is essential to achieve consistent and uniform film quality [84].

6.2.12 Drop Casting

Drop casting is a simple fabrication technique involving the deposition of a liquid dispersion onto a substrate, allowed to dry and forming a thin film. This method is frequently used in proof-of-concept investigations; however, it is challenged related to poor film uniformity and limited scalability. Achieving consistent thickness and homogeneity is complex, making drop casting less suitable for large-scale applications.

Activated carbon (AC) is widely recognized as the predominant electrode material used in SCs commercial applications due to its cost-effective production, often utilizing waste materials such as polyurethane and *Prunus persica* seed stones, and large surface area, and ease of processing [85–89]. The high porosity of AC maximizes the interfacial contact between the electrode and electrolyte facilitating the diffusion of the electrolyte ions through the electrode material, and improving the charge storage capacity. Hence, AC can make a nanocomposite with different materials such as GO [90]. The process of synthesizing AC typically includes the carbonization of organic precursors such as *P. persica* seed stones, coconut shells, wood, or synthetic polymers, followed by activation via chemical treatments, such as H_3PO_4, KOH or $ZnCl_2$ or even via physical methods, including CO_2 or steam. For example, Aryal et al. [89] prepared AC from waste peach stones at 250°C for 4 h to make pre-carbonized char (PCC), followed by impregnation process in H_3PO_4 in a 1:1 ratio and dried at 80°C as shown in Figure 6.5i and ii, subsequent carbonization was performed under inert gas at 400 (HAC-4), 600 (HAC-6), and 800°C (HAC-8). The resulting high-surface-area carbons are integrated with conductive additives such as carbon black (BC) and polymeric binders such as polyvinylidene

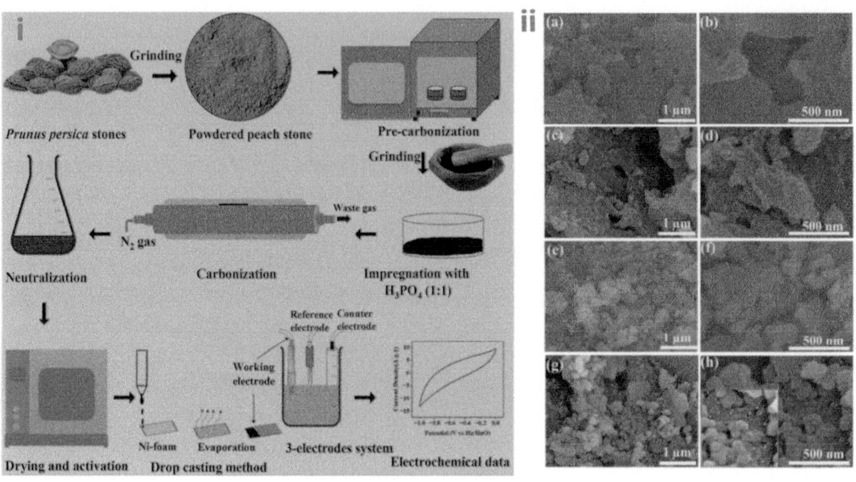

FIGURE 6.5 (i) Schematic illustration of the preparation of AC from *Prunus persica* seed stone, and (ii) structural morphology: FESEM image (a and b) PCC, (c and d) HAC-4 (e and f) HAC-6 and (g and h) HAC-8. (Reprinted with permission from [89]. © Elsevier 2025.)

TABLE 6.1
Comparison between the Different Electrode Materials and Fabrication Techniques and their Corresponding Electrochemical Parameters (C_{Sp}, E, and P) Recently Reported in Literature

Device Materials	Technique (Substrate)	Electrolyte	Cell Type/ (PW[a])	Capacitance ($F\ m^{-2}$)	E ($mWh\ m^{-2}$)/ P ($W\ m^{-2}$)	Capacitance Retention % (Cycle)	Ref.
Co_3O_4@LPEG	Inkjet (Kapton)	PVA/KOH/KI	Symm. (0.4 V)	89 $F\ m^{-2}$ @ 0.2 $A\ m^{-2}$	1.7/0.0388	–	[3]
β-Co_3O_4/LPEG// LPEG	Inkjet (Kapton)	PVA/KOH/KI	Asymm. (1.2 V)	7.3 $F\ m^{-2}$ @ 1 $A\ m^{-2}$	1/0.5	89.76 (2500)	[3]
Fe-MnO_2@PEDOT[(I)]	Inkjet (Kapton)	PVA/LiCl	Symm. (1.0 V)	12 $F\ m^{-2}$ @ 0.3 $A\ m^{-2}$	1.469/0.143	78.7 (5200)	[22]
δ-MnO_2@PEDOT[(I)]	Inkjet (Kapton)	PVA/LiCl	Symm. (0.8 V)	1.44 $F\ m^{-2}$ @ 0.03 $A\ m^{-2}$	0.108/0.0108	78 (3600)	[21]
Graphene	Inkjet (Si Wafer)	PVA/H_3PO_4	Symm. (1.0 V)	1.3 $F\ m^{-2}$ @ 5 $mV\ s^{-1}$	–	88 (1000)	[97]
MnO_2/G	Pulse-current deposition (Kapton)	PVA/LiCl	Symm. (0.8 V)	7@ 5 $mV\ s^{-12}$	–	–	[50]
Fe_2O_3/MnO_2 (I)	Pulse-current deposition (Kapton)	PVA/LiCl	Asymm. (1.6 V)	19 $F\ m^{-2}$ @ 1 $A\ m^{-2}$	6.8/0.045	96 (10,000)	[50]
Graphene (EEG)	Inkjet (Kapton)	PSSH	Symm. (1.0 V)	2 $F\ m^{-2}$ @ 0.03 $A\ m^{-2}$	0.25/0.025	–	[98]
MnO_2 nanoflowers	Inkjet (Paper)	PVA/LiCl	Symm. (1.0 V)	6.8 $F\ m^{-2}$ @ 0.26 $A\ m^{-2}$	0.10/0.011	80.8 (3000)	[99]
$K_2Co_3(P_2O_7)_2$. $2H_2O$ // Graphene (I)	Inkjet (PET)	PVA/KOH	Asymm. (1.07 V)	4.2 $F\ m^{-2}$ @ 0.007 $A\ m^{-2}$	0.67/0.038	94.4 (5000)	[100]
Bis-terpyridyl-based molecular cobalt complexes	Inkjet (Kapton)	PVA/LiCl	Symm. (0.9 V)	10.9 $F\ m^{-2}$ @ 1 $A\ m^{-2}$	–	97 (10,000)	[101]
δ-MnO_2@EEG	Inkjet (Kapton)	PVA/LiCl	Symm. (0.9 V)	17.0 @ 0.1 $A\ m^{-2}$	1.7/0.048	97.65 (2000)	[4]
δ-MnO_2@EEG//EEG	Inkjet (Kapton)	PVA/LiCl	Asymm. (1.5 V)	10.75 $F\ m^{-2}$ @ 0.3 $A\ m^{-2}$	–	–	[4]
Graphene	Aerosol jet printing (Kapton)	EMIM-BF4 organic electrolyte	Symm. (1.0 V)	0.7 $F\ m^{-2}$ @ 0.05 $A\ m^{-2}$	–	66.6 (10,000)	[102]
N-doped sucrosecarbon	Inkjet (Glass)	PVA/H_2SO_4	Symm. (1.0 V)	39 $F\ m^{-2}$	–	96(10,000)	[103]
Hexagonal boron nitride	Screen printing (Kapton)	PVA/ H_3PO_4	Symm. (1.0 V)	10.0 $F\ m^{-2}$	–	–	[104]
GQDs/PANi	Electrodeposition (Kapton)	PVA/H_3PO_4	Symm. (0.8 V)	2.1 @ 0.15 $A\ m^{-2}$	0.29/74.6	85.6 (1500)	[105]
GO@Cu/Au film	In situ reduction and assembly (Kapton)	PVA/H_2SO_4	Symm. (0.8 V)	9.5 @ 1 $mV\ s^{-1}$		98.3% (11,000)	[106]

[a] PW: Potential window.

fluoride (PVDF), applied to current collectors such as nickel foam, and then processed through drying to produce mechanically stable electrodes.

MOFs are materials distinguished by their high porosity and tunable structures [91, 92]. They can be directly synthesized on conductive substrates through solvothermal techniques or undergo carbonization to produce high-surface-area carbon structures suitable for SC applications [93]. Notably, Zhang et al. [94] used the direct laser scribing technique to fabricate an enhanced MOF-carbon structure for micro-SC applications.

6.3 CELL ASSEMBLY AND PACKAGING

After preparation of the electrodes and electrolytes, the final step involves the assembly of the SC cell and its subsequent packaging for practical application [6, 95]. The assembly process requires the precise alignment of electrodes, the careful insertion of separators to minimize the risk of short circuits, and the encapsulation of the device within protective casings to avoid the rapid evaporation of the electrolyte, which could severely compromise electrocapacitive performance [96]. Different device configurations, including coin cells, pouch cells, and cylindrical cells, are utilized based on specific application requirements. Effective sealing and encapsulation are crucial for ensuring long-term stability and safety. Additionally, innovative methods, such as flexible packaging and the integration of micro-SCs, have significantly expanded the potential for future energy storage solutions.

To summarize, Table 6.1 presents a comprehensive comparison of recently reported SC applications in terms of critical design elements, including electrode materials, electrolyte formulations, substrate selection, and device structure, which collectively determine the scalability and commercial viability of SCs technologies.

6.4 CONCLUSION

The development and fabrication of SCs involve a diverse range of advanced techniques and functional materials, each playing a critical role in enhancing the electrochemical performance, mechanical stability, and practical applicability of the devices. The selection of fabrication methods, ranging from traditional approaches to innovative printing technologies, directly impacts factors such as energy and power densities, charge/discharge rates, and cycle life. Moreover, critical design parameters, including electrode materials, electrolyte formulations, substrate selection, and device structure, collectively determine the scalability and commercial viability of SCs. With the growing demand for high-performance, compact, and flexible energy storage systems, particularly in wearable electronics, electric vehicles, and renewable energy platforms, the need for continual innovation in fabrication strategies becomes increasingly important. Achieving cost-effective, environmentally sustainable, and scalable manufacturing processes will be crucial to transitioning from laboratory-scale prototypes to mass production solutions. Additionally, interdisciplinary collaboration across material science, nanotechnology, and engineering will accelerate the optimization of SC systems tailored for next-generation energy storage. Ultimately, sustained research efforts and technological advancements that overcome the current limitations while addressing th evolving needs of modern energy applications will pave the way for the broader commercialization and enhanced performance of SCs.

6.5 AUTHOR AI DECLARATION

The authors hereby declare the use of generative AI and AI-assisted technologies in the writing process. While preparing this work, the authors used the ChatGPT tool/service to rephrase several statements in the manuscript. All AI-generated content was reviewed and edited by the authors as needed. The authors take full responsibility for the content of the publication.

ACKNOWLEDGMENT

The Editors acknowledge the use of Grammarly as a tool to assist in editing this book chapter. The suggestions made by Grammarly were carefully considered and incorporated where appropriate, but the author is solely responsible for the final content.

REFERENCES

1. M. Ezzat, A.E. Rashed, S.A. Sabra, M. Haroun, A. Abd El-Moneim, Fully inkjet-printed graphene/gold nonenzymatic biosensor for glucose detection, Materials Today Communications 37 (2023). https://doi.org/10.1016/j.mtcomm.2023.107549.
2. M.A. Belal, S. Hajra, S. Panda, K.R. Kaja, M.M.M. Abdo, A. Abd El-Moneim, D. Janas, Y.K. Mishra, H.J. Kim, Advances in gas sensors using screen printing, Journal of Materials Chemistry A 13(8) (2025) 5447–5497. https://doi.org/10.1039/d4ta06632d.
3. M.A. Belal, H.H. Khalil, R.L. Mahajan, A.E. Rashed, S.N. Khattab, A.A. El-Moneim, Layered structure design of inkjet-printed graphene/Co_3O_4 for high-performance flexible microsupercapacitors, Journal of Energy Storage 101 (2024). https://doi.org/10.1016/j.est.2024.113900.
4. M.A. Belal, R. Yousry, G. Taulo, A.A. AbdelHamid, A.E. Rashed, A.A. El-Moneim, Layer-by-layer inkjet-printed manganese oxide nanosheets on graphene for high-performance flexible supercapacitors, ACS Applied Materials & Interfaces 15(46) (2023) 53632–53643. https://doi.org/10.1021/acsami.3c07339.
5. S.K. Xu, Y. Dall'Agnese, G.D. Wei, C. Zhang, Y. Gogotsi, W. Han, Screen-printable microscale hybrid device based on MXene and layered double hydroxide electrodes for powering force sensors, Nano Energy 50 (2018) 479–488. https://doi.org/10.1016/j.nanoen.2018.05.064.
6. A.M. Bayoumy, A. Hessein, M.A. Belal, M. Ezzat, M.A. Ibrahim, A. Osman, A. Abd El-Moneim, Microdrop InkJet printed supercapacitors of graphene/graphene oxide ink for flexible electronics, Journal of Power Sources 617 (2024). https://doi.org/10.1016/j.jpowsour.2024.235145.
7. W.J. Hyun, C.M. Thomas, L.E. Chaney, A.C. Mazarin de Moraes, M.C. Hersam, Screen-printable hexagonal boron nitride ionogel electrolytes for mechanically deformable solid-state lithium-ion batteries, Nano Letters 22(13) (2022) 5372–5378. https://doi.org/10.1021/acs.nanolett.2c01364.
8. E.B. Secor, P.L. Prabhumirashi, K. Puntambekar, M.L. Geier, M.C. Hersam, Inkjet printing of high conductivity, flexible graphene patterns, The Journal of Physical Chemistry Letters 4(8) (2013) 1347–1351. https://doi.org/10.1021/jz400644c.

9. L.E. Chaney, W.J. Hyun, M. Khalaj, J. Hui, M.C. Hersam, Fully printed, high-temperature micro-supercapacitor arrays enabled by a hexagonal boron nitride ionogel electrolyte, Advanced Materials 36(52) (2024) e2305161. https://doi.org/10.1002/adma.202305161.
10. M. Saquib, S. Shetty, L. M, A. Rathod, K. Naik, R. Nayak, M. Selvakumar, Challenges in carbon ink formulation and strategies for fabrication of flexible supercapacitors, Carbon Trends 19 (2025). https://doi.org/10.1016/j.cartre.2025.100458.
11. M. Perez-Chavez, H.J. Ojeda-Galvan, C. Rodriguez-Gonzalez, P.E. Martinez-Munoz, A.I. Mtz-Enriquez, M.E. Rodriguez-Garcia, J. Oliva, Synergistic effect between MoS_2/WS_2 composite and ecofriendly electrolytes for the fabrication of graphene supercapacitors with high energy density, Inorganic Chemistry Communications 176 (2025). https://doi.org/10.1016/j.inoche.2025.114212.
12. S. Pise, T. Shaikh, O. Kulkarni, R. Bhosale, D. Narale, M. Vadiyar, K.W. Nam, S. Kolekar, Design and fabrication of gold nanoparticles decorated SSM@$NiCo_2O_4$ as a binder-free electrode for solid-state symmetric supercapacitor application, Journal of Energy Storage 114 (2025). https://doi.org/10.1016/j.est.2025.115965.
13. S.S. Pradeepa, K. Sutharthani, R. Subadevi, M. Sivakumar, Fabrication of 1.8 V hybrid supercapacitor based on molybdenum disulfide-Zinc oxide nanocomposite electrode alongside polyvinyl alcohol (PVA) gel electrolyte, Advanced Powder Technology 35(6) (2024). https://doi.org/10.1016/j.apt.2024.104490.
14. L. Pradhan, B. Mohanty, G. Padhy, R.K. Trivedi, D.P. Das, B. Chakraborty, B.K. Jena, Supercapacitor properties of partially oxidised-MXene quantum dots/graphene hybrids: Fabrication of flexible/wearable micro-supercapacitor devices, Chemical Engineering Journal 497 (2024). https://doi.org/10.1016/j.cej.2024.154587.
15. N. Punnakkal, S. Naneena, C.P.S. Lal, A. Pradeep, T.G.S. Babu, P.V. Suneesh, Nitrogen-doped graphene quantum dot embedded polyaniline for the fabrication of high-performance flexible supercapacitor with enhanced cycling stability, Journal of Energy Storage 100 (2024). https://doi.org/10.1016/j.est.2024.113527.
16. Z.Q. Qiao, N.N. Li, Y.P. Deng, D.B. Ji, D.Q. Ji, D.D. Yuan, W.N. Song, Z.D. Li, H.J. Wu, Fabrication of nano-carbon cages via molten salt CO_2 electrolysis for high-performance symmetrical supercapacitor, Nano Energy 136 (2025). https://doi.org/10.1016/j.nanoen.2025.110704.
17. L. Qin, Y. Cao, Z. Cui, F. Feng, Rapid fabrication of mesoporous N- and P-codoped carbon from corn straw for advanced supercapacitor applications, Journal of Energy Storage 106 (2025). https://doi.org/10.1016/j.est.2024.114849.
18. B.K. Saikia, S.M. Benoy, M. Bora, D. Neog, D. Bhattacharjya, A. Rajbongshi, P. Saikia, Fabrication of pouch cell supercapacitors using abundant coal feedstock and their hybridization with Li-ion battery for e-rickshaw application, Journal of Energy Storage 78 (2024). https://doi.org/10.1016/j.est.2023.110312.
19. Sajida, I. Shaheen, K. Shehzad, M. Ahmad, Reshail, I. Raheem, Samia, P. Rosaiah, M. Altaf, B. Akkinepally, W.U. Arifeen, I. Hussain, Fabrication of binder-free Ni-Mn oxalate electrode for supercapacitor application, Electrochimica Acta 513 (2025). https://doi.org/10.1016/j.electacta.2024.145527.
20. Z.-H. Wang, J.-H. Gong, J.-C. Zhou, S.-Z. Wu, C.-Y. Wei, Y.-Y. Liang, Y.-M. Chen, G.-H. Li, M.-Q. Shi, J. Gao, Facile electrochemical fabrication of PEDOT/hydrated tungsten oxides/W self-supporting electrodes with high performance for supercapacitors, Electrochimica Acta 524 (2025). https://doi.org/10.1016/j.electacta.2025.146056.
21. Y. Wang, Y.Z. Zhang, D. Dubbink, J.E. ten Elshof, Inkjet printing of δ-MnO_2 nanosheets for flexible solid-state micro-supercapacitor, Nano Energy 49 (2018) 481–488. https://doi.org/10.1016/j.nanoen.2018.05.002.

22. Y. Wang, Y.Z. Zhang, Y.Q. Gao, G. Sheng, J.E. ten Elshof, Defect engineering of MnO_2 nanosheets by substitutional doping for printable solid-state micro-supercapacitors, Nano Energy 68 (2020). https://doi.org/10.1016/j.nanoen.2019.104306.
23. J.E. ten Elshof, Y. Wang, Advances in Ink-Jet printing of MnO_2-nanosheet based pseudocapacitors, Small Methods 3(8) (2018). https://doi.org/10.1002/smtd.201800318.
24. J.E. ten Elshof, Y. Wang, M. Mehrali, M. Huijben, Printable 2D materials for thin film batteries and supercapacitors, ECS Meeting Abstracts MA2023-02(4) (2023) 791. https://doi.org/10.1149/MA2023-024791mtgabs.
25. S. Sahoo, A. Sood, R. Kumar, A. Milton, S.M. Choi, Y.J. Jo, I. Hussain, S.S. Han, Rapid fabrication of $MnSnO_3$ perovskite-based nanocomposites via microwave-assisted approach for supercapacitor devices, Journal of Energy Storage 110 (2025). https://doi.org/10.1016/j.est.2025.115368.
26. L. Chen, P. Yao, W. Li, Q. Huang, J. Chen, H. Zhang, X. Wang, 3D-printed flexible and integrable asymmetric microsupercapacitors with high-areal-energy-density, ACS Applied Materials & Interfaces 17(12) (2025) 18666–18676. https://doi.org/10.1021/acsami.5c01164.
27. S. Sain, S. Chowdhury, S. Maity, G. Maity, S.S. Roy, Sputtered thin film deposited laser induced graphene based novel micro-supercapacitor device for energy storage application, Scientific Reports 14(1) (2024) 16289. https://doi.org/10.1038/s41598-024-62192-y.
28. L.J. Du, B. Quan, Z.X. Xu, X. Sun, Y. Luo, J. Travas-Sejdic, B.C. Zhu, Biomass-derived laser-induced graphene doped with nitrogen and sulfur for enhanced supercapacitor performance, Carbon 238 (2025). https://doi.org/10.1016/j.est.2025.115368.
29. M.H. El-Shafei, A.G. El-Deen, A. Abd El-Moneim, A. Hessein, Controlled-synthesis of hierarchical $NiCo_2O_4$ anchored on carbon nanofibers mat for free-standing and highly-performance supercapacitors, Journal of Materials Science: Materials in Electronics 32(12) (2021) 15882–15897. https://doi.org/10.1007/s10854-021-06140-w.
30. A. Ray, B. Saruhan, Application of ionic liquids for batteries and supercapacitors, Materials (Basel) 14(11) (2021). https://doi.org/10.3390/ma14112942.
31. S.S. Shah, M.A. Aziz, Properties of electrode materials and electrolytes in supercapacitor technology, Journal of Chemistry and Environment (2024) 1–45. https://doi.org/10.56946/jce.v3i1.309.
32. S. Sharma, P. Chand, S. Kaushik, A critical review of recent advancements in zinc based metal organic framework nanocomposites and their derivatives for supercapacitor applications with future perspectives and challenges, Sustainable Materials and Technologies 41 (2024). https://doi.org/10.1016/j.susmat.2024.e01045.
33. A. Jadhav, P.K. Jha, M. Salomäki, S. Granroth, P. Damlin, C. Kvarnström, Supercapacitive performance of ionic-liquid-intercalated two-dimensional $Ti_3C_2T_x$ in redox electrolyte, Cell Reports Physical Science 5(2) (2024). https://doi.org/10.1016/j.xcrp.2024.101788.
34. C. Gao, Q. You, J. Huang, J. Sun, X. Yao, M. Zhu, Y. Zhao, T. Deng, Ultraconformable integrated wireless charging micro-supercapacitor skin, Nanomicro Letters 16(1) (2024) 123. https://doi.org/10.1007/s40820-024-01352-1
35. C. Gao, J. Huang, Y. Xiao, G. Zhang, C. Dai, Z. Li, Y. Zhao, L. Jiang, L. Qu, A seamlessly integrated device of micro-supercapacitor and wireless charging with ultrahigh energy density and capacitance, Nature Communications 12(1) (2021) 2647. https://doi.org/10.1038/s41467-021-22912-8.
36. E. Jónsson, Ionic liquids as electrolytes for energy storage applications – A modelling perspective, Energy Storage Materials 25 (2020) 827–835. https://doi.org/10.1016/j.ensm.2019.08.030.

37. Z. Bo, X. Zhang, Z. Huang, Y. Huang, J. Yan, K. Cen, H. Yang, Binary ionic liquids hybrid electrolyte based supercapacitors with high energy & power density, RSC Advances 13(23) (2023) 15762–15771. https://doi.org/10.1039/d3ra01634j.
38. S. Sarma Choudhury, N. Katiyar, R. Saha, S. Bhattacharya, Inkjet-printed flexible planar Zn-MnO_2 battery on paper substrate, Scientific Reports 14(1) (2024) 1597. https://doi.org/10.1038/s41598-024-51871-5.
39. V.K.R. Kondapalli, X. He, M. Khosravifar, S. Khodabakhsh, B. Collins, S. Yarmolenko, Y.P.A. Paz, V. Shanov, CVD synthesis of 3D-shaped 3D graphene using a 3D-printed nickel-PLGA catalyst precursor, ACS Omega 6(43) (2021) 29009–29021. https://doi.org/10.1021/acsomega.1c04072.
40. S. Sahoo, G. Sahoo, S.M. Jeong, C.S. Rout, A review on supercapacitors based on plasma enhanced chemical vapor deposited vertical graphene arrays, Journal of Energy Storage 53 (2022). https://doi.org/10.1016/j.est.2022.105212.
41. A. Hamed, A. Hessein, A.A. El-Moneim, Towards high performance flexible planar supercapacitors: In-situ laser scribing doping and reduction of graphene oxide films, Applied Surface Science 551 (2021). https://doi.org/10.1016/j.apsusc.2021.149457.
42. M. Imtiaz, H.A. Alyousef, B.M. Alotaibi, A.W. Alrowaily, M.F. Alotiby, H.M.T. Farid, A.G. Al-Sehemi, A.M.A. Hanaesh, Fabrication of cerium vanadate-embedded on carbon based graphene material (rGO) with significant performance for supercapacitor electrode, Journal of Energy Storage 101 (2024). DOI: 10.1039/C5EE02284C.
43. S. Kumaran, P.R. Buvaneswari, K.V.M. Shree, V. Rajmohan, Fabrication of rGO-modified ternary metal chalcogenide hybrid nanocomposite (rGO/Cu_2MnSnS_4) for high-performance supercapacitors, Diamond and Related Materials 148 (2024). https://doi.org/10.1016/j.diamond.2024.111384.
44. P. Sengodan, T. Yuanmeng, X. Yang, S. Xiaoxian, J.H.T. Luong, A. Ammasi, Y. Orooji, Design and fabrication of 3D petal-like FeS_2/r-GO nanocomposite as a high-performance electrode in hybrid supercapacitors, Fuel 394 (2025). https://doi.org/10.1016/j.fuel.2025.135114.
45. J.M. Bone, J.L. Jenkins, Understanding polymer electrodeposition and conducting polymer modified electrodes using electrochemistry, spectroscopy, and scanning probe microscopy, Journal of Chemical Education 100(10) (2023) 4062–4071. https://doi.org/10.1021/acs.jchemed.3c00656.
46. A. Samos-Puerto, G. Rodríguez-Gattorno, M.A. Ruiz-Gómez, Fine tuning of ink-jet printability parameters for NiO nanofilms fabrication, Colloids and Surfaces a-Physicochemical and Engineering Aspects 583 (2019). https://doi.org/10.1016/j.colsurfa.2019.123959.
47. R. Brisse, R. Faddoul, T. Bourgeteau, D. Tondelier, J. Leroy, S. Campidelli, T. Berthelot, B. Geffroy, B. Jousselme, Inkjet printing NiO-based p-type dye-sensitized solar cells, ACS Applied Materials & Interfaces 9(3) (2017) 2369–2377. https://doi.org/10.1021/acsami.6b12912.
48. Z.H. Bakr, E.A.A. Aboelazm, C.S. Khe, G.A.M. Ali, K.F. Chong, Recycling the spent lithium-ion battery into nanocubes cobalt oxide supercapacitor electrode, Current Nanoscience 20(6) (2024) 820–829. https://doi.org/10.2174/0115734137265230231020181833.
49. Q.C. He, J.J. Ye, Z.Y. Peng, Y.Y. Guo, L.C. Tan, Y.W. Chen, Electrodeposition of poly(3,4-ethylenedioxythiophene) coated manganese dioxide nanospheres for flexible asymmetric planar supercapacitor with superior energy density, Journal of Power Sources 506 (2021). https://doi.org/10.1016/j.jpowsour.2021.230176.
50. Z. Xia, V. Mishukova, S. Sollami Delekta, J. Sun, J.S. Sanchez, J. Li, V. Palermo, Selective deposition of metal oxide nanoflakes on graphene electrodes to obtain high-performance asymmetric micro-supercapacitors, Nanoscale 13(5) (2021) 3285–3294. https://doi.org/10.1039/d0nr07076a.

51. L. Merabet, K. Rida, N. Boukmouche, Sol-gel synthesis, characterization, and supercapacitor applications of MCo_2O_4 (M = Ni, Mn, Cu, Zn) cobaltite spinels, Ceramics International 44(10) (2018) 11265–11273. https://doi.org/10.1016/j.ceramint.2018.03.171.
52. S. Yadav, A. Sharma, Importance and challenges of hydrothermal technique for synthesis of transition metal oxides and composites as supercapacitor electrode materials, Journal of Energy Storage 44 (2021). https://doi.org/10.1016/j.est.2021.103295.
53. A. Nandagudi, S.H. Nagarajarao, M.S. Santosh, B.M. Basavaraja, S.J. Malode, R.J. Mascarenhas, N.P. Shetti, Hydrothermal synthesis of transition metal oxides, transition metal oxide/carbonaceous material nanocomposites for supercapacitor applications, Materials Today Sustainability 19 (2022). https://doi.org/10.1016/j.mtsust.2022.100214.
54. M.R. Islam, S. Afroj, J. Yin, K.S. Novoselov, J. Chen, N. Karim, Advances in printed electronic textiles, Advanced Science (Weinheim) 11(6) (2024) e2304140. https://doi.org/10.1002/advs.202304140.
55. E.B. Secor, S. Lim, H. Zhang, C.D. Frisbie, L.F. Francis, M.C. Hersam, Gravure printing of graphene for large-area flexible electronics, Advanced Materials 26(26) (2014) 4533–4538. https://doi.org/10.1002/adma.201401052.
56. J.S. Dhatterwal, K.S. Kaswan, N. Kumar, Self-powered integrated sensor system for printable fabrication, Self-Powered Sensors (2025) 187–235. https://doi.org/10.1016/b978-0-443-13792-1.00008-0.
57. M.D. Stoller, S. Park, Y. Zhu, J. An, R.S. Ruoff, Graphene-based ultracapacitors, Nano Letters 8(10) (2008) 3498–502. https://doi.org/10.1021/nl802558y.
58. W.F. Ji, M.M.M. Ahmed, A. Bibi, Y.C. Lee, J.M. Yeh, Graphene-based PANI composite coatings with fine-controllable 3D hierarchical structures prepared from bio-inspired photo-/colloidal-lithography technique for flexible supercapacitor application, Electrochimica Acta 390 (2021). https://doi.org/10.1016/j.electacta.2021.138890.
59. R. Khan, C. Ryu, H.M. Do, J.B. In, Fabrication of functional laser-induced graphene using gelatin-2,6-diaminoanthraquinone coating for high-performance micro-supercapacitor, Journal of Power Sources 621 (2024). https://doi.org/10.1016/j.jpowsour.2024.235305.
60. E. Ghoniem, A.A. El-Moneim, Flexible interdigital in-plane supercapacitor based on laser reduced graphene oxide (LRGO), Key Engineering Materials 705 (2016) 138–144. https://doi.org/10.4028/www.scientific.net/KEM.705.138.
61. E. Ghoniem, S. Mori, A. Abdel-Moniem, Low-cost flexible supercapacitors based on laser reduced graphene oxide supported on polyethylene terephthalate substrate, Journal of Power Sources 324 (2016) 272–281. https://doi.org/10.1016/j.jpowsour.2016.05.069.
62. M.H. El-Shafei, E. Ghoniem, A. Hassanin, A.A. El-Moneim, Novel Technique for Producing Porous Carbon Nanofiber Mate for Supercapacitors Application, Key Engineering Materials 735 (2017) 199–204. https://doi.org/10.4028/www.scientific.net/KEM.735.199.
63. M.H. El-Shafei, A.H. Hassanin, N.M. Shaalan, T. Sharshar, A.A. El-Moneim, Free-standing interconnected carbon nanofiber electrodes: new structural designs for supercapacitor application, Nanotechnology 31(18) (2020) 185403. https://doi.org/10.1088/1361-6528/ab6d22.
64. M. Hussein El-Shafei, M.S. Abdel-Latif, A. Hessein, A. Abd El-Moneim, Laser-induced carbonization of carbon nanofibers free-standing electrodes for flexible interdigitated micro-supercapacitors, FlatChem 42 (2023). https://doi.org/10.1016/j.flatc.2023.100570.
65. W.O. Makinde, M.A. Hassan, Y. Pan, G.Q. Guan, N. López-Salas, A.S.G. Khalil, Sulfur and nitrogen co-doping of peanut shell-derived biochar for sustainable supercapacitor applications, Journal of Alloys and Compounds 991 (2024). https://doi.org/10.1016/j.jallcom.2024.174452.

66. S. Chen, Y. Chang, A. Liu, Y. Wang, Z. Feng, J. Li, L. Jiao, Laser repeat scanning preparation of nitrogen sulfur doped graphene supercapacitor, Surfaces and Interfaces 56 (2025). https://doi.org/10.1016/j.surfin.2024.105581.
67. Z.H. Tang, Y. Lei, Y.F. Deng, J. Chen, K.W. Liu, C.R. Zhang, T.Q. Wang, Y.N. Lei, MOF-derived MoS_2/nitrogen-doped graphene aerogel for supercapacitor electrodes, Diamond and Related Materials 153 (2025). https://doi.org/10.1016/j.diamond.2025.112098.
68. Z. Peng, R. Ye, J.A. Mann, D. Zakhidov, Y. Li, P.R. Smalley, J. Lin, J.M. Tour, Flexible boron-doped laser-induced graphene microsupercapacitors, ACS Nano 9(6) (2015) 5868–5875. https://doi.org/10.1021/acsnano.5b00436
69. M. Khandelwal, C.V. Tran, J.B. In, Nitrogen and phosphorous Co-doped laser-induced graphene: A high-performance electrode material for supercapacitor applications, Applied Surface Science 576 (2022). https://doi.org/10.1016/j.apsusc.2021.151714.
70. R. Xu, Z. Wang, L. Gao, S. Wang, J. Zhao, Effective design of MnO_2 nanoparticles embedded in laser-induced graphene as shape-controllable electrodes for flexible planar microsupercapacitors, Applied Surface Science 571 (2022). https://doi.org/10.1016/j.apsusc.2021.151385.
71. C. Zhu, X. Dong, X. Mei, M. Gao, K. Wang, D. Zhao, General fabrication of metal oxide nanoparticles modified graphene for supercapacitors by laser ablation, Applied Surface Science 568 (2021). https://doi.org/10.1016/j.apsusc.2021.150978.
72. A. Saraei, A. Esfandiar, N. Taghavinia, Highly efficient photo-supercapacitor based on Mg-doped NiOx/SnO_2 p-n heterojunction, Journal of Alloys and Compounds 1015 (2025). https://doi.org/10.1016/j.jallcom.2025.178920.
73. M. Shariq, D. Alhashmialameer, H. Adawi, M.R. Alrahili, M.Y.A. Almashnowi, A. Alzahrani, M. Sharma, S.K. Ali, Y. Slimani, Advancements in transition metal sulfide supercapacitors: A focused review on high-performance energy storage, Journal of Industrial and Engineering Chemistry 144 (2025) 269–291. https://doi.org/10.1016/j.jiec.2024.11.012.
74. O. Öztürk, E. Gür, Layered transition metal sulfides for supercapacitor applications, ChemElectroChem 11(11) (2024). https://doi.org/10.1002/celc.202300575.
75. K. Samdhyan, P. Chand, H. Anand, Fabrication of high-performance asymmetric supercapacitor device based on CuS with marigold flower like framework, Journal of Energy Storage 106 (2025). https://doi.org/10.1016/j.est.2024.114719.
76. N.S. Akiiga, A.M. Rashad Fath El-Bab, M. Yoshihisa, A.A. El-Moneim, Enzyme-Free glucose detection in sweat using 2D inkjet-printed cobalt sulfide anchored on graphene in a paper-based microfluidic device, Journal of Colloid and Interface Science 688 (2025) 490–504. https://doi.org/10.1016/j.jcis.2025.02.129.
77. M. Girirajan, S. Yarramsetti, S. Kalluri, S. Sakkarapani, S. Sangaraju, P.S. Maram, A novel 2D bismuthene-molybdenum disulfide nanocomposite for high energy density supercapacitors and fabrication scaled to pouch cell, Journal of Energy Storage 85 (2024). https://doi.org/10.1016/j.est.2024.111042.
78. G. Ghanashyam, H.K. Jeong, Electrochemical performance of poly sodium 4-styrenesulfonate intercalated thermally reduced graphite oxide for supercapacitors, Chemical Physics Letters 722 (2019) 39–43. https://doi.org/10.1016/j.cplett.2019.03.008.
79. Y.C. Chen, H.C. Yang, Z.J. Han, Z. Bo, J.H. Yan, K.F. Cen, K.K. Ostrikov, MXene-based electrodes for supercapacitor energy storage, Energy & Fuels 36(5) (2022) 2390–2406. https://doi.org/10.1021/acs.energyfuels.1c04104.
80. S. Kumar, M.A. Rehman, S. Lee, M. Kim, H. Hong, J.Y. Park, Y. Seo, Supercapacitors based on $Ti_3C_2T_x$ MXene extracted from supernatant and current collectors passivated by CVD-graphene, Scientific Reports 11(1) (2021) 649. https://doi.org/10.1038/s41598-020-80799-9.

81. R. Akhter, S.S. Maktedar, MXenes: A comprehensive review of synthesis, properties, and progress in supercapacitor applications, Journal of Materiomics 9(6) (2023) 1196–1241. https://doi.org/10.1016/j.jmat.2023.08.011.
82. S. Iravani, R.S. Varma, MXene-based wearable supercapacitors and their transformative impact on healthcare, Materials Advances 4(19) (2023) 4317–4332. https://doi.org/10.1039/d3ma00365e.
83. Z. Xiong, X. Yun, L. Qiu, Y. Sun, B. Tang, Z. He, J. Xiao, D. Chung, T.W. Ng, H. Yan, R. Zhang, X. Wang, D. Li, A dynamic graphene oxide network enables spray printing of colloidal gels for high-performance micro-supercapacitors, Advanced Materials 31(16) (2019) e1804434. https://doi.org/10.1002/adma.201804434.
84. M.G. Say, R. Brooke, J. Edberg, A. Grimoldi, D. Belaineh, I. Engquist, M. Berggren, Spray-coated paper supercapacitors, NPJ Flexible Electronics 4(1) (2020). https://doi.org/10.1038/s41528-020-0079-8.
85. A.G. El-Deen, M. Hussein El-Shafei, A. Hessein, A.H. Hassanin, N.M. Shaalan, A.A. El-Moneim, High-performance asymmetric supercapacitor based hierarchical $NiCo_2O_4$@ carbon nanofibers//Activated multichannel carbon nanofibers, Nanotechnology 31(36) (2020) 365404. https://doi.org/10.1088/1361-6528/ab97d6.
86. A.C. Nwanya, A. Musheghyan-Avetisyan, E. György, A.P. del Pino, Fabrication of asymmetric supercapacitors by laser processing of activated carbon-based electrodes produced from rice husk waste, Surfaces and Interfaces 54 (2024). https://doi.org/10.1016/j.surfin.2024.105200.
87. D.V. Rupnar, H.M. Yadav, A.V.S.L. Sai Bharadwaj, P.B. Sarawade, N.S. Ramgir, G.S. Ghodake, B.G. Pawar, Fabrication and evaluation of carbon black of waste Tyre-derived activated carbon electrodes for enhanced supercapacitor performance, Carbon Trends 19 (2025). https://doi.org/10.1016/j.cartre.2025.100489.
88. X.R. Liu, Z.X. Geng, Y.K. Sun, W. Sun, Polyurethane-converted activated carbons with tailorable porosity and surface functionality for supercapacitors, Journal of Energy Storage 115 (2025). https://doi.org/10.1016/j.est.2025.115988.
89. S. Aryal, K.R. Shrestha, T. Shrestha, H.B. Oli, I. Pathak, R.L. Shrestha, D.P. Bhattarai, Activated carbon from *Prunus persica* seed stones as a negatrode material for high-performance supercapacitors, Journal of Molecular Structure 1323 (2025). https://doi.org/10.1016/j.molstruc.2024.140810.
90. M. Rathnayaka, H. Sitinamaluwa, Graphene oxide/Activated carbon nano composite with hierarchical pore structure for supercapacitor applications, Electrochimica Acta 517 (2025). https://doi.org/10.1016/j.electacta.2025.145752.
91. B. Diao, C. Cong, F. Sun, H. Li, H. Zhang, X. Wang, M. Jin, S. Lim, X. Li, S.H. Kim, Printed flexible supercapacitors utilizing composites of MWCNT/MOF ultrathin nanosheets: Exploring 1D/2D structural adjustment, Chemical Engineering Journal 500 (2024). https://doi.org/10.1016/j.cej.2024.156993.
92. B. Ramulu, J.S. Yu, Recent advances in metal-organic frameworks-based materials for supercapacitor applications, Applied Science and Convergence Technology 33(3) (2024) 53–61. https://doi.org/10.5757/asct.2024.33.3.53.
93. J.L. Hou, Y. Ning, K.N. Guo, W.X. Jiao, C. Chen, B.W. Zhang, X.F. Wu, J.L. Zhao, D.H. Lin, S.G. Sun, Application of metal-organic framework materials in supercapacitors, Journal of Energy Storage 113 (2025). https://doi.org/10.1016/j.est.2025.115535.
94. W. Zhang, R. Li, H. Zheng, J.S. Bao, Y.J. Tang, K. Zhou, Laser-assisted printing of electrodes using metal-organic frameworks for micro-supercapacitors, Advanced Functional Materials 31(14) (2021). https://doi.org/10.1002/adfm.202009057.
95. E. Aboelazm, C.S. Khe, S. Grätz, K.F. Chong, M.S.M. Saheed, M.B.Z. Hegazy, Layered construction of integrated sulfur-bridged CoNi-S/rGO architecture for enhanced electrochemical energy storage, Chemical Engineering Journal 490 (2024). https://doi.org/10.1016/j.cej.2024.151557.

96. E. Aboelazm, C.S. Khe, K.F. Chong, M.S. Mohamed Saheed, M.B.Z. Hegazy, Interconnected CoNi-Se hollow flakes through reduced graphene oxide sheets as a cathode material for hybrid supercapacitors, ACS Applied Materials & Interfaces 16(12) (2024) 15011–15022. https://doi.org/10.1021/acsami.3c17615.
97. J. Li, V. Mishukova, M. Östling, All-solid-state micro-supercapacitors based on inkjet printed graphene electrodes, Applied Physics Letters 109(12) (2016). https://doi.org/10.1063/1.4962728.
98. J. Li, S. Sollami Delekta, P. Zhang, S. Yang, M.R. Lohe, X. Zhuang, X. Feng, M. Ostling, Scalable fabrication and integration of graphene microsupercapacitors through full inkjet printing, ACS Nano 11(8) (2017) 8249–8256. https://doi.org/10.1021/acsnano.7b03354.
99. A. Sajedi-Moghaddam, M. Gholami, N. Naseri, Inkjet printing of MnO_2 nanoflowers on surface-modified A4 paper for flexible all-solid-state microsupercapacitors, ACS Applied Materials & Interfaces 15(3) (2023) 3894–3903. https://doi.org/10.1021/acsami.2c08939.
100. H. Pang, Y. Zhang, W.-Y. Lai, Z. Hu, W. Huang, Lamellar $K_2Co_3(P_2O_7)_2{\cdot}2H_2O$ nanocrystal whiskers: High-performance flexible all-solid-state asymmetric micro-supercapacitors via inkjet printing, Nano Energy 15 (2015) 303–312. https://doi.org/10.1016/j.nanoen.2015.04.034.
101. G. Li, Z. Meng, J. Qian, C.-L. Ho, S.P. Lau, W.-Y. Wong, F. Yan, Inkjet printed pseudocapacitive electrodes on laser-induced graphene for electrochemical energy storage, Materials Today Energy 12 (2019) 155–160. https://doi.org/10.1016/j.mtener.2019.01.005.
102. A.P.S. Gaur, W. Xiang, A. Nepal, J.P. Wright, P. Chen, T. Nagaraja, S. Sigdel, B. LaCroix, C.M. Sorensen, S.R. Das, Graphene aerosol gel ink for printing micro-supercapacitors, ACS Applied Energy Materials 4(8) (2021) 7632–7641. https://doi.org/10.1021/acsaem.1c00919.
103. Y. Bräuniger, S. Lochmann, J. Grothe, M. Hantusch, S. Kaskel, Piezoelectric inkjet printing of nanoporous carbons for micro-supercapacitor devices, ACS Applied Energy Materials 4(2) (2021) 1560–1567. https://doi.org/10.1021/acsaem.0c02745.
104. L.E. Chaney, W.J. Hyun, M. Khalaj, J. Hui, M.C. Hersam, Fully printed, high-temperature micro-supercapacitor arrays enabled by a hexagonal boron nitride ionogel electrolyte, Advanced Materials (2023) e2305161. https://doi.org/10.1002/adma.202305161.
105. A. Khodabandehlo, A. Noori, M.S. Rahmanifar, M.F. El-Kady, R.B. Kaner, M.F. Mousavi, Laser-scribed graphene–polyaniline microsupercapacitor for internet-of-things applications, Advanced Functional Materials 32(39) (2022). https://doi.org/10.1002/adfm.202204555.
106. Z.-K. Wu, Z. Lin, L. Li, B. Song, K.-s. Moon, S.-L. Bai, C.-P. Wong, Flexible micro-supercapacitor based on in-situ assembled graphene on metal template at room temperature, Nano Energy 10 (2014) 222–228. https://doi.org/10.1016/j.nanoen.2014.09.019.

7 Characterization Techniques for Supercapacitors

Ubaid Sidiqi and Dinesh Kumar

7.1 INTRODUCTION

Supercapacitors have become a key technology that bridges the gap between traditional capacitors and batteries in the search for cutting-edge and environmentally friendly energy storage solutions. Supercapacitors, known for their high-power density, quick charge and discharge times, and remarkable cycle stability, are becoming more and more important for a variety of applications, from grid energy management to portable electronics and electric cars.

A comprehensive grasp of the electrode materials and the intricate electrochemical processes at the electrode–electrolyte interface is necessary to realize their full potential. The foundation of supercapacitor research and development is characterization techniques, which provide an in-depth understanding of material characteristics and device performance.

Techniques for material characterization, such as X-ray diffraction (XRD) and X-ray photoelectron spectroscopy (XPS), are crucial for identifying the structural and compositional features necessary to customize electrode functions. Scanning electron microscopy (SEM) and transmission electron microscopy (TEM) offer insights into surface texture and nanostructure, which play a role in affecting charge storage and ion transport. Optical techniques like Fourier Transform Infrared (FTIR) Spectroscopy and Raman Spectroscopy (RS) elucidate chemical bonding and structural dynamics. Surface area measurements using the Brunauer–Emmett–Teller (BET) method quantify the available active sites, while thermogravimetric analysis (TGA) and zeta potential (ζ-potential) assessments offer insights into thermal stability and colloidal dispersion, respectively. Complementing these material-focused analyses, electrochemical characterization techniques directly evaluate the supercapacitor's performance under operational conditions. Methods such as cyclic voltammetry (CV), galvanostatic charge–discharge (GCD), and electrochemical impedance spectroscopy (EIS) quantify crucial parameters, including specific capacitance, energy and power densities, rate capability, charge transfer resistance, and long-term stability. Emerging approaches like self-discharge testing and in situ/operando electrochemical analyses further enhance understanding by monitoring dynamic changes during real-time operation.

This chapter comprehensively addresses both material and electrochemical characterization techniques, emphasizing their role in the rational design and optimization of supercapacitor systems. It guides readers through the principles, practical

DOI: 10.1201/9781003535836-7

considerations, and interpretation of experimental data, fostering a deeper comprehension essential for advancing next-generation energy storage technologies.

7.2 ANALYTICAL TECHNIQUES

Analytical techniques are the fundamental and most important tools used to investigate the compositional and structural characteristics of materials at both the atomic and molecular levels. In the supercapacitor field, these techniques provide valuable insights into the crystalline nature, phase purity, elemental composition, and chemical states of electrode materials, which significantly influence their electrochemical properties.

In this context, two widely used analytical techniques are XRD and XPS.

7.2.1 X-Ray Diffraction (XRD)

XRD is the main analytical technique, which is mainly used to investigate the crystal structure, phase purity, and crystallite dimensions of the electrode material, as well as other materials. This technique gives valuable insights into lattice spacing and types of crystalline phases due to the diffraction of X-rays by the atomic planes. These diffraction peaks arise due to the constructive interference of the scattered X-rays, which follows Bragg's law as stated in Equation 7.1.

$$n\lambda = 2dsin\theta \tag{7.1}$$

where n defines the order of diffraction, λ is the wavelength of the X-rays, d is the spacing between crystal planes, and θ is the angle of incidence.

The average crystallite size is calculated using Debye-Scherrer's equation, as shown in Equation 7.2, which correlates the broadening of the most prominent diffraction peak to the size of coherently diffracting crystalline domains [1].

$$D = \frac{K\lambda}{\beta cos\theta} \tag{7.2}$$

where D is the average crystallite size, K is the Scherrer constant (typically around 0.9), λ is the wavelength of the X-ray source, β is the full width at half maximum (FWHM) of the diffraction peak (in radians), and θ is the Bragg angle corresponding to the peak position.

This provides an understanding of the grain size at the nanoscale, which can affect surface area and electrochemical activity. One significant advantage of XRD is its ability to analyze bulk powder samples, offering statistically representative results for the materials' overall crystallinity and phase purity. Phase identification is achieved by comparing the 2θ peak positions and intensities with standard reference patterns from databases such as the International Centre for Diffraction Data (ICDD), formerly recognized as "Joint Committee on Powder Diffraction Standards" (JCPDS). This helps confirm the formation of desired phases and detect impurities or secondary phases. However, XRD has limitations in detecting amorphous materials or extremely small particles (below ~3 nm), where diffraction peaks become broad and less distinct. The synthesis method can substantially influence the crystallinity and crystallite size of electrode materials. For instance, Hareendra K. Kumar et al. [2] studied $LiNiVO_4$ electrodes synthesized via different routes, demonstrating that hydrothermal methods tend to produce smaller crystallites with broader peaks compared to those produced by mechanochemical synthesis. Smaller crystallite sizes often shorten ion diffusion paths and reduce charge-transfer resistance, enhancing electrochemical performance. Furthermore, XRD peak broadening can result from crystallite size and lattice strain within the material. This interplay impacts the material's structural stability and charge storage capabilities. It is also noteworthy that crystallite sizes calculated from XRD can differ from particle sizes observed by TEM, as XRD measures coherent domains while TEM captures whole particles, which may consist of multiple crystallites. In addition to providing structural information, XRD can indicate preferential crystal growth directions, which influence surface energy and material morphology. For example, Yan et al. [3] synthesized AuNPs that showed preferential growth along the (111) plane, correlating with the strongest XRD peak and the lowest surface energy orientation. This suggests that growth will occur in the direction with the maximum surface energy. Overall, XRD remains an indispensable tool for understanding and optimizing the crystal structure, phase purity, and nanoscale features of supercapacitor electrode materials, enabling better design of high-performance energy storage devices.

7.2.2 X-Ray Photoelectron Spectroscopy (XPS)

XPS is one of the important techniques used to analyze the surface chemistry of electrode material. Since the charge storage mechanism of pseudocapacitors relies on Faradaic redox reactions, which occur on or near the electrode's surface, the understanding of the structural and chemical state of surface atoms is essential. This surface information is provided by XPS, which identifies the elemental composition, oxidation states, and electronic environments of elements within the material's top few nanometers.

The principle of XPS is based on the photoelectric effect. The irradiated monochromatic X-rays, typically Al Kα radiation with an energy of 1486.6 eV, eject core electrons from the surface atoms. The kinetic energy of these emitted photoelectrons is measured, and their corresponding binding energy is evaluated using Equation 7.3.

$$E_b = h\nu - E_k - \phi \tag{7.3}$$

where E_b is the binding energy, $h\nu$ is the photon energy, E_k is the kinetic energy of the emitted electron, and ϕ is the work function of the spectrometer.

This binding energy, which is characteristic of specific elements and their chemical states, makes XPS a highly informative tool for surface analysis. XPS has the feature to detect all the elements other than hydrogen and helium within the top 5–10 nm of the material's surface, depending on the electron escape depth. For thinner layers, angle-resolved XPS (ARXPS) can enhance depth sensitivity by varying the emission angle of the photoelectrons [4, 5].

XPS has been used in supercapacitor research to trigger a number of important phenomena. Toupin et al. [6] have employed XPS to track changes in the valence state of MnO_2 electrodes over the charge–discharge cycle. Their study revealed the Mn oxidation state shifted from III to IV, driven by Na^+ ion interactions, confirming the Faradaic nature of the charge storage mechanism.

Similarly, Wu et al. [7] utilized XPS to verify the successful doping of nitrogen and boron atoms in graphene nanosheets, demonstrating homogeneous distribution of heteroatoms that can enhance electronic conductivity and active sites for redox reactions.

XPS is also highly effective in confirming the formation of core-shell structures, functional surface groups, and ligand exchange reactions.

Sullivan et al. [8] employed XPS to track surface functionalization of glassy carbon electrodes and observed changes in functional groups before and after redox cycling. However, they noted that due to the limited sampling depth (~5 nm or ~50 Å), XPS alone may not capture the full chemical structure of thicker surface layers, requiring complementary techniques such as FTIR for bulk surface characterization.

In an advanced application, Tõnisoo et al. [9] employed in situ XPS with a symmetric two-electrode supercapacitor setup to study electrochemical behavior at the electrode–electrolyte interface. By analyzing C 1s and N 1s photoelectron spectra under varying cell potentials, they observed peak shifts and intensity changes correlating with surface-bound functional groups and redox activity. This approach provided direct insights into the real-time charge storage processes at the interface.

Overall, XPS is an essential tool for understanding surface chemistry, oxidation states, and chemical changes in supercapacitor materials, helping researchers link surface makeup with electrochemical performance.

7.3 MORPHOLOGICAL TECHNIQUES

Morphological techniques are employed to investigate electrode materials' surface topography, particle size, shape, and structural features, which directly impact ion transport, surface area, and charge behavior in supercapacitors.

Two key morphological techniques discussed below are SEM and TEM.

7.3.1 Scanning Electron Microscopy (SEM)

The surface morphology of the material is investigated by SEM. The high-resolution images are produced by the detection of secondary electrons, which are emitted when the high-energy electrons interact with the surface of the sample. These images reveal the actual topography of the sample, such as particle size, porosity, and surface texture, which are important and critical in determining ion diffusion pathways and charge storage behavior [10].

In supercapacitor research, SEM is essential for evaluating hierarchical and porous architecture formation. For instance, 3D-printed or template-assisted electrode materials often require SEM to confirm uniformity and structural integrity. Using a stereolithographic approach, Xue et al. [11] demonstrated this by fabricating 3D hierarchical porous reduced graphene oxide (rGO) on metal lattices. SEM images revealed well-developed pores formed during the reduction process, facilitating better ion accessibility and improved capacitance as shown in Figure 7.1, which illustrates the surface morphology of the synthesized rGO, as observed through both low- and high-magnification FESEM imaging. In Figure 7.1(a), a well-defined porous architecture is evident on the metallic lattice framework. This porous network originates from the evolution of hydrogen gas bubbles, which are generated during the reduction of hydrogen ions (H^+) in HCl solution as graphene oxide (GO) is transformed into rGO. This unique hierarchical porous structure is advantageous, as it promotes better utilization of the active material, thereby enhancing the overall capacitance.

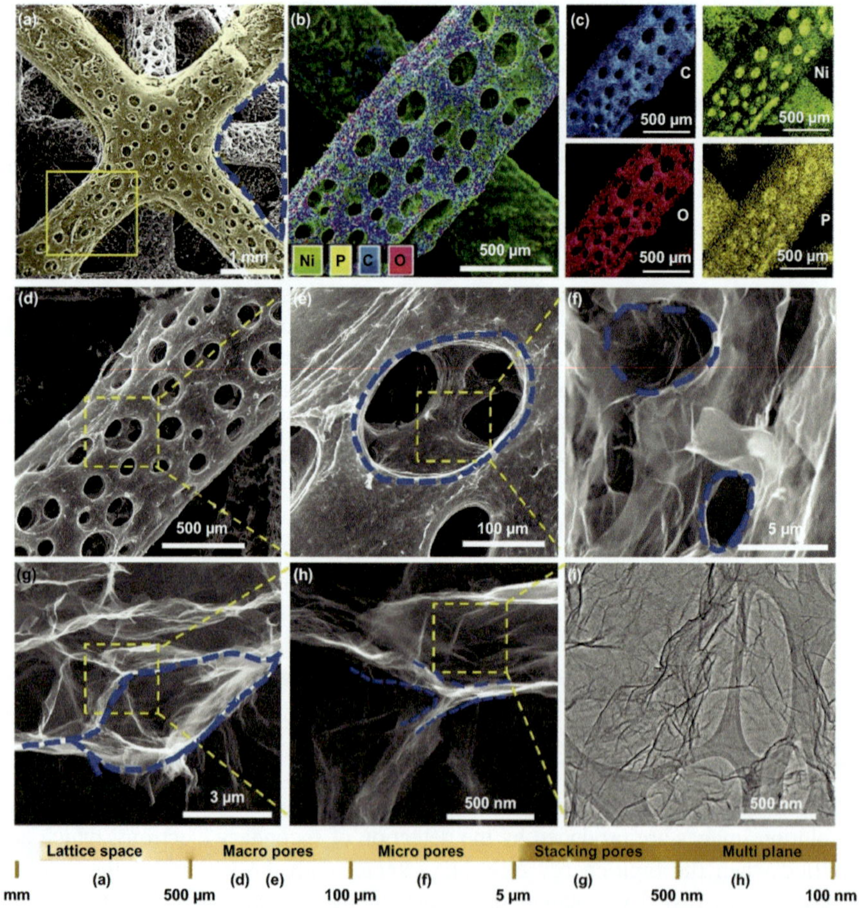

FIGURE 7.1 (a) FESEM image showcasing an individual porous lattice unit, (b–d) SEM–EDX elemental mapping and corresponding FESEM image highlighting the morphology of a single porous structure, (e) magnified view revealing the continuous porous architecture, (f) micropores observed along the sidewalls of the rGO, (g) pores generated due to the stacking of graphene sheets, (h) pores formed by the multilayer arrangement of graphene, (i) TEM image displaying the graphene sheets, exhibiting characteristic wrinkles and a wave-like morphology. (Reproduced from the ref. [11]. © Springer Nature 2019.)

Figure 7.1(b) and (c) presents SEM-EDX elemental mapping results, confirming the composition and uniform distribution of elements within the porous framework. Each segment of the lattice consists of rGO with a diameter of approximately 1 mm and a consistent pore size of about 200 μm, as shown in Figure 7.1(d) and (e). This uniform porosity plays a critical role in facilitating ion transport. Furthermore, Figure 7.1(f) highlights the presence of microcavities around 5 μm in diameter on the rGO sidewalls. In Figure 7.1(g) and (h), nanopores are visible, formed by the stacking of porous, multilayered graphene sheets. The wrinkled and wave-like morphology of the graphene nanosheets, Figure 7.1(i), enhances electrochemical properties by providing multiple diffusion pathways created by the lattices and porous rGO.

In conclusion, SEM validates the successful synthesis of electrode nanostructures and provides insights into how surface engineering influences supercapacitor efficiency. Its compatibility with EDX further allows simultaneous structural and compositional evaluation, making SEM an indispensable tool in supercapacitor material development.

7.3.2 Transmission Electron Microscopy (TEM)

TEM is one of the most powerful and widely utilized characterization techniques for investigating nanostructured materials, especially in energy storage systems like supercapacitors. It transmits a focused beam of high-energy electrons through an ultra-thin sample (typically less than 100 nm thick). As electrons interact with the sample, some are transmitted while others are scattered either elastically or inelastically, depending on the specimen's density, thickness, and atomic number. The transmitted electrons are then used to form a high-resolution image, revealing fine structural and morphological details of the material at the nanoscale [12].

TEM is particularly advantageous for evaluating nanoparticle morphology, crystallinity, grain boundaries, and size distributions. Its direct imaging capability enables a precise assessment of particle shape and dispersion, offering more accurate insights than light microscopy or SEM for materials at the nanometer scale. For instance, in nanomaterials used for supercapacitor electrodes, TEM can distinguish between layered and porous structures, allowing researchers to correlate morphology with electrochemical performance. Figure 7.1(i) shows TEM image displaying the graphene sheets, exhibiting characteristic wrinkles and a wave-like morphology. However, one of the critical limitations of TEM is its labor-intensive sample preparation. Improper handling can lead to aggregation or deformation of nanoparticles, often due to drying artifacts or residual solvents. This can result in an inaccurate interpretation of structural features. Additionally, TEM has a limited field of view, making it challenging to analyze statistically significant numbers of particles. Therefore, TEM is often complemented by techniques such as XRD, which offer more bulk-sensitive and statistically representative measurements [13].

An advanced variation of TEM, High-Resolution TEM (HR-TEM), provides even greater insight by using phase-contrast imaging. In this mode, both transmitted and scattered electrons are collected to form images with lattice-level resolution. HR-TEM allows visualization of atomic arrangements, crystallographic defects, and interplanar spacings, which are crucial for identifying the crystal structure and validating the formation of composite or heterostructured materials. Such atomic-level characterization is critical in supercapacitor research, where the alignment, defect density, and interface quality between different components significantly affect charge transport and storage behavior [14, 15].

7.4 OPTICAL TECHNIQUES

Optical techniques analyze the molecular structure, chemical bonding, and functional groups present in electrode materials, providing vital information about material composition and surface chemistry relevant to supercapacitor performance. Two widely used optical characterization methods discussed below are FTIR spectroscopy and RS.

7.4.1 Fourier Transform Infrared (FTIR) Spectroscopy

FTIR is a widely used technique for functional group identification and understanding the molecular structure of materials. Due to the infrared radiation, the absorption of the sample is measured, which causes characteristic molecular vibrations such as stretching or bending chemical bonds. The vibrations appear as peaks in the IR spectrum, within the range of 4000–400 cm^{-1}, providing a unique fingerprint for each material [16, 17].

In the case of supercapacitors, FTIR helps in detecting surface functional groups such as hydroxyl, carbonyl, and carboxyl groups that are involved in influencing electron transfer and interaction with electrodes. This method is used for all the forms of matter and efficiently evaluates both organic and inorganic components. The mass of the atoms and the bond strength between them greatly influence the vibrational frequency. The molecules absorb IR radiation only when the energy matches their natural vibrational frequencies, leading to a change in the dipole moment and making them IR active.

FTIR not only reveals the types of bonds but also gives information about molecular interactions and surface modifications. Recent studies have demonstrated its utility in identifying functional groups in layered double hydroxides (LDHs) used for energy storage. Wu et al. [18] confirmed the presence of hydroxyl groups in NiCo-LDHs, contributing to their excellent interaction with the electrolyte. Similarly, Chai et al. [19] used FTIR to verify functional group incorporation in CoGa-LDHs with varying metal ratios.

In summary, FTIR is a powerful tool for determining surface chemistry and verifying chemical structures. Thus, it is relevant for optimizing and developing advanced electrode materials.

7.4.2 Raman Spectroscopy (RS)

RS is also a very important, powerful, non-destructive tool for investigating electrode materials. It is based on the inelastic scattering of monochromatic light, known as the Raman effect. When a high-intensity laser interacts with the material, a small amount of scattered photons exhibits a shift in energy corresponding to molecular vibrations. This shift, known as the Raman shift, provides a fingerprint for material identification [20]. The technique is sensitive to crystalline symmetry, strain, and bond disorder changes, making it especially useful for characterizing structural evolution during electrochemical cycling. It requires no sample preparation and yields sharp spectral features corresponding to specific functional groups and structures within organic and certain inorganic materials. In-situ RS has proven valuable in relation to structural changes with electrochemical performance. For example, Sato et al. [21] employed in situ Raman analysis to evaluate the electrochemical behavior of air-oxidized single-walled carbon nanotubes (AO-SWCNTs) in H_2SO_4 electrolyte. A notable downshift of the G^+ band under negative current application indicated increased defect density and diameter distribution due to oxidation. These changes enhanced ion accessibility and contributed to improved specific capacitance (50 F/g) and charge storage (60 C/g) compared to the pristine nanotubes (38 F/g and 46 C/g, respectively). The larger G^+ upshift under positive current further confirmed the altered phonon energy and weakened C–C bonds in the oxidized nanotubes.

7.5 SURFACE AREA MEASUREMENTS

Surface area characterization is essential in evaluating supercapacitor materials, as the available surface directly influences charge storage capacity, particularly in EDLC. The BET method is one of the most widely used techniques.

7.5.1 Brunauer–Emmett–Teller (BET) Method

The electrochemical performance of supercapacitor electrodes, whether through EDLC or pseudocapacitive mechanisms, is highly influenced by surface-driven processes such as charge adsorption/desorption and redox reactions at the electrode–electrolyte interface. A key factor affecting these processes is the electrode materials' specific surface area and pore architecture, as they determine ion accessibility and charge storage capacity.

Based on the physical adsorption of inert gases such as argon (87.3 K) or nitrogen (77.4 K), the BET method is a commonly used approach to determine the specific surface area of porous materials [22, 23]. The surface area information is the result of the adsorption–desorption isotherm. Further, the non-local density functional theory (NLDFT) model can be used for determining pore size distribution. The geometry of the pore is usually assumed to be cylindrical, slit shaped, or combination of both, depending on the nature of the material.

The nature of the material contributes differently to electrochemical behavior, e.g., micropores (<2 nm) enhance EDLC performance due to high surface area, as they may restrict ion mobility if the pore size is too narrow, while mesopores (2–50 nm) facilitate efficient electrolyte diffusion and can improve rate capability. However, all pores are not equally accessible to electrolyte ions, making it essential to characterize the surface area and pore size distribution accurately.

The ultra-micropore (<1 nm) investigation is carried out by adsorption of carbon dioxide (CO_2) at 273 K, as CO_2 has high diffusivity in confined spaces. For instance, Zhu et al. [24] analyzed pore structure using NLDFT by utilizing both CO_2 and N_2

adsorption isotherms. This study demonstrated that the adsorption of CO2 provides more accurate insights into ultra-micropores. In contrast, nitrogen isotherms effectively captured the presence of micropores larger than 1 nm and mesopores up to 4 nm in diameter. These combined results confirmed the hierarchical pore structure essential for optimizing charge storage and ion transport.

7.6 OTHER TECHNIQUES

7.6.1 Thermogravimetric Analysis (TGA)

TGA is responsible for detecting the thermal behavior and compositional stability of the electrode material. TGA provides information about thermal decomposition of the material by monitoring mass loss as a function of temperature, unlike FTIR which provides information about functional groups. Sample is heated at controlled rate in an inert or oxidative atmosphere. On increasing temperature, decomposition starts at different temperature range for different components such as adsorbed moisture, surface ligands or structural elements. The changes result in mass loss, which is recorded as a function of temperature. These results are used to estimate the composition and thermal stability of the material by identifying key degradation steps. In the case of supercapacitors, TGA evaluates organic content in carbon-based materials, confirming the thermal endurance and identifying the optimal process temperature. Gogotsi et al. [25] investigated the removal of sulfur from the sulfur-loaded MXene composites using TGA. Their results showed complete removal of sulfur at approximately 350°C under an argon atmosphere. Due to this, the porous MXene structures were formed due to the breaking of interlayer interactions. These porous structures are an important source for ion transport in supercapacitor materials. This structural transformation was later verified by XRD analysis.

Similarly, Wei et al. [26] applied TGA to evaluate the composition and stability of NiMn-LDH/CNT core/shell electrodes. TGA helped in determining Ni/Mn feeding ratio and thermal stability, both of which are essential for achieving controlled thickness and optimal performance in supercapacitors.

7.6.2 Zeta Potential (ζ-Potential)

The electrostatic stability of colloidal dispersions is indicated by the ζ-potential. It is used to determine the electrical potential of the suspended particles, which is influenced by some factors such as pH, ionic concentration, and surface chemistry. Those particles with high positive or negative ζ-potential repel each other, which results in stable dispersions with less aggregation. When the ζ-potential approaches zero (near isoelectric point), the particles agglomerate due to the dominating attractive forces, which leads to the destabilization of the suspension. The electrostatically stable dispersions have ζ-potential $\pm$ 20 mV. The ζ-potential value can be altered by functionalizing the surface or by adding the dopants or solvents, thereby enhancing dispersion stability through improved electrostatic repulsion. Thus, modifying the surface chemistry acts as an efficient method to control the colloidal behavior of nanomaterials used in supercapacitors. ζ-potential is often complemented with dynamic light scattering (DLS) to calculate the particle size distribution and dispersion quality. Erdem et al. [27] analyzed the ζ-potential of Mn:PANI:N-CDot composite electrodes using a Zetasizer. The incorporation of nitrogen-doped carbon dots (N-CDots) slightly reduces the potential from ~13 to ~10.93 mV, indicating successful integration into the composite. However, while increasing Mn content, the rise in ζ-potential is seen, which suggests the improved incorporation of Mn^{2+} ions into the structure. Chandra et al. [28], on the other hand, used ζ-potential, showing increased electrochemical activity of hollow nanostructures as compared to solid counterparts, which is due to the more surface-active sites in the hollow structure, resulting in better ion diffusion, therefore enhancing capacitive performance.

In cuprous oxide (Cu_2O), which exhibits a negative ζ-potential. Upon dispersion, the hydroxyl ions (OH^-) get accumulated at the surface, leading to the formation of a stable nanolayer. The lowering pH introduces protons (H^+), which shifts the ζ-potential toward a positive regime and increases the electropositivity of the hollow nanostructures.

Thus, zeta potential is not only essential for understanding the colloidal behavior of electrode materials but also serves as a powerful tool in optimizing surface interactions and ion accessibility in supercapacitor systems.

7.7 ELECTROCHEMICAL TECHNIQUES

The performance and behavior of supercapacitor materials are evaluated by using electrochemical techniques such as CV, GCD, and EIS. These methods provide detailed insights into charge storage mechanisms, capacitance, resistance, and overall device stability.

7.7.1 Cyclic Voltammetry (CV)

CV is a basis and fundamental electrochemical technique which is widely used to characterize electrode material. This method evaluates electrochemical properties such as capacitive behavior, redox activity, charge storage mechanism, and system reversibility. Due to its accuracy, simple process, fast data acquisition, and capability to differentiate between faradic and non-faradic processes, it remains cornerstone in supercapacitor research.

In the field of supercapacitors, CV provides critical insights into whether charge storage has been arrived from EDLC, pseudocapacitive faradic reactions or the combination of both of them. These insights are derived from the shape and area of cyclic voltammogram, which is a plot between current (I) and applied potential (V) as shown in Figure 7.2.

Based on faraday's law, the total charge (Q) involved in an electrochemical reaction is directly proportional to the number of electrons exchanged (n) and the faraday constant (F), as mentioned in Equation 7.4.

$$Q = n \; X \; F \tag{7.4}$$

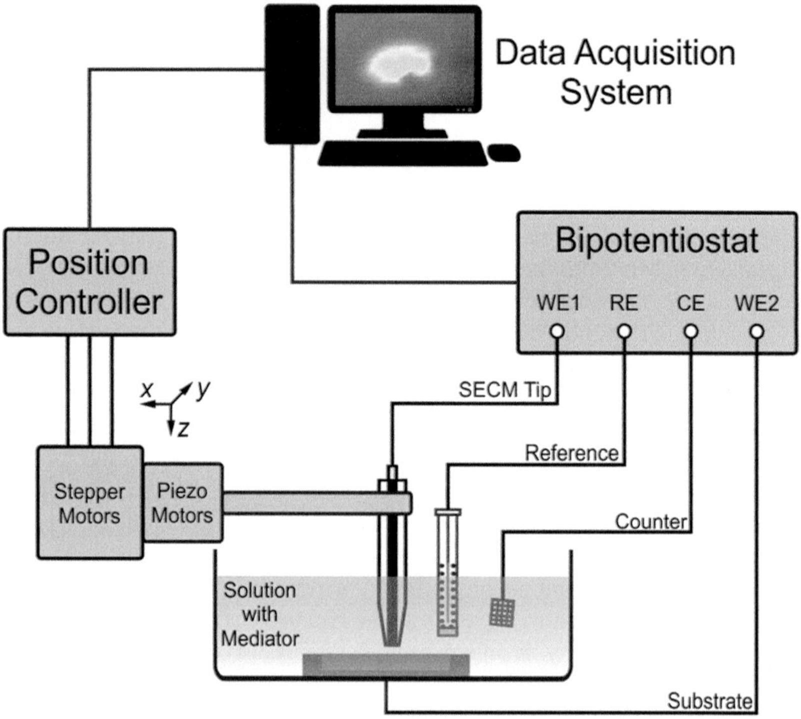

FIGURE 7.2 Schematic of a scanning electrochemical microscopy instrument (SECM). (Adapted with permission [10]. © ACS 2016.)

The capacitance (C) of a supercapacitor can be determined from the current (i) response obtained during CV, as described by Equation 7.5.

$$C = \frac{\Delta Q}{\Delta V} \tag{7.5}$$

During the charge–discharge cycles, the current (i) is associated with the capacitance (C) and the rate of change of voltage (dV/dt), as expressed in Equation 7.6.

$$i = C\ X\ \frac{dV}{dt} \tag{7.6}$$

At elevated scan rates, the current increases as it is directly influenced by both the ionic flux toward the electrode and the diffusion of ions at the electrode interface. Considering the number of electrons transferred (n), faraday's constant (F), the active surface area of the working electrode (A), the initial analyte concentration (C_0), the diffusion coefficient (D), and time (t), this current response (i) is effectively described by the Cottrell equation, as presented in Equation 7.7.

$$i = nfAC_o\sqrt{\frac{D}{\pi\ X\ t}} \tag{7.7}$$

The Randles–Sevcik and Cottrell equations each characterize diffusion-controlled currents in electrochemical systems, though they are applicable under different experimental conditions. In both cases, the presence of the diffusion coefficient (D) allows verification of diffusion-limited behavior using either method.

7.7.1.1 Essential Components of the CV Experimental Setup

Although CV can be performed using a two-electrode cell configuration, the three-electrode setup is more frequently utilized in fundamental research (Figure 7.2). This configuration, often called a half-cell, allows for the isolated examination of a single electrode by reducing interference from reactions occurring at the other electrodes. The following are the primary components used in CV characterization of hybrid SCs:

7.7.1.1.1 Working Electrode (WE)

The WE is a fundamental component in the three-electrode configuration used for CV measurements, as it functions as the primary site where the targeted electrochemical reactions take place. During the CV experiment, the potential applied to the WE is systematically varied, inducing oxidation and reduction processes of the analyte or material under investigation. The current generated at the WE during these electrochemical reactions is measured, offering important insights into the material's electrochemical properties, including its charge storage capacity, electrical conductivity, and kinetics of response.

The composition and surface characteristics of the WE are essential factors that significantly influence the accuracy and consistency of CV measurements. Typical materials used for working electrodes include platinum, gold, glassy carbon, and carbon-based composites, which are chosen for their high electrical conductivity, chemical stability, and minimal interference with the reactions being studied. Furthermore, the surface area and morphology of the working electrode can significantly impact the current response, necessitating careful consideration of these parameters in experimental design.

7.7.1.1.2 Reference Electrode (RE)

The RE is an essential component in a three-electrode system, providing a stable and known reference potential against which the WE's potential is measured. It does not participate in the electrochemical reactions occurring at the WE. Common reference electrodes include the silver/silver chloride (Ag/AgCl) and saturated calomel electrode (SCE), which maintain a constant potential due to the immobility of their ionic solutions. The RE ensures accurate potential control and helps minimize measurement variations due to fluctuations in the electrochemical environment.

7.7.1.1.3 Counter Electrode (CE)

The CE completes the electrical circuit in a three-electrode electrochemical cell by allowing current to flow between the WE and itself when potential is applied to the WE. It does not participate in the electrochemical reaction but records the current to maintain charge balance. Commonly used CEs include large-area platinum plates or mesh-type platinum electrodes, chosen for their high conductivity and chemical stability.

7.7.1.1.4 Supporting Electrolyte (SE)

The SE, which can be aqueous or organic depending on the electrode materials used, plays a vital role in maintaining ionic conductivity and minimizing solution resistance in electrochemical systems. It must demonstrate high solubility in the selected solvent, remain chemically and electrochemically stable under the experimental conditions, and be suitable for purification. Additionally, the solvent should effectively dissolve both the electrolyte and the target compound. The concentration of redox species at the electrode interface is described by the Nernst equation [29], which relates the electrode potential to the ionic concentrations, as presented in Equation 7.8.

$$E_T = E^o - \frac{RT}{nF} ln \frac{C_{red}}{C_{ox}} \tag{7.8}$$

where E_T = reversible cell voltage at an arbitrary temperature T, E^o = standard state reversible cell voltage, F = faraday's constant, R = gas constant, C_{red} = concentration of reductive species, C_{ox} = concentration of oxidative species.

The peak current associated with a reversible redox reaction is defined by the Randles–Sevcik equation. This equation illustrates the dependence of peak current on the concentration of the electroactive species, as represented in Equation 7.9.

$$i_p = \left(2.69\ X\ 10^5\right) n^{3/2} A D^{1/2} C v^{1/2} \tag{7.9}$$

where i_p = peak current (A), n = number of electrons transferred, A = electrode area (cm^2), D = diffusion coefficient of the species (cm^2/s), C = concentration of the species (mol/cm^3), v = scan rate (V/s).

The contributions from diffusion-controlled and capacitive processes in an electrochemical system can be assessed by evaluating the b-value. This parameter is crucial for analyzing the electrochemical kinetics of nano-engineered electrode materials. Examining the b-value and the CV curves' shape makes it possible to differentiate between EDLCs, pseudocapacitive materials, and battery-type electrodes. For electrodes based on EDLC, the CV curve generally exhibits an ideal rectangular shape (Figure 7.3a) [30, 31], with a characteristic b-value of precisely 1, and a linear voltage response (a triangular shaped curve) during constant current charging/discharging, Figure 7.3 (c). A CV curve that retains its rectangular shape across varying scan rates, without notable shifts of anodic peaks toward higher potentials or cathodic peaks toward lower potentials, signifies a strong capacitive behavior. On the other hand, pseudocapacitive materials exhibit CV curves that are close to rectangular but often include inflection points, reflecting subtle redox activity (Figure 7.3b) [32]. For these materials, the b-value remains close to 1, suggesting a hybrid storage mechanism combining both capacitive and faradaic processes. Capacitance, measured in farads (F), represents the ability of an electrode to store electrical charge. When it comes to battery-type electrodes, the CV curves show well-defined redox peaks, indicative of a quasi-reversible charge transfer mechanism. For these systems, the b-value is typically around 0.5, as shown in Figure 7.3(g-i). Various configurations for combining EDLC, pseudocapacitive, and battery-type electrodes to form asymmetric and hybrid supercapacitors are illustrated in Figure 7.3(d-f)

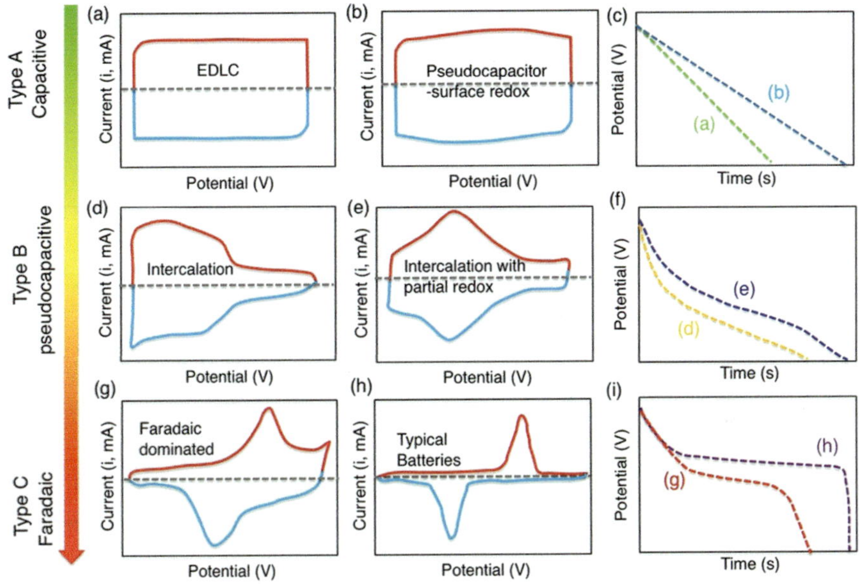

FIGURE 7.3 (a, b, d, e, g, h) Schematic cyclic voltammograms and (c, f, i) corresponding galvanostatic discharge curves for various kinds of energy-storage materials. (Adapted with permission [34]. © ACS 2018.)

[33, 34]. Experiments are generally performed at lower scan rates for such materials, since high scan rates can amplify ohmic resistance effects, distorting the CV results. Furthermore, CV data can extract important electrochemical parameters, including specific capacitance (C), energy density, and power density. Cs typically exhibits higher values at lower scan rates, as the extended time allows electrolyte ions to more efficiently adsorb and desorb on the electrode surface. The Cs, expressed in F/g, can be calculated from the CV curves using Equation 7.10.

$$Cs = \frac{\int_{V-}^{V+} I(V)\,dV}{mv\,(V_+ - V_-)} \tag{7.10}$$

where I(V) denotes the current as a function of voltage, while V+ and V– represent the upper and lower voltage limits, respectively. The variable m refers to the mass of the active electrode material (in grams), and *v* corresponds to the scan rate of the potential (measured in V/s).

CV is a versatile electroanalytical technique with wide-ranging applications across science and engineering, as shown in Figure 7.4. It is widely utilized for the electrochemical fabrication of conductive polymers and the precise deposition of nanoparticle films on electrode surfaces. CV is essential for evaluating the reversibility of redox reactions and analyzing electron transfer kinetics, making it a fundamental tool in mechanistic electrochemistry. It also plays a key role in investigating the corrosion behavior of metals. In the field of energy storage and conversion, CV contributes to the development and performance assessment of lithium-ion batteries,

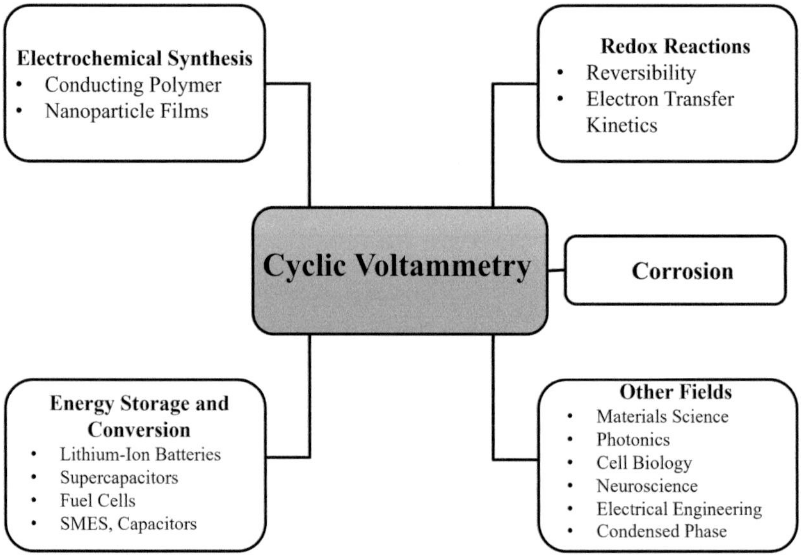

FIGURE 7.4 Applications of CV across multiple disciplines.

supercapacitors, fuel cells, superconducting magnetic energy storage (SMES) systems, and conventional capacitors. Beyond energy-related fields, CV is extensively applied in materials science, photonics, cell biology, neuroscience, electrical engineering, and condensed-phase physics to investigate charge transport, molecular interactions, and interfacial processes.

7.7.2 Galvanostatic Charging–Discharging (GCD)/ Chronopotentiometry (CP)

Alongside CV, GCD analysis is routinely used to evaluate essential electrochemical parameters, including specific capacitance, energy density, and power density. These assessments are generally conducted across a range of current densities (e.g., 1A/g, 2A/g, etc.), where the change in potential over time is recorded. The cyclic stability of the electrode material can be assessed by examining the consistency of the GCD curves over numerous charge–discharge cycles, particularly by tracking the duration of these cycles. There is an inverse correlation between current density and specific capacitance [35], whereby increasing the current density typically decreases capacitance. For EDLCs [36], the GCD curves display a linear profile, (Figure 7.3 (c)) and the b-value typically equals 1, an indicator of ideal capacitive performance. In contrast, pseudocapacitive materials [37] show non-linear charge–discharge behavior, (Figure 7.3(f)) resulting from quasi-reversible faradaic processes, which lead to asymmetric GCD curves due to the time-dependent nature of the charge/voltage ratio. In the case of battery-type electrodes, GCD profiles exhibit distinct voltage plateaus, signifying phase transformations or faradaic reactions occurring at specific potentials (Figure 7.3(i)). The particular capacitance obtained from GCD measurements can be determined using the established formula in Equation 7.11, offering valuable insight into the efficiency and electrochemical performance of the electrode material.

$$C_{CD} = \frac{2I_m}{V^2} \int_{V^-}^{V^+} Vdt \tag{7.11}$$

where C_{CD} = charging–discharging capacitance, V = operational voltage window, I_m is current density in A/g, $\int_{V^-}^{V^+} Vdt$ is current integral area.

Or, it can also be written as in Equation 7.12.

$$C_s = \frac{I \times \Delta t}{\Delta V \times m} \tag{7.12}$$

where I, Δt, ΔV, m correspond to discharge current (A), discharge time (s), potential window (V), and the mass of active materials (g), respectively.

Enhancing the electrochemical performance of a supercapacitor requires rapid charge–discharge capability and minimal voltage drop, the latter reflecting low internal resistance (IR). Typically, as current density increases, power density decreases, whereas energy density tends to rise. The specific energy density E, measured in watt-hours per kilogram (Wh/kg), is defined by Equation 7.13.

$$E = \frac{1}{2} C_s (\Delta V)^2 \tag{7.13}$$

where ΔV denotes the operational voltage window, and Cs represents the specific capacitance of the fabricated electrode. The specific power density P, expressed in watts per kilogram (W/kg), is defined by Equation 7.14.

$$P = \frac{I \ X \ \Delta V}{m} \tag{7.14}$$

where I = applied constant current, and m = total mass of supercapacitor in grams.

Similarly, GCD [38, 39] measurements can be used to determine the specific capacitance, energy, and power densities of a symmetric two-electrode supercapacitor system. When connected in series, a pair of such supercapacitors can power a white LED [40].

7.7.2.1 Electrochemical Impedance Spectroscopy (EIS)

EIS is a powerful analytical technique widely employed to investigate the electrochemical behavior of electrode materials, especially in energy storage systems like supercapacitors. This method enables the extraction of critical parameters, including series resistance (R_s), charge transfer resistance (R_{ct}), double-layer capacitance (C_{dl}), and Warburg impedance (Z_w), which arises from ionic diffusion processes, as illustrated in Figure 7.5 [41,42]. The impedance (Z) of an electrochemical system can be modeled using basic electrical elements: resistors (Z = R), inductors ($Z_l = i\omega L$), and capacitors $\left(Z = \frac{1}{i\omega C} \right)$, where ω is the angular frequency. Most commercial EIS instruments have a measurable impedance range up to approximately 10^{10} Ω.

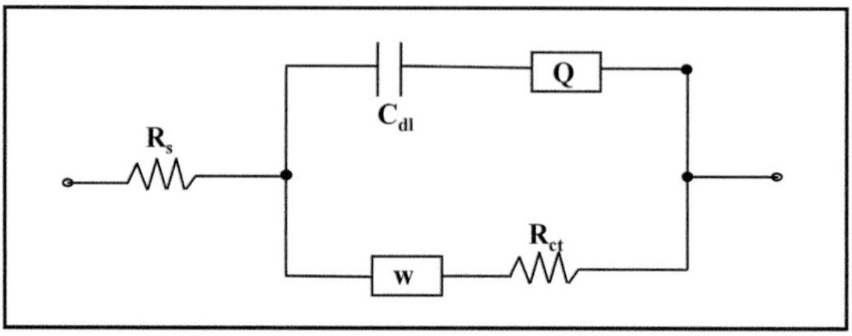

FIGURE 7.5 General diagram for EIS.

EIS measurements are usually performed across a broad frequency range, typically spanning from 0.01 Hz to 1 mHz, to thoroughly evaluate the dynamic behavior of the electrochemical system [43, 44]. Impedance data are typically represented using Nyquist and Bode plots. In a Nyquist plot (Figure 7.6), the real part of the impedance is shown on the x-axis, while the imaginary part is displayed on the y-axis. The phase angle (θ), representing the phase difference between voltage and current, is given by θ = arg (Z) with Z(w) expressed as $Z(w) = Z_o(\cos\theta + i\ \sin\theta)$ [45].

Nyquist plots often feature semicircles at high frequencies, indicating electron transfer or redox processes at the electrode surface. A linear segment with a slope of 45° typically appears at intermediate frequencies, corresponding to the Warburg region, which reflects diffusion-controlled processes. At low frequencies, a vertical line suggests the buildup of the electrical double layer at the electrode–electrolyte interface. The intersection of the semicircle with the real axis represents the solution resistance (R_s), while the charge transfer resistance (R_{ct}) is inferred from the low-frequency semicircle diameter or extension, as shown in Figure 7.5. Charge transport across the electrode–electrolyte interface

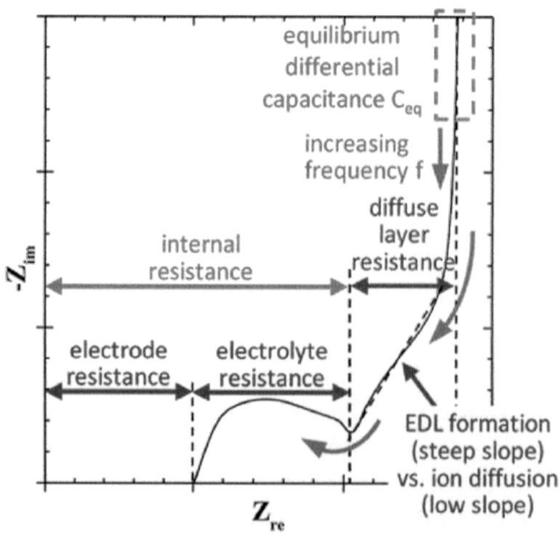

FIGURE 7.6 Schematic representation of the Nyquist plots generation mechanism. (Adapted with permission [43]. © ACS 2018.)

is more efficient when R_{ct} is small, indicating a faster charge transfer mechanism. The Bode plot complements the Nyquist representation, plotting frequency (ω) along the x-axis and either the phase angle (θ) or log |Z| on the y-axis. The total faradaic impedance (R_f), accounting for both charge transfer and diffusional contributions, is expressed in Equation 7.15.

$$R_f = R_{ct} + R_w \tag{7.15}$$

where R_w is the Warburg resistance, calculated by Equation 7.16.

$$R_w = \frac{\sigma}{w^{1/2}} \tag{7.16}$$

and σ is the Warburg coefficient $\left(\Omega\ rad^{\frac{1}{2}} s^{-1}\right)$.

The Warburg impedance (Z_f) is calculated using Equation 7.17.

$$Z_f = \left(\frac{2}{\omega}\right)^{1/2} \tag{7.17}$$

The phase angle in the Bode plot is theoretically determined by Equation 7.18.

$$\theta = tan^{-1}\left(\frac{\sigma\omega^{-1/2}}{R_{ct} + \sigma\omega^{-1/2}}\right) \tag{7.18}$$

This relation indicates that in a fully reversible process ($R_{ct} = 0$), the phase angle is ≤ 45°, while in a quasi-reversible system ($R_{ct} > 0$), it falls below 45°. The double-layer capacitance (C_{dl}) is determined by Equation 7.19.

$$C_{dl} = \frac{1}{\sigma\omega^{1/2}} \tag{7.19}$$

In Nyquist plots, Warburg impedance appears as a 45° line, while in Bode plots, it manifests as a 45° phase shift. For optimal performance in supercapacitor applications, both R_{ct} and R_w should be minimized to reduce the IR drop, thereby enhancing energy and power output [43, 46].

EIS is thus an essential diagnostic tool for evaluating charge transfer characteristics and optimizing the design of electrode materials for energy storage technologies.

7.8 FUTURE PERSPECTIVES AND CHALLENGES

At this stage of research, critical considerations must guide future advancements in supercapacitor technology. A primary concern is selecting cost-effective, environmentally friendly, non-toxic, and sustainable synthesis methods and materials essential for meeting global energy demands. Emerging materials like MOFs, metal nitrides, phosphorene, and MXenes show promise for next-generation supercapacitor applications due to their exceptional physicochemical properties.

Innovative fabrication techniques should be explored to produce electrode materials with hierarchical porosity, hollow architectures, reduced ion diffusion pathways, and high structural density. Furthermore, to enhance electrochemical performance, the use of redox-active electrolytes (RAEs) in combination with aqueous systems is recommended. These additives can introduce additional redox sites at the electrode–electrolyte interface, leading to improved potential windows, increased surface area, higher specific capacitance, enhanced energy density, and superior cyclic stability.

Finally, increased attention should be given to the development of asymmetric supercapacitor configurations, which offer the advantage of simultaneously achieving high energy and power densities, making them attractive for practical energy storage systems.

7.9 AUTHOR AI DECLARATION

The authors hereby declare the use of generative AI and AI-assisted technologies in the writing process. While preparing this work, the authors used the ChatGPT tool/service to rephrase several statements in the manuscript. All AI-generated content was reviewed and edited by the authors as needed. The authors take full responsibility for the content of the publication.

ACKNOWLEDGMENT

The Editors acknowledge the use of Grammarly as a tool to assist in editing this book chapter. The suggestions made by Grammarly were carefully considered and incorporated where appropriate, but the author is solely responsible for the final content.

REFERENCES

1. Sidiqi, U., Ubaidullah, M., Kumar, A., Kumar, D., Muzammil, K., and Imran, M. "Progress on cupric oxide-based nanomaterials: Exploring advancements in their synthesis, applications, and prospects." *Materials Science and Engineering: B* 308 (2024): 117598. doi:10.1016/j.mseb.2024.117598
2. Hareendra Krishna Kumar, H., Chulliyote, R., and Joseph, M. G. "Effect of crystallite size on the intercalation pseudocapacitance of lithium nickel vanadate in aqueous electrolyte." *Journal of Solid State Electrochemistry* 22 (2018): 1–9. doi:10.1007/s10008-017-3712-2
3. Yan, W., Petkov, V., Mahurin, S. M., Overbury, S. H., and Dai, S. "Powder XRD analysis and catalysis characterization of ultra-small gold nanoparticles deposited on titania-modified SBA-15." *Catalysis Communications* 6 (2005): 404–408. doi:10.1016/j.catcom.2005.04.004
4. Sarma, D., Santra, P. K., Mukherjee, S., and Nag, A. "X-ray photoelectron spectroscopy: A unique tool to determine the internal heterostructure of nanoparticles." *Chemistry of Materials* 25 (2013): 1222–1232. doi:10.1021/cm303567d
5. Greczynski, G., and Hultman, L. "X-ray photoelectron spectroscopy: Towards reliable binding energy referencing." *Progress in Materials Science* 107 (2020): 100591. doi:10.1016/j.pmatsci.2019.100591

6. Toupin, M., Brousse, T., and Bélanger, D. "Charge storage mechanism of MnO_2 electrode used in aqueous electrochemical capacitor." *Chemistry of Materials* 16 (2004): 3184–3190. doi:10.1021/cm049649j
7. Wu, Z. S., Winter, A., Chen, L., Sun, Y., Turchanin, A., Feng, X., and Müllen, K. "Three dimensional nitrogen and boron co-doped graphene for high-performance all-solid-state supercapacitors." *Advanced Materials* 24 (2012): 5130–5135. doi:10.1002/adma.201201948
8. Sullivan, M., Schnyder, B., Bärtsch, M., Alliata, D., Barbero, C., Imhof, R., and Kötz, R. "Electrochemically modified glassy carbon for capacitor electrodes: Characterization of thick anodic layers by cyclic voltammetry, differential electrochemical mass spectrometry, spectroscopic ellipsometry, X-ray photoelectron spectroscopy, FTIR, and AFM." *Journal of The Electrochemical Society* 147 (2000): 2636. doi:10.1149/1.1393582
9. Tõnisoo, A., Kruusma, J., Pärna, R., Kikas, A., Hirsimäki, M., Nõmmiste, E., and Lust, E. "In situ XPS studies of electrochemically negatively polarized molybdenum carbide derived carbon double layer capacitor electrode." *Journal of The Electrochemical Society* 160 (2013): A1084–A1093. doi:10.1149/2.042308jes
10. Polcari, D., Dauphin-Ducharme, P., and Mauzeroll, J. "Scanning electrochemical microscopy: A comprehensive review of experimental parameters from 1989 to 2015." *Chemical Reviews* 116 (2016): 13234–13278. doi:10.1021/acs.chemrev.6b00067
11. Xue, J., Gao, L., Hu, X., Cao, K., Zhou, W., Wang, W., and Lu, Y. "Stereolithographic 3D printing-based hierarchically cellular lattices for high-performance quasi-solid supercapacitor." *Nano-Micro Letters* 11 (2019): 46. doi:10.1007/s40820-019-0280-2
12. Zhang, L., Shi, W., and Zhang, B. "A review of electrocatalyst characterization by transmission electron microscopy." *Journal of Energy Chemistry* 26 (2017): 1117–1135. doi:10.1016/j.jechem.2017.10.016
13. Song, Z., and Xie, Z. H. "A literature review of in situ transmission electron microscopy technique in corrosion studies." *Micron* 112 (2018): 69–83. doi:10.1016/j.micron.2018.04.001
14. Zhou, W., and Thomas, J. M. "HRTEM surface profile imaging of solids." *Current Opinion in Solid State and Materials Science* 5 (2001): 75–83. doi:10.1016/S1359-0286(00)00037-1
15. Greer, H. F., and Zhou, W. "Electron diffraction and HRTEM imaging of beam-sensitive materials." *Crystallography Reviews* 17 (2011): 163–185. https://doi.org/10.1080/0889311X.2010.535525
16. Dendisová, M., Jeništová, A., Parchaňská-Kokaislová, A., Matějka, P., Prokopec, V., and Švecová, M. "The use of infrared spectroscopic techniques to characterize nanomaterials and nanostructures: A review." *Analytica Chimica Acta* 1031 (2018): 1–14. doi:10.1016/j.aca.2018.05.046
17. Petit, T., and Puskar, L. "FTIR spectroscopy of nanodiamonds: Methods and interpretation." *Diamond and Related Materials* 89 (2018): 52–66. doi:10.1016/j.diamond.2018.08.005
18. Chen, H., Hu, L., Chen, M., Yan, Y., and Wu, L. "Nickel–cobalt layered double hydroxide nanosheets for high-performance supercapacitor electrode materials." *Advanced Functional Materials* 24 (2014): 934–942. doi:10.1002/adfm.201301747
19. Chen, X., Chai, H., Cao, Y., Zhou, W., Li, Y., and Yang, Y. "Hierarchical CoGa layered double hydroxides grown on nickel foam as high energy density hybrid supercapacitor." *Chemical Engineering Journal* 381 (2020): 122620. doi:10.1016/j.cej.2019.122620
20. Kumar, C. S. *Raman Spectroscopy for Nanomaterials Characterization*. Dordrecht: Springer Science & Business Media, 2012.

21. Ogino, S., Itoh, T., Mabuchi, D., Yokoyama, K., Motomiya, K., Tohji, K., and Sato, Y. "In situ electrochemical raman spectroscopy of air-oxidized semiconducting single-walled carbon nanotube bundles in aqueous sulfuric acid solution." *Journal of Physical Chemistry C* 120 (2016): 7133–7143. doi:10.1021/acs.jpcc.5b12057
22. Sinha, P., Datar, A., Jeong, C., Deng, X., Chung, Y.-G., and Lin, L.-C. "Surface area determination of porous materials using the Brunauer–Emmett–Teller (BET) method: Limitations and improvements." *Journal of Physical Chemistry C* 123 (2019): 20195–20209. doi:10.1021/acs.jpcc.9b02116
23. Mohan, V. B., Jayaraman, K., and Bhattacharyya, D. "Brunauer–Emmett–Teller (BET) specific surface area analysis of different graphene materials: A comparison to their structural regularity and electrical properties." *Solid State Communications* 319–320 (2020): 114004. doi:10.1016/j.ssc.2020.114004
24. Zhu, Y., Murali, S., Stoller, M. D., Ganesh, K., Cai, W., Ferreira, P. J., Pirkle, A., Wallace, R. M., Cychosz, K. A., and Thommes, M. "Carbon-based supercapacitors produced by activation of graphene." *Science* 332 (2011): 1537–1541. doi:10.1126/science.1200770
25. Xie, X., Kretschmer, K., Anasori, B., Sun, B., Wang, G., and Gogotsi, Y. "Porous $Ti_3C_2T_x$ MXene for ultrahigh-rate sodium-ion storage with long cycle life." *ACS Applied Nano Materials* 1 (2018): 505–511. doi:10.1021/acsanm.8b00045
26. Zhao, J., Chen, J., Xu, S., Shao, M., Zhang, Q., Wei, F., Ma, J., Wei, M., Evans, D. G., and Duan, X. "Hierarchical NiMn layered double hydroxide/carbon nanotubes architecture with superb energy density for flexible supercapacitors." *Advanced Functional Materials* 24 (2014): 2938–2946. doi:10.1002/adfm.201303638
27. Alaş, M. O., Güngör, A., Genç, R., and Erdem, E. "Feeling the power: Robust supercapacitors from nanostructured conductive polymers fostered with Mn^{2+} and carbon dots." *Nanoscale* 11 (2019): 12804–12816. doi:10.1039/C9NR03544C
28. Sharma, V., Singh, I., and Chandra, A. "Hollow nanostructures of metal oxides as next generation electrode materials for supercapacitors." *Scientific Reports* 8 (2018): 1307. doi:10.1038/s41598-018-19815-y
29. Weber, K., and Creager, S. E. "Voltammetry of redox-active groups irreversibly adsorbed onto electrodes. Treatment using the Marcus relation between rate and overpotential." *Analytical Chemistry* 66, no. 19 (1994): 3164–3172. doi:10.1021/ac00091a027
30. Lämmel, C., Schneider, M., Weiser, M., and Michaelis, A. "Investigations of electrochemical double layer capacitor (EDLC) materials—A comparison of test methods." *Materialwissenschaft und Werkstofftechnik* 44, no. 7 (2013): 641–649. doi:10.1002/mawe.201300122
31. Wang, J., Polleux, J., Lim, J., and Dunn, B. "Pseudocapacitive contributions to electrochemical energy storage in TiO_2 (anatase) nanoparticles." *The Journal of Physical Chemistry C* 111, no. 40 (2007): 14925–14931. doi:10.1021/jp074464w.
32. Costentin, C., and Savéant, J.-M. "Energy storage: Pseudocapacitance in prospect." *Chemical Science* 10, no. 22 (2019): 5656–5666. doi:10.1039/C9SC01662G.
33. Sharma, S., and Chand, P. "Supercapacitor and electrochemical techniques: A brief review." *Results in Chemistry* 5 (2023): 100885. doi:10.1016/j.rechem.2023.100885.
34. Gogotsi, Y., and Penner, R. M. "Energy storage in nanomaterials–capacitive, pseudocapacitive, or battery-like?" *ACS Nano* 12, no. 3 (2018): 2081–2083. doi:10.1021/acsnano.8b01914.
35. Lu, W., Hartman, R., Qu, L., and Dai, L. "Nanocomposite electrodes for high-performance supercapacitors." *The Journal of Physical Chemistry Letters* 2, no. 6 (2011): 655–660. doi:10.1021/jz200104n.
36. Stoller, M. D., and Ruoff, R. S. "Best Practice methods for determining an electrode material's performance for ultracapacitors." *Energy & Environmental Science* 3, no. 9 (2010): 1294–1301. doi:10.1039/C0EE00074D.

37. Kim, B.-K., Chabot, V., and Yu, A. "Carbon nanomaterials supported $Ni(OH)_2$/NiO hybrid flower structure for supercapacitor." *Electrochimica Acta* 109 (2013): 370–380. doi:10.1016/j.electacta.2013.07.119.
38. Zeng, L., Wu, T., Ye, T., Mo, T., Qiao, R., and Feng, G. "Modeling galvanostatic charge–discharge of nanoporous supercapacitors." *Nature Computational Science* 1, no. 11 (2021): 725–731. doi:10.1038/s43588-021-00153-5.
39. Noori, A., El-Kady, M. F., Rahmanifar, M. S., Kaner, R. B., and Mousavi, M. F. "Towards establishing standard performance metrics for batteries, supercapacitors and beyond." *Chemical Society Reviews* 48, no. 5 (2019): 1272–1341. doi:10.1039/C8CS00581H.
40. Sundriyal, S., Shrivastav, V., Kaur, H., Mishra, S., and Deep, A. "High-performance symmetrical supercapacitor with a combination of a ZIF-67/rGO composite electrode and a redox additive electrolyte." *ACS Omega* 3, no. 12 (2018): 17348–17358. doi:10.1021/acsomega.8b02065.
41. Lasia, A. *Electrochemical Impedance Spectroscopy and its Applications.* In *Modern Aspects of Electrochemistry*, edited by B.E. Conway, J.O. Bockris, and R.E. White, vol. 32. Boston: Springer, 2002.
42. Panchal, K., Bhakar, K., Sharma, K. S., Kumar, D., and Prasad, S. "Review on electrochemical impedance spectroscopy: A technique applied to hollow structured materials for supercapacitor and sensing applications." *Applied Spectroscopy Reviews* 60, no. 1 (2025): 30–55. doi:10.1080/05704928.2024.2361082.
43. Mei, B. A., Munteshari, O., Lau, J., Dunn, B., and Pilon, L. "Physical interpretations of Nyquist plots for EDLC electrodes and devices." *The Journal of Physical Chemistry C* 122, no. 1 (2018): 194–206. doi:10.1021/acs.jpcc.7b10582.
44. Barsoukov, E., & Macdonald, J. R., editors. *Impedance Spectroscopy: Theory, Experiment, and Applications.* Hoboken: Wiley-Interscience, 2005.
45. Sedlakova, V., Sikula, J., Majzner, J., Sedlak, P., Kuparowitz, T., Buergler, B., and Vasina, P. "Supercapacitor equivalent electrical circuit model based on charges redistribution by diffusion." *Journal of Power Sources* 286 (2015): 58–65. doi:10.1016/j.jpowsour.2015.03.122.
46. Jović, V. D. *Determination of the Correct Value of Cdl from the Impedance Results Fitted by the Commercially Available Software.* Gamry Instruments, 2003. https://www.gamry.com/assets/Uploads/the-correct-value-of-Cdl.pdf.

8 Applications of Supercapacitors in Energy Storage

Anusha Srinivas and Sapna Nehra

8.1 INTRODUCTION

The global energy landscape has been undergoing rapid changes, driven by the need for energy solutions that are both sustainable and efficient. The rising trend in energy demands is also increasing the need for better energy storage technologies that can effectively integrate renewable sources, reduce the gap between supply and demand, and provide the highest performance for electronic appliances and vehicles. Supercapacitors have emerged as a promising technology in the energy sector [1]. They can store small amounts of energy quickly and tolerate frequent cycles of energy influx and consumption as well as operate at various temperatures. Supercapacitors are being rapidly incorporated into various storage applications and markets for use [2].

Supercapacitors have an advantage in storing a large amount of energy, which sets them apart from other devices. One of their main benefits is that they have an electric double layer (EDL), which can be a double or triple layer, differing from batteries, which are electrochemical cells that can last for very long periods. With the supercapacitor, which is "nearly as fast" as the discharge of a capacitor, it is reliably expected that energy storage and release will also become faster [3–5].

The chapter presents a variety of applications of supercapacitors, including renewable energy systems, portable electronics, industrial applications, and other examples. It also addresses the challenges associated with supercapacitors, such as their lower energy density compared to batteries, and highlights recent improvements in materials and technologies that aim to overcome these limitations.

8.2 APPLICATIONS IN RENEWABLE ENERGY INTEGRATION

Solar energy is the most widely accepted form of renewable energy, with the installation of photovoltaic (PV) systems enabling the direct conversion of solar energy into electricity [6]. However, solar energy production is inherently intermittent and depends on sunlight availability, which varies throughout the day and depends on local climate conditions. To control the energy flow, these systems are used in solar power setups to store excess energy during peak sunlight times [7–9]. Batteries are devices used to store and release electrical energy for future use. Supercapacitors are the preferred choice when short-term operation involves repetitive charging and discharging, such as in processes with highly variable solar energy generation.

DOI: 10.1201/9781003535836-8

Supercapacitors can quickly store excess energy when there is strong sunlight, causing the PV cells to generate surplus energy that can later be saved for lighting or running appliances during periods without sunlight or at night, i.e., when energy is more consistent and reliable [10, 11]. Besides this, supercapacitors contribute to additional supply utilization if used as energy storage devices [12–14]. In energy harvesting systems, solar cells utilize supercapacitors to capture and store energy, which can then power ultra-low-power applications.

8.2.1 Wind Energy Systems

The other main source of renewable energy is wind power, where wind energy is converted into electrical energy by wind turbines [15]. Supercapacitors are used in wind energy systems to fairly distribute the electricity generated by wind turbines [16, 17]. During high winds, the excess power exceeds the grid's capacity, so supercapacitors can instantly store this extra energy without any loss through the conversion process [18]. As a result, when wind speeds are low, the stored energy is released to maintain a stable and continuous power supply [19–22], as shown in Figure 8.1.

FIGURE 8.1 Wind energy system.

Supercapacitors are connected to the grid on a large scale to ensure fast frequency regulation and voltage stabilization [23]. They respond to grid fluctuations almost instantaneously, making it possible to avert the blackouts and correct the stability of the power supply [24].

8.3 SUPERCAPACITORS IN ELECTRIC VEHICLES

Supercapacitor adoption in electric vehicles (EVs) and hybrid electric vehicles (HEVs), as well as how they can be integrated into vehicle systems, has been a game changer in the automotive industry [25]. Supercapacitors have the potential to release power from their stored energy, and accordingly, they are able to take up energy quickly, which is key for regenerative braking; this fact makes supercapacitors essential for the new vehicles' powertrains [26]. In this section, we examine the various functions of supercapacitors—whether to power or store energy—to improve the performance, efficiency, and sustainability of EVs.

8.3.1 Enhancing Acceleration and Power Delivery

Serving the high-power needs that come with the rapid acceleration of EVs is a key role fulfilled by supercapacitors. However, while lithium-ion batteries are efficient at storing energy, they are not as fast at delivering high power [27]. When needed, the increased power density of supercapacitors is used to assist batteries in supplying the required power during peak demand, acceleration, or climbing [28]. This tandem action allows the EV's speed to become more stable and makes the vehicle more responsive to driver input, resulting in a better driving experience [29]. Additionally, supercapacitors can take over the task of providing high-power short bursts from the vehicle's battery, which in the long run extends its service life [30]. Therefore, by handling short, high-power loads, the battery can operate more effectively and with less thermal stress, leading to improved reliability and durability of the entire powertrain.

Regenerative braking is a key feature in EVs and HEVs, serving as a method to enhance energy efficiency through kinetic energy recovery. Unlike traditional brakes that convert and dissipate kinetic energy as heat, EVs and HEVs benefit from this system. Supercapacitors are ideal for regenerative braking systems because they can charge and discharge rapidly, among other advantages. When a driver presses the brake pedal, the supercapacitor captures the energy and stores it for future use, such as during the next acceleration. This process not only boosts the overall efficiency of the vehicle but also reduces wear on brake components, which lowers maintenance costs and extends the lifespan of the braking system.

Supercapacitors have gained popularity in the latest EVs for energy harvesting through regenerative braking, complementing the batteries. Although batteries are more effective for large-scale energy storage, supercapacitors are the only way to store electricity using a similar neo-mega power source, where energy is delivered in extremely high volumes over very short periods, such as during braking [31]. The combined support of electrochemical supercapacitors and batteries creates a healthy environment in the vehicle, enabling optimal regenerative energy recovery and achieving high energy savings along with reducing environmental impact at various levels [32].

8.3.2 Improving Battery Longevity and Reducing Energy Losses

Supercapacitors are important and are primarily used to enhance the performance and lifespan of EV batteries by reducing the need for frequent deep discharges, thereby extending battery life. In most cases, infrequent use of EV batteries leads to a rapid loss of energy cell power, which can result from normal and high-speed operation [33, 34].

Supercapacitors, apart from storage and conversion processes, can reduce energy loss [35]. They are linked to improved energy management and higher efficiency in conversion processes, which affects the driving range and capabilities of an EV [36]. In cold weather, the performance of lithium-ion batteries is impacted because they are more prone to experiencing a decrease in energy at low temperatures, leading to reduced power retention [37]. However, supercapacitors perform very well at higher temperatures, making them suitable for EVs used in colder regions [38].

In cold temperatures, supercapacitors are very useful. While electric cars rely on supercapacitors to start, these devices remain effective even in very cold weather [39–41].

8.3.3 Fast Charging and Energy Recapture

The rapid recharge is essential for EVs, which has made it achievable worldwide. The main advantages users find are that the quick charging reduces idle time and helps them stay on schedule. Bi-directional charging of the vehicle, whether inside or outside the building, may reduce network congestion and help balance the intermittent features of the grid [42–44].

With the help of supercapacitors, it is possible to provide more efficient energy recovery from the regenerative energy of braking or energy harvesting systems [45]. The rapid transfer of energy to the storage system from various sources remains one of the key factors in the vehicle's energy management system, which is both accessible and robust, ultimately optimizing energy use throughout the entire system [46]. It is an accepted fact that supercapacitors will play a larger role in the EV sector when new materials and technologies emerge. Currently, trials are underway for new technological advancements, such as improving the sustainability of supercapacitors and increasing energy density using materials like graphene and carbon nanotubes, which are used in supercapacitors [47, 48].

Supercapacitors are becoming increasingly important in EVs. Supercapacitors help EVs accelerate and decelerate. They also extend battery life and enable faster charging. This improves the overall performance and efficiency of EVs [49, 50]. It is a fact that supercapacitors are predicted to become an essential component of electrified transportation in the future.

8.4 PORTABLE ELECTRONICS AND WEARABLE DEVICES

The rapid growth of portable electronics and wearables has created a need for smaller, improved, and more reliable energy storage solutions. Supercapacitors have emerged as an exciting new technology in this field, and they are expected to address issues such as severe power-density limitations, long charging times, and low durability [51]. This section will focus on the unique applications of supercapacitors in portable electronics and wearable devices, highlighting their impact as well as the benefits in device operation, such as extended battery life and the development of new functions.

8.4.1 Powering Portable Electronics

Supercapacitors are used when a device needs to be powered quickly. One concrete example is smartphones, where supercapacitors can be depended upon to supply high energy–consuming functions such as the camera flash, and other similar functions such as GPS and speaker [52–55].

8.4.2 Enabling New Functionalities in Wearables

The inclusion of supercapacitors in wearable devices can also enable new and surprisingly complex functions with traditional batteries alone. Wearables and supercapacitors are combined with thin-film batteries to develop hybrid storage systems on a larger scale [56]. It is not only a device that enhances user convenience and maintains power due to its operational characteristics but also a task focused on power saving, which conserves energy and handles tasks that require high power occasionally [57, 58].

Furthermore, self-powered wearable devices that use supercapacitors are an innovation that greatly benefits the fight against climate change. With supercapacitors, energy can be absorbed from the environment or the user's actions, such as sunlight, and then stored for later use [59]. The device becomes independent of external charging, and the extended battery lifespan enables continuous operation anywhere [60].

Furthermore, the inherent flexibility of a supercapacitor has been demonstrated through its use in the production of not only flexible but also stretchable wearables. These portable devices function as more advanced batteries and can adapt to the user's needs. Supercapacitors are typically made from advanced materials such as carbon nanotubes or graphene, which provide the necessary flexibility and ensure the energy storage system remains operational even when the device is bent or stretched during use [61–63]. The most significant advantage of supercapacitors for portable and wearable electronics is their quick recharging ability, unlike other types of batteries [64]. While traditional batteries take a long time to charge fully, supercapacitors can often be recharged in just a few minutes, sometimes even in seconds.

Supercapacitors are valued for more than just quick recharging; they are delivered to consumers with exceptional durability. They can be cycled tens of thousands of times with little to no significant loss in energy capacity, making them ideal for demanding rechargeable devices [65]. The expectation that a wearable product can last for years without needing a battery replacement is the main focus here, and the durability of supercapacitors adds a new dimension [66–68]. Their ability to deliver large amounts of energy in a short period, their fast charge rate, and their resistance to harsh conditions drive the development of the next generation of portable and wearable technology [69].

8.5 INDUSTRIAL AND GRID STABILIZATION

Supercapacitors are used in industrial operations and grid stabilization. They can generate energy emissions, retain high-power capacity, and withstand numerous processes, making them an ideal solution for addressing issues in these economic sectors [70, 71]. Here, the various applications of supercapacitors in industrial power systems and grid stabilization are presented, along with their roles in enhancing energy efficiency, reliability, and resilience, as depicted in Figure 8.2.

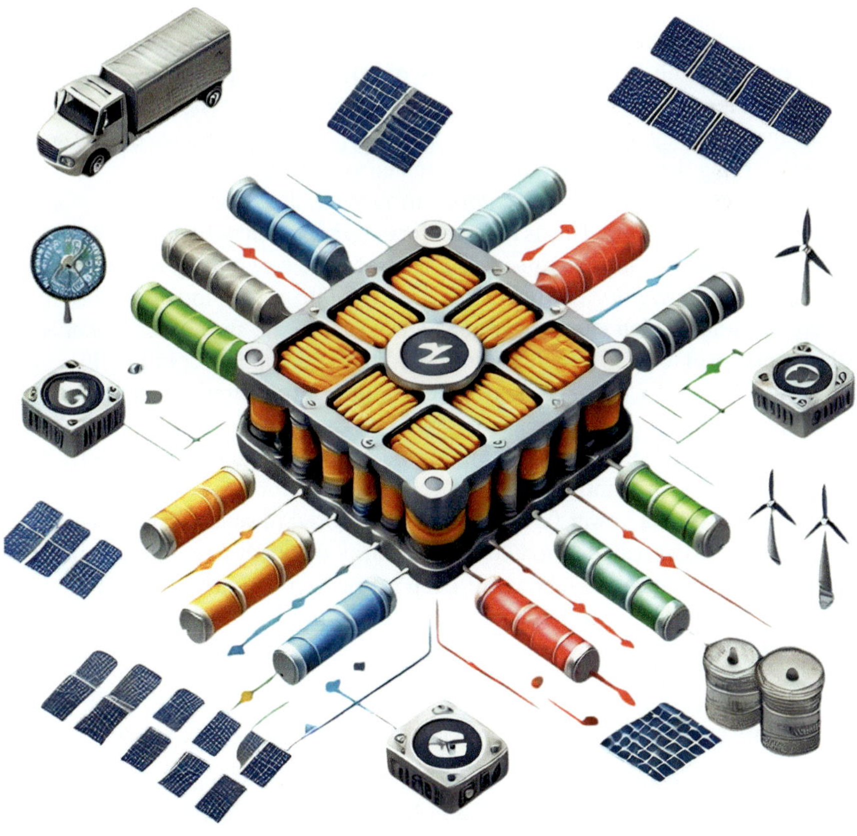

FIGURE 8.2 Industrial and grid stabilization.

8.5.1 Industrial Power Quality Management

In industry, consistent power quality is a crucial factor for the uninterrupted operation of machines and processes. Power surges, voltage sags, and sudden power fluctuations are hazards that must be avoided to prevent significant material loss, which can lead to system shutdowns, machine malfunctions, work stoppages, or financial loss. Supercapacitors, which are the first to detect these inline disturbances, also assist in power management [72, 73].

Supercapacitors are used in UPS systems and power conditioning equipment to reduce power quality issues, with the goal of lowering the supercapacitor level [74]. During voltage sags or short-term power shortages, supercapacitors can instantly release stored energy to stabilize the voltage and keep critical systems running without shutdowns [75].

The implementation of supercapacitors in heavy machinery industries has brought valuable solutions. For instance, power surges due to the frequent start and stop of heavy machinery can be smoothed out. Therefore, stress on the power infrastructure will be decreased, and the equipment will be protected from damage by supercapacitors.

8.5.2 Enhancing Reliability of Industrial Automation Systems

Industrial automation heavily depends on a continuous and stable power supply to operate smoothly. The negative effects of power interruptions or irregularities include equipment failures, shutdowns, and safety hazards. The idea here is that supercapacitors play a key role in energy storage, providing a reliable energy source for maintaining the stability of these automated systems [76, 77].

In robotic systems, supercapacitors can provide the power needed for quick, high-energy movements. This is relevant for activities such as lifting heavy objects or producing rapid changes. Speed and accuracy are very important in manufacturing, so the capability of supercapacitors is quite useful today. Additionally, supercapacitors, which can deliver constant power to the system, may enable the robotic platform to operate continuously without interruption, resulting in a more efficient device, less damage, and fewer equipment failures. Furthermore, the actual power output of supercapacitors and their ability to store data in programmable logic controllers in the automation industry, used to control various other devices, is a major advantage.

Thus, in case of a power failure, the energy stored in supercapacitors—which can also be used as backup—could provide immediate power to either shut down the system in a controlled manner or keep it functioning until the main power is restored. This feature of supercapacitors helps minimize data loss and equipment damage [78–80].

8.5.3 Supporting Grid Stabilization and Frequency Regulation

The power grid's renewable energy sources are increasing, with the addition of solar and wind making it easier for the grid to become unstable and face a new, simple frequency regulation problem at the same time [81]. Renewables such as wind and solar are inherently unstable energy sources, which is a major factor in power generation issues leading to fluctuations that directly affect grid frequency and jeopardize its stability. The use of supercapacitors for occasional energy storage is becoming more common due to their ability to provide quick, transient energy storage for instant responses to power changes [82].

Supercapacitors play a key role in maintaining grid frequency stability because of their ability to charge and discharge very quickly. When power demand exceeds supply, supercapacitors are used to absorb or supply power to the grid, which helps keep the system's frequency steady and maintains overall system security. High-quality, efficient supercapacitors can charge and discharge instantly, enabling them to support the grid during unsteady power conditions. At the same time, these devices contribute to the security and stability of the grid [83, 84].

8.5.4 Integration with Renewable Energy Systems

The increasing adoption of renewable energy sources highlights the need for energy storage solutions to handle their intermittency and variability. Supercapacitors are utilized in both solar and wind power systems to improve the stability and energy conversion efficiency of the overall system [85, 86].

In solar energy systems, supercapacitors can store energy generated during peak sunlight hours and then use this energy during periods of low sunlight [87]. These power bursts can also assist in establishing the energy resource, helping to stabilize the system's output [88].

Power generation in industrial applications with supercapacitors not only helps decrease costs but also reduces the amount of electricity needed in the system. Additionally, the use of peak shaving helps keep the environment cleaner by cutting down the demand for additional power sources during times of peak operation [89–92].

Super capacitors are increasingly in demand in both industrial and power grid stabilization sectors. Their ability to store energy quickly and deliver high power makes them ideal for managing power quality, supporting grid stability, and enhancing the efficiency of renewable energy systems.

8.6 CONSUMER APPLIANCES AND SMART GRIDS

Supercapacitors have recently been innovative, powering areas such as consumer electronics and smart grids, while consistently maintaining energy storage technology in these fields with improved performance, efficiency, and reliability [93, 94]. The section below introduces supercapacitors and their functions in the mentioned applications, indicating they are gradually transforming our daily lives. It also highlights the key interfaces between them and smart energy systems capable of generating energy anywhere, anytime.

8.6.1 Enhancing Consumer Appliances

Home appliances designed to serve consumer needs, such as vacuum cleaners and ovens, are becoming increasingly sophisticated with features that prioritize instant power, energy efficiency, and reliable performance over time. Supercapacitors are an ideal solution for these needs because they offer high-power density, fast charge–discharge capabilities, and durability. These qualities not only meet the user's power demands but also fulfill the requirement for energy efficiency [95, 96].

Supercapacitors are important in devices that face sudden power fluctuations, as they can provide the energy needed for household and electric tools more efficiently [97].

8.6.2 Supporting Energy-Efficient Appliances

Supercapacitors not only enable energy to be stored but also ensure that any energy wasted during the operation of regenerative systems is saved. A perfect example of this is when spinning the washing machine; the work done is stored in a supercapacitor and then used to power the next cycle or other functions of the machine [98–103]. The machine's thus reduced overall energy consumption to a minimum, and the energy star rating consequently improved.

Simultaneously, supercapacitors offer the benefit of efficient power management in the appliances where they are used by providing short-term energy storage to stabilize the power during energy supply periods. Just as a smart refrigerator with supercapacitors can accumulate and release stored energy on demand to maintain smooth operation even when the power supply is more stable but of lower voltage. This reduces the power supply effort of the equipment and helps the energy operation remain stable and efficient.

8.6.3 Revolutionizing Smart Grids

The future of energy distribution is represented by smart grids, which implement modern technologies to produce electricity, distribute it, and improve consumption efficiency [104]. Supercapacitors play a crucial role in this transition by enabling fast response times and high-power density necessary for smart grid operation [105].

Supercapacitors play an important role in the smart grid by providing quick energy storage that can quickly match the movements in power supply and demand [106]. As supercapacitors are used in the distributed storage segment of the energy network and thus the smart grid, they act as energy storage systems. These devices can be installed at various points in the grid, store energy, and release it to the local area as needed, making it easier for small, decentralized power generation to replace centralized sources. In conclusion, a distributed-energy power system improves the grid's resilience to interruptions and enhances its capacity to incorporate more renewable energy sources [107–111].

8.7 RENEWABLE ENERGY-POWERED IOT DEVICES

The Internet of Things (IoT) is undoubtedly one of the most important trends in technology, with billions of connected devices used across many industries. It is predicted that a significant portion of the electronics are powered by solar energy generated on-site or from solar panels [112–120]. Supercapacitors in these systems provide green and reliable energy storage solutions that not only consume minimal energy but can also be configured to meet the specific power needs of IoT applications [121].

It's common to see IoT devices in areas lacking traditional power sources or where such sources are minimal. In these locations, PV solar panels are responsible for maximum power generation [122]. Supercapacitors play a crucial role in this setting by allowing energy to be stored and used when needed, ensuring the continuous operation of IoT devices. These battery-less energy storage units can also include an onboard energy harvesting system, such as a PV cell, to utilize local renewable energy [123]. The durability and high charge–discharge cycle capability of supercapacitors make them ideal for IoT applications where maintenance or battery replacement is difficult and costs are high. Integrating supercapacitors into IoT devices can significantly extend their lifespan, reducing maintenance costs and enhancing reliability. Energy harvesting technology involves capturing available stray energy and converting it into electrical power, which is already commonly used to supply IoT devices [124, 125]. Overall, supercapacitors, as part of energy storage technology, are closely related to energy harvesting systems, as they store the collected energy and can provide it when the load demands it. The interoperability of these two components likely increases the efficiency and renewability of IoT devices, enabling longer periods of autonomous operation [126].

8.8 CHALLENGES AND LIMITATIONS

Although supercapacitors have many good features, some challenges make wide adoption difficult. One issue is that their energy density is lower than batteries, which makes them less suitable for long-term energy storage. Also, the materials and manufacturing processes for supercapacitors can be expensive, especially with advanced materials like graphene or metal oxides, limiting their use in cost-sensitive sectors.

The substances (activated carbon, electrolytes, and metal oxides) used in making supercapacitors are generally harmful to the environment if not handled properly. The extraction, processing, and disposal of these materials must be carefully managed to reduce their environmental impact. However, developing sustainable and eco-friendly materials for supercapacitors remains a key area of ongoing research.

8.9 FUTURE PROSPECTS AND ADVANCEMENTS

Supercapacitors are poised to revolutionize the way energy is stored globally. Many exciting discoveries are anticipated, which could lead to new possibilities by addressing the current limitations of supercapacitors and building upon them to introduce new applications. An example of such innovative, high-performance, low-cost materials is the development of new nanostructured carbons and advanced metal oxides by the research community. These materials will not only significantly enhance supercapacitor performance but also reduce production time and costs. The concept of hybrid systems that combine supercapacitors with other storage technologies like batteries aims to overcome the shortcomings of each type and find a balanced solution. Advances in flexible and wearable supercapacitors are opening new opportunities for use in wearable electronics and portable devices, previously unseen, which enhances user convenience and broadens potential applications.

8.10 CONCLUSION

Supercapacitors have gained prominence as an energy storage solution over the past decade. They serve as a link between traditional batteries and standard capacitors. Supercapacitors are incredibly quick to charge and discharge; therefore, they are unrivaled for many applications in batteries, power systems, and similar fields.

Although the main challenges with supercapacitors, such as high costs and low energy density, may seem minor, ongoing research makes significant progress in improving them. The impact of introducing new materials (like graphene) is clearly visible and contributes positively to developments in this field. It is certain that in the next few years, the role of supercapacitors in energy storage will not only persist but also expand, especially when used alongside other emerging storage technologies like batteries. The new and improved energy storage devices, i.e., supercapacitors, are expected to be environmentally friendly, helping to mitigate the energy crisis in power distribution systems, vehicle transport, and even the IoT concepts.

In conclusion, supercapacitors are set to revolutionize the energy storage industry by providing a stable, eco-friendly solution and being essential to meeting future energy needs. Undoubtedly, their ongoing development will be the turning point for a more sustainable and environmentally friendly energy system worldwide.

8.11 AUTHOR AI DECLARATION

The authors hereby declare the use of generative Artificial Intelligence (AI) and AI-assisted technologies in the writing process. While preparing this work, the authors used the ChatGPT tool/service to rephrase several statements in the manuscript. All AI-generated content was reviewed and edited by the authors as needed. The authors take full responsibility for the content of the publication.

ACKNOWLEDGMENT

The Editors acknowledge the use of Grammarly as a tool to assist in editing this book chapter. The suggestions made by Grammarly were carefully considered and incorporated where appropriate, but the author is solely responsible for the final content.

REFERENCES

1. Majumdar, D., M. Mandal, and S. K. Bhattacharya. "Journey from supercapacitors to supercapatteries: recent advancements in electrochemical energy storage systems". *Emergent Materials* 3 (2020): 347–367. https://doi.org/10.1007/s42247-020-00090-5.
2. Adedoja, O. S., E. R. Sadiku, and Y. Hamam. "An overview of the emerging technologies and composite materials for supercapacitors in energy storage applications". *Polymers* 15 (2023): 2272. doi:10.3390/polym15102272.
3. Smdani, G., M. R. Islam, A. N. Ahmad Yahaya, and S. I. Bin Safie. "Performance evaluation of advanced energy storage systems: a review". *Energy & Environment* 34 (2023): 1094–1141. https://doi.org/10.1177/0958305X221074729.
4. Saikia, B. K., S. M. Benoy, M. Bora, J. Tamuly, M. Pandey, and D. Bhattacharya. "A brief review on supercapacitor energy storage devices and utilization of natural carbon resources as their electrode materials". *Fuel* 282 (2020): 118796. doi:10.1016/j.fuel.2020.118796.
5. Gür, T. M. "Review of electrical energy storage technologies, materials and systems: challenges and prospects for large-scale grid storage". *Energy & Environmental Science* 11 (2018): 2696–2767. doi:10.1039/C8EE01419A.
6. Zhou, T., and W. Sun. "Optimization of battery–supercapacitor hybrid energy storage station in wind/solar generation system". *IEEE Transactions on Sustainable Energy* 5 (2014): 408–415. doi:10.1109/TSTE.2013.2288804.
7. González, L. G., R. Chacon, B. Delgado, D. Benavides, and J. Espinoza. "Study of energy compensation techniques in photovoltaic solar systems with the use of supercapacitors in low-voltage networks". *Energies* 13 (2020): 3755. doi:10.3390/en13153755.
8. Elmorshedy, M. F., M. R. Elkadeem, K. M. Kotb, I. B. Taha, and D. Mazzeo. "Optimal design and energy management of an isolated fully renewable energy system integrating batteries and supercapacitors". *Energy Conversion and Management* 245 (2021): 114584. doi:10.1016/j.enconman.2021.114584.
9. Nordin, N. A., M. N. M. Ansari, S. M. Nomanbhay, N. A. Hamid, N. M. Tan, Z. Yahya, and I. Abdullah. "Integrating photovoltaic (PV) solar cells and supercapacitors for sustainable energy devices: a review". *Energies* 14 (2021): 7211. doi:10.3390/en14217211.
10. Hill, C. A., M. C. Such, D. Chen, J. Gonzalez, and W. M. Grady. "Battery energy storage for enabling integration of distributed solar power generation". *IEEE Transactions on Smart Grid* 3 (2012): 850–857. doi:10.1109/TSG.2012.2190113.
11. Zhou, Y., H. Qi, J. Yang, Z. Bo, F. Huang, M. S. Islam, et al. "Two-birds-one-stone: multifunctional supercapacitors beyond traditional energy storage". *Energy & Environmental Science* 14 (2021): 1854–1896. https://doi.org/10.1039/D0EE03167D.

12. Zhao, X., B. M. Sánchez, P. J. Dobson, and P. S. Grant. "The role of nanomaterials in redox-based supercapacitors for next generation energy storage devices". *Nanoscale* 3 (2011): 839–855. https://doi.org/10.1039/C0NR00594K.
13. Khalid, M. "A review on the selected applications of battery-supercapacitor hybrid energy storage systems for microgrids". *Energies* 12 (2019): 4559. doi:10.3390/en12234559.
14. Wang, X., B. Ha, G. Y. Lee, H. Kim, J. Yu, H. Rhee, et al. "Low-cost far-field wireless electrical load monitoring system applied in an off-grid rural area of Tanzania". *Sustainable Cities and Society* 59 (2020): 102209. doi:10.1016/j.scs.2020.102209.
15. Anvari, M., G. Lohmann, M. Wächter, P. Milan, E. Lorenz, D. Heinemann, et al. "Short term fluctuations of wind and solar power systems". *New Journal of Physics* 18 (2016): 063027. doi:10.1088/1367-2630/18/6/063027.
16. Mendis, N., K. M. Muttaqi, and S. Perera. "Management of battery-supercapacitor hybrid energy storage and synchronous condenser for isolated operation of PMSG based variable-speed wind turbine generating systems". *IEEE Transactions on Smart Grid* 5 (2014): 944–953. doi:10.1109/TSG.2013.2287874.
17. Hassan, Q., M. Jaszczur, A. K. Al-Jiboory, A. Hasan, and A. Mohamad. "Optimizing of hybrid renewable photovoltaic/wind turbine/super capacitor for improving self-sustainability". *Energy Harvesting and Systems* 9 (2022): 151–164. doi:https://doi.org/10.1515/ehs-2021-0095.
18. Hadjipaschalis, I., A. Poullikkas, and V. Efthimiou. "Overview of current and future energy storage technologies for electric power applications". *Renewable and Sustainable Energy Reviews* 13 (2009): 1513–1522. doi:10.1016/j.rser.2008.09.028.
19. Pan, T. L., H. S. Wan, and Z. C. Ji. "Stand-alone wind power system with battery/supercapacitor hybrid energy storage". *International Journal of Sustainable Engineering* 7 (2014): 103–110. https://doi.org/10.1080/19397038.2013.779327.
20. Panhwar, I. H., K. Ahmed, M. Seyedmahmoudian, A. Stojcevski, B. Horan, S. Mekhilef, et al. "Mitigating power fluctuations for energy storage in wind energy conversion system using supercapacitors". *IEEE Access* 8 (2020): 189747–189760. doi:10.1109/ACCESS.2020.3029442.
21. Datta, R., and V. T. Ranganathan. "A method of tracking the peak power points for a variable speed wind energy conversion system". *IEEE Transactions on Energy Conversion* 18 (2003): 163–168. doi:10.1109/TEC.2002.808346.
22. Rocabert, J., R. Capo-Misut, and R. S. Muñoz-Aguilar, J. I. Candela, and P. Rodriguez. "Control of energy storage system integrating electrochemical batteries and supercapacitors for grid-connected applications". *IEEE Transactions on Industry Applications* 55, no. 2 (2018): 1853–1862. doi:10.1109/TIA.2018.2873534.
23. Qi, H., X. Wang, L. M. Tolbert, F. Li, F. Z. Peng, P. Ning, and M. Amin. "A resilient real-time system design for a secure and reconfigurable power grid". *IEEE Transactions on Smart Grid* 2, no. 4 (2011): 770–781. doi:10.1109/TSG.2011.2159819.
24. Mustafi, N. N. "An overview of hybrid electric vehicle technology". *Engines and Fuels for Future Transport* (2022): 73–102. doi:10.1109/VPPC.2009.5289703.
25. Tzortzis, G., A. Amargianos, S. Piperidis, E. Koutroulis, and N. C. Tsourveloudis. "Development of a compact regenerative braking system for electric vehicles." In *Proceedings of the 23rd Mediterranean Conference on Control and Automation (MED)*, 102–108 (2015). doi:10.1109/MED.2015.7158751.
26. Wen, J., D. Zhao, and C. Zhang. "An overview of electricity powered vehicles: lithium-ion battery energy storage density and energy conversion efficiency". *Renewable Energy* 162 (2020): 1629–1648. doi:10.1016/j.renene.2020.09.008.
27. Li, F., J. Xie, Y. Fan, and J. Qiu. "Potential of different forms of gravity energy storage". *Sustainable Energy Technologies and Assessments* 64 (2024): 103728. https://doi.org/10.1016/j.seta.2024.103728.

28. Pandey, D., K. Sambath Kumar, L. N. Henderson, G. Suarez, P. Vega, H. R. Salvador, et al. "Energized composites for electric vehicles: a dual function energy-storing supercapacitor-based carbon fiber composite for the body panels". *Small* 18, no. 9 (2022): 2107053. doi:10.1002/smll.202107053.
29. Rashidi, S., N. Karimi, B. Sunden, K. C. Kim, A. G. Olabi, and O. Mahian. "Progress and challenges on the thermal management of electrochemical energy conversion and storage technologies: fuel cells, electrolysers, and supercapacitors". *Progress in Energy and Combustion Science* 88 (2022): 100966. doi:10.1016/j.pecs.2021.100966.
30. Gao, Y., L. Chen, and M. Ehsani. "Investigation of the effectiveness of regenerative braking for EV and HEV." In *SAE Transactions*, 3184–3190 (1999). doi:10.4271/1999-01-2910.
31. Shchurov, N. I., S. I. Dedov, B. V. Malozyomov, A. A. Shtang, N. V. Martyushev, R. V. Klyuev, and S. N. Andriashin. "Degradation of lithium-ion batteries in an electric transport complex". *Energies* 14, no. 23 (2021): 8072. doi:10.3390/en14238072.
32. Li, J., S. He, Q. Yang, Z. Wei, Y. Li, and H. He. "A comprehensive review of second life batteries toward sustainable mechanisms: potential, challenges, and future prospects". *IEEE Transactions on Transportation Electrification* 9, no. 4 (2022): 4824–4845. doi:10.1109/TTE.2022.3220411.
33. Hou, J., Y. Shao, M. W. Ellis, R. B. Moore, and B. Yi. "Graphene-based electrochemical energy conversion and storage: fuel cells, supercapacitors and lithium ion batteries". *Physical Chemistry Chemical Physics* 13, no. 34 (2011): 15384–15402. https://doi.org/10.1039/C1CP21915D.
34. Farajpour, Y., H. Chaoui, M. Khayamy, S. Kelouwani, and M. Alzayed. "Novel energy management strategy for electric vehicles to improve driving range". *IEEE Transactions on Vehicular Technology* 72, no. 2 (2022): 1735–1747. doi:10.1109/TVT.2022.3213472.
35. Gering, K. L. "Low-temperature performance limitations of lithium-ion batteries". *ECS Transactions* 1, no. 26 (2006): 119. doi:10.1149/1.2355880.
36. Rashidi, S., N. Karimi, B. Sunden, K. C. Kim, A. G. Olabi, and O. Mahian. "Progress and challenges on the thermal management of electrochemical energy conversion and storage technologies: fuel cells, electrolysers, and supercapacitors". *Progress in Energy and Combustion Science* 88 (2022): 100966. doi:10.1016/j.pecs.2021.100966.
37. Bonges, H. A. III, and A. C. Lusk. "Addressing electric vehicle (EV) sales and range anxiety through parking layout, policy and regulation". *Transportation Research Part A: Policy and Practice* 83 (2016): 63–73. doi:10.1016/j.tra.2015.09.011.
38. Chodankar, N. R., S. J. Patil, S.-K. Hwang, S. V. Karekar, K. Jayaramulu, W. Zhang, et al. "Supercapacitors operated at extremely low environmental temperatures". *Journal of Materials Chemistry A* 9, no. 47 (2021): 26603–26627. https://doi.org/10.1039/D1TA08538G.
39. Potdar, V., S. Batool, and A. Krishna. "Risks and challenges of adopting electric vehicles in smart cities". *Smart Cities: Development and Governance Frameworks* (2018): 207–240. doi.org/10.1007/978-3-319-76669-0_9.
40. Ghasemi-Marzbali, A. "Fast-charging station for electric vehicles, challenges and issues: a comprehensive review". *Journal of Energy Storage* 49 (2022): 104136. doi:10.1016/j.est.2022.104136.
41. Hepel, M. "Advances in micro-supercapacitors (MSCs) with high energy density and fast charge-discharge capabilities for flexible bioelectronic devices—a review". *Electrochemical Science Advances* 3, no. 3 (2023): e2100222. doi:10.1002/elsa.202100222.
42. Collin, R., Y. Miao, A. Yokochi, P. Enjeti, and A. Von Jouanne. "Advanced electric vehicle fast-charging technologies". *Energies* 12, no. 10 (2019): 1839. doi:10.3390/en12101839.

43. Hosseini, S. M., M. Soleymani, S. Kelouwani, and A. Amamou. "Energy recovery and energy harvesting in electric and fuel cell vehicles, a review of recent advances". *IEEE Access* 11 (2023): 83107–83135. doi:10.1109/ACCESS.2023.3301329.
44. Khalid, M. "A review on the selected applications of battery-supercapacitor hybrid energy storage systems for microgrids". *Energies* 12, no. 23 (2019): 4559. doi:10.3390/en12234559.
45. Lemian, D., and F. Bode. "Battery-supercapacitor energy storage systems for electrical vehicles: a review". *Energies* 15, no. 15 (2022): 5683. doi:10.3390/en15155683.
46. Wong, S. I., J. Sunarso, B.-T. Wong, H. Lin, A. Yu, and B. Jia. "Towards enhanced energy density of graphene-based supercapacitors: current status, approaches, and future directions". *Journal of Power Sources* 396 (2018): 182–206. doi:10.1016/j.jpowsour.2018.06.024.
47. Jyoti, J., T. K. Gupta, B. P. Singh, M. Sandhu, and S. K. Tripathi. "Recent advancement in three dimensional graphene-carbon nanotubes hybrid materials for energy storage and conversion applications". *Journal of Energy Storage* 50 (2022): 104235. doi:10.1016/j.est.2022.104235.
48. Iqbal, M. Z., M. M. Faisal, and S. R. Ali. "Integration of supercapacitors and batteries towards high-performance hybrid energy storage devices". *International Journal of Energy Research* 45, no. 2 (2021): 1449–1479. doi:10.1002/er.5830.
49. Benzigar, M. R., V. D. Dasireddy, X. Guan, T. Wu, and G. Liu. "Advances on emerging materials for flexible supercapacitors: current trends and beyond". *Advanced Functional Materials* 30, no. 40 (2020): 2002993. doi:10.1002/adfm.202002993.
50. Mushroor, S., S. Haque, and A. A. Riyadh. "The impact of smart phones and mobile devices on human health and life". *International Journal of Community Medicine and Public Health* 1 (2020): 9–15.
51. Singh, S., *Ride Technology Wave for Career Success: Continuous Learning Newer Technologies, Innovating and Integrating are Keys to Success*. Notion Press, Chennai (2022).
52. Abraham, T. R., K. Sunil, M. Shah, N. Ashok, and S. Thomas. "Energy storage devices: batteries and supercapacitors." In *Nanobiohybrids for Advanced Wastewater Treatment and Energy Recovery*, 61 (2023) IWA Publishings, London.
53. Riaz, A., M. R. Sarker, M. H. M. Saad, and R. Mohamed. "Review on comparison of different energy storage technologies used in micro-energy harvesting, WSNs, low-cost microelectronic devices: challenges and recommendations". *Sensors* 21, no. 15 (2021): 5041. doi:10.3390/s21155041.
54. Kumar, N. M., N. K. Singh, and V. K. Peddiny. "Wearable smart glass: Features, applications, current progress and challenges." In *2018 Second International Conference on Green Computing and Internet of Things (ICGCIoT)*, 577–582. IEEE (2018). doi:10.1109/ICGCIoT.2018.8753045.
55. Khan, H. A., M. Tawalbeh, B. Aljawrneh, W. Abuwatfa, A. Al-Othman, H. Sadeghifar, and A. G. Olabi. "A comprehensive review on supercapacitors: their promise to flexibility, high temperature, materials, design, and challenges". *Energy* 2024: 131043. https://doi.org/10.1016/j.energy.2024.131043.
56. Govindan, S., D. Wang, A. Sivasubramaniam, and B. Urgaonkar. "Leveraging stored energy for handling power emergencies in aggressively provisioned datacenters." In *Proceedings of the Seventeenth International Conference on Architectural Support for Programming Languages and Operating Systems*, 75–86. ACM (2012). https://doi.org/10.1145/2150976.2150985.
57. Chen, X., N. S. Villa, Y. Zhuang, L. Chen, T. Wang, Z. Li, and T. Kong. "Stretchable supercapacitors as emergent energy storage units for health monitoring bioelectronics". *Advanced Energy Materials* 10, no. 4 (2020): 1902769. doi:10.1002/aenm.201902769.
58. Varma, J., K. Sambath Kumar, S. Seal, S. Rajaraman, and J. Thomas. "Fiber-type solar cells, nanogenerators, batteries, and supercapacitors for wearable applications". *Advanced Science* 5, no. 9 (2018): 1800340. doi:10.1002/advs.201800340.

59. Lopez, X., K. Afrin, and B. S. Nepal. "Examining the design, manufacturing and analytics of smart wearables". *Medical Devices & Sensors* 3, no. 3 (2020): e10087. doi:10.1002/mds3.10087.
60. Yousaf, M., H. T. H. Shi, Y. Wang, Y. Chen, Z. Ma, A. Cao, …, and R. P. Han. "Novel pliable electrodes for flexible electrochemical energy storage devices: recent progress and challenges". *Advanced Energy Materials* 6, no. 17 (2016): 1600490. doi:10.1002/aenm.201600490.
61. Liu, L., Z. Niu, and J. Chen. "Unconventional supercapacitors from nanocarbon-based electrode materials to device configurations". *Chemical Society Reviews* 45, no. 15 (2016): 4340–4363. https://doi.org/10.1039/C6CS00041J.
62. El-Kady, M. F., Y. L. Shao, and R. B. Kaner. "Graphene for batteries, supercapacitors and beyond". *Nature Reviews Materials* 1, no. 7 (2016): 1–14. doi:10.1038/natrevmats.2016.16.
63. Shchurov, N. I., S. I. Dedov, B. V. Malozyomov, A. A. Shtang, N. V. Martyushev, R. V. Klyuev, and S. N. Andriashin. "Degradation of lithium-ion batteries in an electric transport complex". *Energies* 14, no. 23 (2021): 8072. doi:10.3390/en14238072.
64. Hooshmand, M., D. Zordan, D. D. Testa, E. Grisan, and M. Rossi. "Boosting the battery life of wearables for health monitoring through the compression of biosignals". *IEEE Internet of Things Journal* 4, no. 5 (2017): 1647–1662. doi:10.1109/JIOT.2017.2689164.
65. Chodankar, N. R., S. J. Patil, S. K. Hwang, P. A. Shinde, S. V. Karekar, S. R. Raju, …, and Y. K. Han. "Refurbished carbon materials from waste supercapacitors as industrial-grade electrodes: empowering electronic waste". *Energy Storage Materials* 49 (2022): 564–574. doi:10.1016/j.ensm.2022.04.004.
66. Zhang, L., D. Liu, Z. S. Wu, and W. Lei. "Micro-supercapacitors powered integrated system for flexible electronics". *Energy Storage Materials* 32 (2020): 402–417. doi:10.1016/j.ensm.2020.07.027.
67. Mokhtari, F., Z. Cheng, R. Raad, J. Xi, and J. Foroughi. "Piezofibers to smart textiles: a review on recent advances and future outlook for wearable technology". *Journal of Materials Chemistry A* 8, no. 19 (2020): 9496–9522. https://doi.org/10.1039/D0TA00227E.
68. Navarro, G., J. Torres, M. Blanco, J. Nájera, M. Santos-Herran, and M. Lafoz. "Present and future of supercapacitor technology applied to powertrains, renewable generation and grid connection applications". *Energies* 14, no. 11 (2021): 3060. doi:10.3390/en14113060.
69. Riaz, A., M. R. Sarker, M. H. M. Saad, and R. Mohamed. "Review on comparison of different energy storage technologies used in micro-energy harvesting, WSNs, low-cost microelectronic devices: challenges and recommendations". *Sensors* 21, no. 15 (2021): 5041. doi:10.3390/s21155041.
70. Afzal, M. M., M. A. Khan, S. Hassan, A. Wadood, W. Uddin, S. Hussain, and S. B. Rhee. "A comparative study of supercapacitor-based STATCOM in a grid-connected photovoltaic system for regulating power quality issues". *Sustainability* 12, no. 17 (2020): 6781. doi:10.3390/su12176781.
71. Panhwar, I. H., K. Ahmed, M. Seyedmahmoudian, A. Stojcevski, B. Horan, S. Mekhilef, …, and M. Asghar. "Mitigating power fluctuations for energy storage in wind energy conversion system using supercapacitors". *IEEE Access* 8 (2020): 189747–189760. doi:10.1109/ACCESS.2020.3031446.
72. Ismail, N. M., and M. K. Mishra. "A multi-objective control scheme of a voltage source converter with battery–supercapacitor energy storage system used for power quality improvement". *International Journal of Electrical Power & Energy Systems* 142 (2022): 108253. doi:10.1016/j.ijepes.2022.108253.
73. Sahay, K., and B. Dwivedi. "Supercapacitors energy storage system for power quality improvement: an overview". *Journal of Power Sources* 10, no. 10 (2009): 1–8.

74. Yadlapalli, R. T., R. R. Alla, R. Kandipati, and A. Kotapati. "Super capacitors for energy storage: progress, applications and challenges". *Journal of Energy Storage* 49 (2022): 104194. doi:10.1016/j.est.2022.104194.
75. Subasinghage, K., K. Gunawardane, N. Padmawansa, N. Kularatna, and M. Moradian. "Modern supercapacitors technologies and their applicability in mature electrical engineering applications". *Energies* 15, no. 20 (2022): 7752. doi:10.3390/en15207752.
76. Saravanakumar, Y. N., M. T. H. Sultan, F. S. Shahar, W. Giernacki, A. Łukaszewicz, M. Nowakowski, …, and S. Stępień. "Power sources for Unmanned Aerial Vehicles: a state-of-the-art". *Applied Sciences* 13, no. 21 (2023): 11932. doi:10.3390/app132111932.
77. Mohammed, N., "Control and monitoring of battery energy storage system using PLC." Master thesis, Universiti Teknologi Malaysia (2017).
78. Gao, W., H. Haitjema, F. Z. Fang, R. K. Leach, C. F. Cheung, E. Savio, and J. M. Linares. "On-machine and in-process surface metrology for precision manufacturing". *CIRP Annals* 68, no. 2 (2019): 843–866. doi:10.1016/j.cirp.2019.04.006.
79. Anees, A. S. "Grid integration of renewable energy sources: Challenges, issues and possible solutions." In *2012 IEEE 5th India International Conference on Power Electronics (IICPE)*, 1–6. IEEE (2012). doi:10.1109/IICPE.2012.6450474.
80. Parra, D., M. Swierczynski, D. I. Stroe, S. A. Norman, A. Abdon, J. Worlitschek, …, and M. K. Patel. "An interdisciplinary review of energy storage for communities: challenges and perspectives". *Renewable and Sustainable Energy Reviews* 79 (2017): 730–749. doi:10.1016/j.rser.2017.05.003.
81. Akram, U., M. Nadarajah, R. Shah, and F. Milano. "A review on rapid responsive energy storage technologies for frequency regulation in modern power systems". *Renewable and Sustainable Energy Reviews* 120 (2020): 109626. doi:10.1016/j.rser.2019.109626.
82. Hadjipaschalis, I., A. Poullikkas, and V. Efthimiou. "Overview of current and future energy storage technologies for electric power applications". *Renewable and Sustainable Energy Reviews* 13, no. 6–7 (2009): 1513–1522. doi:10.1016/j.rser.2008.09.028.
83. Worku, M. Y. "Recent advances in energy storage systems for renewable source grid integration: a comprehensive review". *Sustainability* 14, no. 10 (2022): 5985. doi:10.3390/su14105985.
84. Nordin, N. A., M. N. M. Ansari, S. M. Nomanbhay, N. A. Hamid, N. M. Tan, Z. Yahya, and I. Abdullah. "Integrating photovoltaic (PV) solar cells and supercapacitors for sustainable energy devices: a review". *Energies* 14, no. 21 (2021): 7211. doi:10.3390/en14217211.
85. Karabacak, K., "Towards sustainable solar energy solutions: harnessing supercapacitors in PV systems". *Energy Storage and Conversion* 2, no. 1 (2024): 436–436. https://doi.org/10.59400/esc.v2i1.436.
86. Jain, S., and V. Agarwal. "An integrated hybrid power supply for distributed generation applications fed by nonconventional energy sources". *IEEE Transactions on Energy Conversion* 23, no. 2 (2008): 622–631. doi:10.1109/TEC.2007.914172.
87. Bandari, N., K. V. G. Raghavendra, S. Heo, S. V. S. V. P. D. Kumar, W. Uddin, L. K. R. Sama, …, and H. J. Kim. "Efficient electricity management system for optimal peak/off-peak hour pricing". *Electronics* 9, no. 8 (2020): 1189. doi:10.3390/electronics9081189.
88. Rahman, F., S. Rehman, and M. A. Abdul-Majeed. "Overview of energy storage systems for storing electricity from renewable energy sources in Saudi Arabia". *Renewable and Sustainable Energy Reviews* 16, no. 1 (2012): 274–283. doi:10.1016/j.rser.2011.07.058.
89. Yadlapalli, R. T., R. R. Alla, R. Kandipati, and A. Kotapati. "Super capacitors for energy storage: progress, applications and challenges". *Journal of Energy Storage* 49 (2022): 104194. doi:10.1016/j.est.2022.104194.

90. Chotia, I., "Electrical performance and economic feasibility analyses of hybrid and battery storage devices used in remote area islanded renewable energy systems." M.S. thesis (2016).
91. Molina, M. G. "Energy storage and power electronics technologies: a strong combination to empower the transformation to the smart grid". *Proceedings of the IEEE* 105, no. 11 (2017): 2191–2219. doi:10.1109/JPROC.2017.2702627.
92. Sharma, P., S. R. Salkuti, and S. C. Kim. "Advancements in energy storage technologies for smart grid development". *International Journal of Electrical and Computer Engineering* 12, no. 4 (2022): 3421–3429. doi:10.11591/ijece.v12i4.pp3421-3429.
93. Sahoo, S., R. Kumar, E. Joanni, R. K. Singh, and J. J. Shim. "Advances in pseudocapacitive and battery-like electrode materials for high performance supercapacitors". *Journal of Materials Chemistry A* 10, no. 25 (2022): 13190–13240. doi:10.1039/D2TA02357A.
94. Bu, F., W. Zhou, Y. Xu, Y. Du, C. Guan, and W. Huang. "Recent developments of advanced micro-supercapacitors: design, fabrication and applications". *npj Flexible Electronics* 4, no. 1 (2020): 31. doi:10.1038/s41528-020-00093-6.
95. Ge, Z., "Energy-efficient wastewater treatment by microbial fuel cells: Scaling up and optimization." Ph.D. dissertation, Virginia Polytechnic Institute and State University (2015).
96. Kim, S., and P. H. Chou. "Energy harvesting with supercapacitor-based energy storage." In *Smart Sensors and Systems*, 215–241. Springer, Cham (2015). doi:10.1007/978-3-319-14711-6_10.
97. Herring, H., and R. Roy. "Technological innovation, energy efficient design and the rebound effect". *Technovation* 27, no. 4 (2007): 194–203. doi:10.1016/j.technovation.2006.11.004.
98. Laheäär, A., P. Przygocki, Q. Abbas, and F. Béguin. "Appropriate methods for evaluating the efficiency and capacitive behavior of different types of supercapacitors". *Electrochemistry Communications* 60 (2015): 21–25. doi:10.1016/j.elecom.2015.07.022.
99. Gao, D., Z. Luo, C. Liu, and S. Fan. "A survey of hybrid energy devices based on supercapacitors". *Green Energy & Environment* 8, no. 4 (2023): 972–988. doi:10.1016/j.gee.2022.03.005.
100. Armaroli, N., and V. Balzani. "Towards an electricity-powered world". *Energy & Environmental Science* 4, no. 9 (2011): 3193–3222. https://doi.org/10.1039/C1EE01249E.
101. Shi, J., S. Liu, L. Zhang, B. Yang, L. Shu, Y. Yang, …, and X. Tao. "Smart textile-integrated microelectronic systems for wearable applications". *Advanced Materials* 32, no. 5 (2020): 1901958. doi:10.1002/adma.201901958.
102. Khalid, M. "A review on the selected applications of battery-supercapacitor hybrid energy storage systems for microgrids". *Energies* 12, no. 23 (2019): 4559. doi:10.3390/en12234559.
103. Biczel, P., "Energy storage systems." In *Power Electronics in Smart Electrical Energy Networks*, 269–302. Springer, London (2008).
104. Pinnangudi, B., M. Kuykendal, and S. Bhadra. "Smart grid energy storage." In *The Power Grid*, 93–135. Academic Press, London (2017).
105. Gandoman, F. H., A. Ahmadi, A. M. Sharaf, P. Siano, J. Pou, B. Hredzak, and V. G. Agelidis. "Review of FACTS technologies and applications for power quality in smart grids with renewable energy systems". *Renewable and Sustainable Energy Reviews* 82 (2018): 502–514. doi:10.1016/j.rser.2017.09.062.
106. Molina, M. G. "Distributed energy storage systems for applications in future smart grids." In *2012 Sixth IEEE/PES Transmission and Distribution: Latin America Conference and Exposition (T&D-LA)*, 1–7. IEEE (2012). doi:10.1109/TDC-LA.2012.6319081.

107. Mohd, A., E. Ortjohann, A. Schmelter, N. Hamsic, and D. Morton. "Challenges in integrating distributed energy storage systems into future smart grid." In *2008 IEEE International Symposium on Industrial Electronics*, 1627–1632. IEEE (2008). doi:10.1109/ISIE.2008.4677097.
108. Kolokotsa, D., N. Kampelis, A. Mavrigiannaki, M. Gentilozzi, F. Paredes, F. Montagnino, and L. Venezia. "On the integration of the energy storage in smart grids: Technologies and applications". *Energy Storage* 1, no. 1 (2019): e50. doi:10.1002/est2.50.
109. Zhang, L., X. Hu, Z. Wang, J. Ruan, C. Ma, Z. Song, ..., and M. G. Pecht. "Hybrid electrochemical energy storage systems: an overview for smart grid and electrified vehicle applications". *Renewable and Sustainable Energy Reviews* 139 (2021): 110581. doi:10.1016/j.rser.2020.110581.
110. Yang, Y., Y. Han, W. Jiang, Y. Zhang, Y. Xu, and A. M. Ahmed. "Application of the supercapacitor for energy storage in China: role and strategy". *Applied Sciences* 12, no. 1 (2021): 354. doi:10.3390/app12010354.
111. Shaqsi, A. Z. A., K. Sopian, and A. Al-Hinai. "Review of energy storage services, applications, limitations, and benefits". *Energy Reports* 6 (2020): 288–306. doi:10.1016/j.egyr.2020.11.002.
112. Luo, X., J. Wang, M. Dooner, and J. Clarke. "Overview of current development in electrical energy storage technologies and the application potential in power system operation". *Applied Energy* 137 (2015): 511–536. doi:10.1016/j.apenergy.2014.09.081.
113. Huang, X., W. Yan, H. Cao, S. Chen, G. Tao, and J. Zhang. "Prospects for purely electric construction machinery: mechanical components, control strategies and typical machines". *Automation in Construction* 164 (2024): 105477. doi:10.1016/j.autcon.2024.105477.
114. Pourkheirollah, H., *Printed Supercapacitors for Energy Storage and Functional Applications: Modeling, Analysis, and Integration* (2023).
115. Elahi, H., K. Munir, M. Eugeni, S. Atek, and P. Gaudenzi. "Energy harvesting towards self-powered IoT devices". *Energies* 13, no. 21 (2020): 5528. doi:10.3390/en13215528.
116. Tawiah, B., R. K. Seidu, B. K. Asinyo, and B. Fei. "A review of fiber-based supercapacitors and sensors for energy-autonomous systems". *Journal of Power Sources* 595 (2024): 234069. https://doi.org/10.1016/j.jpowsour.2024.234069.
117. Saleem, Y., N. Crespi, M. H. Rehmani, and R. Copeland. "Internet of things-aided smart grid: technologies, architectures, applications, prototypes, and future research directions". *IEEE Access* 7 (2019): 62962–63003. doi:10.1109/ACCESS.2019.2913982.
118. Abdul-Qawy, A. S., P. J. Pramod, E. Magesh, and T. Srinivasulu. "The internet of things (IoT): an overview". *International Journal of Engineering Research and Applications* 5, no. 12 (2015): 71–82.
119. Mi, Y., Y. Lu, X. Wang, Z. Zhao, X. Cao, and N. Wang. "From triboelectric nanogenerator to uninterrupted power supply system: the key role of electrochemical batteries and supercapacitors". *Batteries* 8, no. 11 (2022): 215. doi:10.3390/batteries8110215.
120. Liu, L., Y. Feng, and W. Wu. "Recent progress in printed flexible solid-state supercapacitors for portable and wearable energy storage". *Journal of Power Sources* 410 (2019): 69–77. doi:10.1016/j.jpowsour.2018.11.089.
121. He, J., L. Cao, J. Cui, G. Fu, R. Jiang, X. Xu, and C. Guan. "Flexible energy storage devices to power the future". *Advanced Materials* 36, no. 4 (2024): 2306090. doi:10.1002/adma.202306090.
122. Riaz, A., M. R. Sarker, M. H. M. Saad, and R. Mohamed. "Review on comparison of different energy storage technologies used in micro-energy harvesting, WSNs, low-cost microelectronic devices: challenges and recommendations". *Sensors* 21, no. 15 (2021): 5041. doi:10.3390/s21155041.

123. Sanislav, T., G. D. Mois, S. Zeadally, and S. C. Folea. "Energy harvesting techniques for internet of things (IoT)". *IEEE Access* 9 (2021): 39530–39549. doi:10.1109/ACCESS.2021.3064066.
124. Sun, H., Y. Zhang, J. Zhang, X. Sun, and H. Peng. "Energy harvesting and storage in 1D devices". *Nature Reviews Materials* 2, no. 6 (2017): 17023. doi:10.1038/natrevmats.2017.23.
125. Wang, Z. L., and W. Wu. "Nanotechnology-enabled energy harvesting for self-powered micro-/nanosystems". *Angewandte Chemie International Edition* 51, no. 47 (2012): 11700–11721. doi:10.1002/anie.201201656.
126. Yoon, H., Lee, J., and Jang, J. "Conducting-polymer nanomaterials for high-performance sensor applications: issues and challenges". *Advanced Materials* 28, no. 22 (2016): 4217–4246. https://doi.org/10.1002/adfm.200801141.

9 Applications of Supercapacitors in Power Electronics

Kerim Karabacak

9.1 INTRODUCTION

As global demand for sustainable energy rises, power electronics is crucial in managing energy systems. It enables efficient conversion, control, and distribution of electrical energy in applications like renewable energy, electric vehicles (EVs), smart grids, and industrial automation. A significant recent development is the integration of supercapacitors (SCs), which bridge the gap between conventional capacitors and batteries. Known for delivering rapid energy bursts and recharging quickly, SCs are a versatile energy storage solution. Unlike traditional batteries relying on chemical reactions, SCs store energy electrostatically, allowing faster cycling and thousands to millions of charge–discharge cycles without significant degradation, making them ideal for high power–density applications and long-term reliability. SCs offer unique advantages in power electronics over other storage technologies, especially when quick energy response and efficiency are crucial. SCs buffer fluctuations, stabilize output, and enhance energy quality in renewable systems like wind and solar. They support EVs in acceleration and regenerative braking, prolonging the primary battery's life and efficiency.

Additionally, their importance in uninterruptible power supply (UPS) and voltage stabilization highlights their versatility and transformative potential across sectors. This chapter explores the applications of SCs in power electronics, including renewable energy integration, EV systems, industrial power management, and grid stabilization. Each section discusses principles, performance benefits, and challenges of SC implementation. By analyzing SCs' current state and prospects, this chapter offers a comprehensive view of their role in advancing power electronics for a sustainable energy future.

9.2 FUNDAMENTALS OF SCs IN POWER ELECTRONICS

SCs, or ultracapacitors, are electrochemical energy storage devices that distinguish themselves by their high-power density, rapid charge–discharge capabilities, and long cycle life. In power electronics, understanding the principles of SCs is essential to harness their full potential. This section discusses the charge–discharge dynamics, lifespan considerations for SCs, and modeling of SCs in power electronics applications.

DOI: 10.1201/9781003535836-9

9.2.1 Charge and Discharge Dynamics

SCs' charge and discharge behavior is a defining characteristic that impacts their role in power electronics. SCs exhibit a nearly linear discharge curve, which differs significantly from the flat voltage output seen in batteries. This characteristic makes them suitable for applications where variable power requirements need a highly responsive energy source [1]. Figure 9.1 shows charge and discharge curves of a SC for different currents. The charging curves initially rise quickly, reaching peak voltages of around 1.5 V, but the rate of increase varies with the applied current. Higher currents (15 and 25 mA) reach their maximum voltage more rapidly than the lower current (5 mA), indicating faster charging times. However, during the discharge phase, the higher currents also show a steeper decline in voltage, suggesting faster energy depletion, while the 5 mA discharge curve shows a slower decline, indicating a longer discharge time and thus greater energy retention at lower currents. This behavior illustrates the trade-off between charge/discharge speed and energy retention in SCs, depending on the current applied.

The effectiveness of SCs in power electronics hinges on capacitance (C), equivalent series resistance (ESR), and cycle efficiency. Capacitance indicates the capacity to store electric charge, with increased capacitance allowing for enhanced energy storage. ESR, representing the SC's internal resistance, plays a significant role in energy efficiency and heat generation during charge–discharge cycles, necessitating a low ESR for high-power applications. Cycle efficiency is equally important, as SCs demonstrate exceptional energy efficiency owing to their minimal internal resistance. Their ability to handle rapid charge–discharge cycles makes them suitable for applications that require frequent energy bursts, like voltage stabilization and regenerative braking in EVs. By fine-tuning these parameters, SCs improve reliability and efficiency across various power electronic systems.

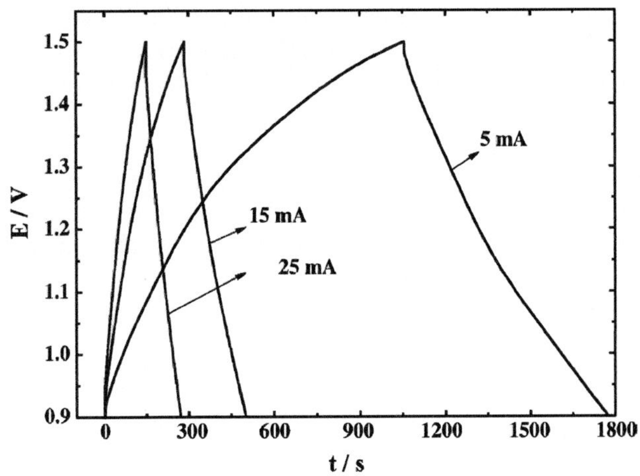

FIGURE 9.1 Charge and discharge characteristics of the supercapacitor under varying current levels. (Reprinted with permission from [2]. © Elsevier 2006.)

9.2.2 Lifespan and Efficiency Factors in Power Electronics

SCs are known for their durability and long life, capable of completing millions of cycles without significant capacity loss. However, their lifespan and efficiency are affected by several factors, including temperature, voltage balancing, and self-discharge. High temperatures can accelerate the degradation of SCs. Managing thermal effects through cooling and insulation is crucial in power electronics to maintain performance and extend the lifespan. Overcharging or imbalanced voltage across series-connected SCs can lead to premature failure. To address this, voltage balancing circuits are often integrated into power electronic systems to ensure uniform voltage distribution across each cell. SCs exhibit higher self-discharge rates than batteries, which can limit their use in applications requiring prolonged energy retention.

Power electronics systems must be designed to account for these factors to maximize lifespan and efficiency, using temperature control, voltage regulation, and periodic monitoring to sustain optimal performance over time. Figure 9.2(a) and (b) shows the cyclic voltammetry (CV) curve of the NiCoP@C@LDHs electrode, reflecting a larger integrated area, which implies that the polished electrode can store more energy after phosphorization and carbon layer coating. The galvanostatic charge–discharge (GCD) curves (Figure 9.2(c) and (d)) indicate a battery-type capacitor behavior in these electrodes. The intersection of the curves on the x-axis in the high-frequency region indicates electrolyte resistance (Rs), which includes the ionic resistance of the electrolyte, and the intrinsic and interface resistances, affecting their electrochemical performance. For NiCoP@C@LDHs, Rs is 1.225 Ω, lower than other electrodes, indicating high conductivity of NiCoP@C and accessible ionic conductivity of the layered double hydroxides (LDHs) shell (Figure 9.2(e)). Figure 9.2(f) illustrates the capacity retention of four different SC materials (NiCoP NWs, NiCoP@C, LDHs, and NiCoP@C@LDHs) as a function of cycle number, up to 4500 cycles. The y-axis represents capacity retention in percentage, while the x-axis represents the number of charge cycles. The NiCoP@C@LDHs (stars) perform best, retaining over 90% of their initial capacity even after 4500 cycles, indicating excellent cycling stability. LDHs (triangles) and NiCoP@C (circles) show moderate retention, with LDHs maintaining around 80% and NiCoP@C around 75% at the end of the cycles. In contrast, NiCoP NWs (squares) exhibit the lowest retention, dropping below 60% by 4500 cycles. This suggests that NiCoP@C@LDHs has the most stable electrochemical performance among the tested materials.

9.2.3 Modeling SCs for Power Electronic Applications

Accurate modeling of SCs is crucial in designing power electronic systems that capitalize on their unique properties. Models commonly used for SCs include electrical equivalent circuit models, which depict the SC as a combination of capacitance, ESR, and leakage resistance, providing insight into their behavior under varying

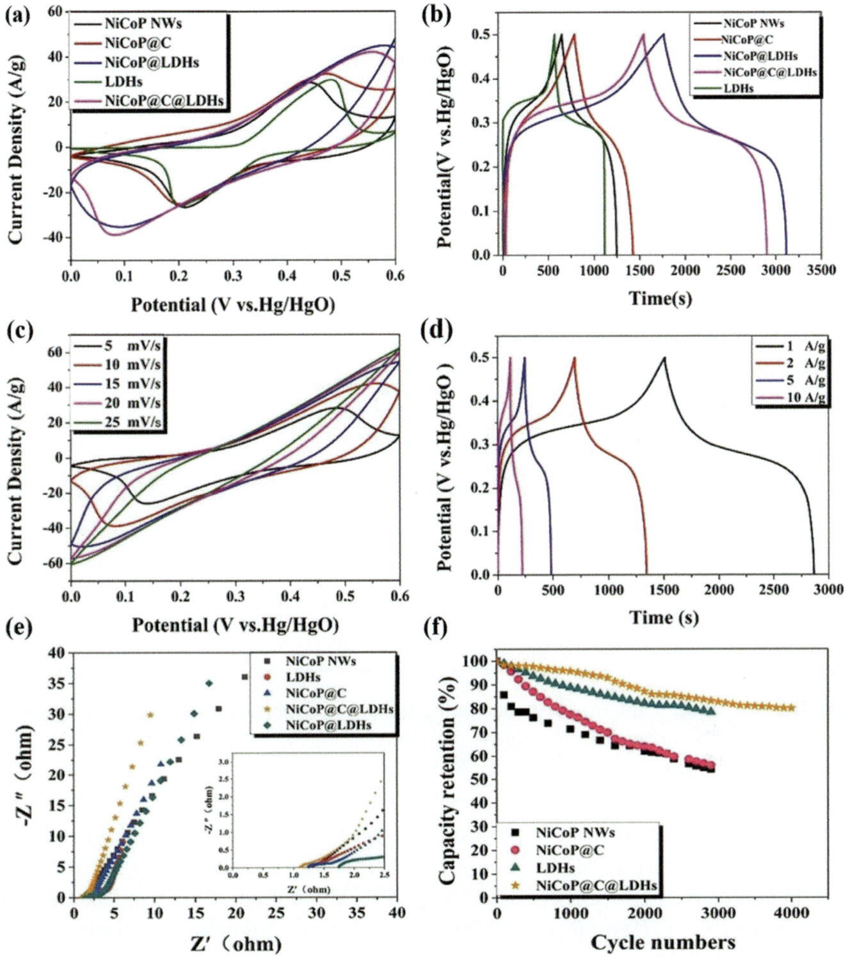

FIGURE 9.2 (a) CV curves at 10 mV/s, (b) CV curves at various scan rates, (c) GCD plots at 1 A/g, (d) charge–discharge plots at varying discharge current densities, (e) Nyquist plots at open circuit potential with an inset view, and (f) cycling performance at 10 A/g. Reprinted with permission from [3]. (© Elsevier 2020.)

power demands. Figure 9.3 depicts the physical phenomena and components of the proposed equivalent electrical circuit model. During the SC charging process, the immediate capacitance C_1 affects the SC capacitance. It is believed that the charges in the electrolyte drift within an electric field close to the electrode/electrolyte interface, resulting in charge redistribution in that area. This increase in capacitance is represented in the model by the charging of parallel capacitor C_2. Over time, the diffuse current diminishes as the gradient of charge concentration decreases. In the model, resistor $R_2(t)$ indicates the declining likelihood of additional charge carriers being transported by diffusion.

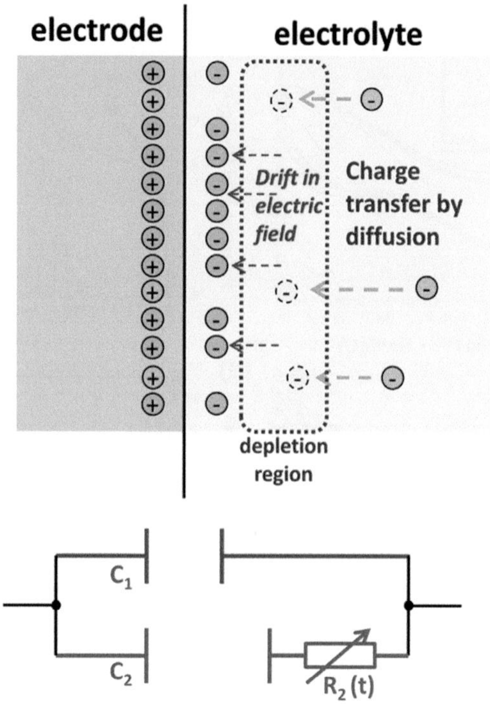

FIGURE 9.3 Schematic drawing illustrating the physical phenomena during the charging and the elements involved in the equivalent electrical circuit model for SC. Modified and adapted with permission [4]. (© Elsevier 2015.)

Dynamic models are more sophisticated because they account for charge redistribution, temperature effects, and non-linear characteristics, enhancing the accuracy of simulations in real-world applications [5–10]. These models are used extensively in simulating the performance of SCs in power systems, allowing engineers to optimize energy efficiency, predict lifespan, and minimize heat generation.

9.3 APPLICATIONS IN RENEWABLE ENERGY SYSTEMS

The deployment of renewable energy sources like wind and solar has created energy storage and management challenges. Their intermittent nature can lead to instability in power output, making efficient energy storage essential. SCs, capable of handling rapid fluctuations and providing high-power density, have become crucial in renewable energy applications. This section explores SCs' role in enhancing renewable energy systems, focusing on wind, solar, and hybrid setups.

9.3.1 Role of SCs in Renewable Energy Systems

In renewable energy systems, SCs enhance stability and efficiency by smoothing power fluctuations from environmental changes, ensuring a consistent energy supply.

They stabilize voltage and frequency, compensating for rapid generation changes and improving grid reliability. SCs buffer energy and level loads, crucial for intermittent generation, effectively managing peak loads and generation dips. These characteristics make SCs invaluable in renewable applications, contributing to system performance and sustainability, while also being ideal for integrating renewable systems with advanced converters and control strategies.

9.3.1.1 SCs in Wind Energy Systems

In renewable energy systems, SCs enhance stability and efficiency. They smooth power fluctuations by absorbing short-term output variations due to changing conditions, ensuring a consistent energy supply [11–14]. SCs stabilize voltage and frequency by compensating for rapid generation changes, improving grid compatibility and reliability. Their energy buffering and load-leveling capabilities are essential for systems with intermittent power generation, managing peak loads and generation dips effectively. These characteristics make SCs invaluable in renewable energy applications, enhancing system performance and sustainability [15–18]. During low or no wind periods, SCs can store and provide energy to restart the turbine, facilitating a smooth transition back to generation mode [19–21].

9.3.1.2 SCs in Solar Energy Systems

Solar power systems experience intermittent generation because of changing sunlight levels during the day and across seasons. SCs improve these systems by providing energy storage and release during peak demand. They can quickly store and release energy at times of maximum solar production, working alongside slower energy sources such as batteries [22–26]. Maintaining stable solar output during cloud cover: As clouds move overhead, solar energy production can significantly decrease. SCs mitigate these reductions by supplying immediate power to ensure consistent energy output [27–30]. Integrating solar microgrids: SCs are ideal for solar microgrid systems as they help balance short-term energy demands and enhance power quality by absorbing or releasing energy when necessary [31–35].

In conjunction with solar batteries, SCs reduce the burden on batteries by handling high-power transients, thus prolonging battery life and increasing overall system efficiency.

9.3.1.3 SCs in Hybrid Renewable Energy Systems

Hybrid renewable energy systems combine wind, solar, and other sources, creating complex power management challenges. SCs are crucial in power electronics by balancing energy sources: They quickly absorb or release energy to maintain balance during transitions [36]. Improving energy management: SCs respond to fluctuations, maintaining stable power output and ensuring efficient power flow from various sources to the grid or storage [37–42]. Reducing reliance on traditional storage: Hybrid systems use batteries for long-term storage, but SCs meet transient demands, extending battery lifespan and lowering maintenance costs [43].

In hybrid setups, SCs can work with energy management systems (EMS) that coordinate the charging and discharging schedules across different energy sources, creating a seamless, stable energy supply even when renewable input varies. Figure 9.4

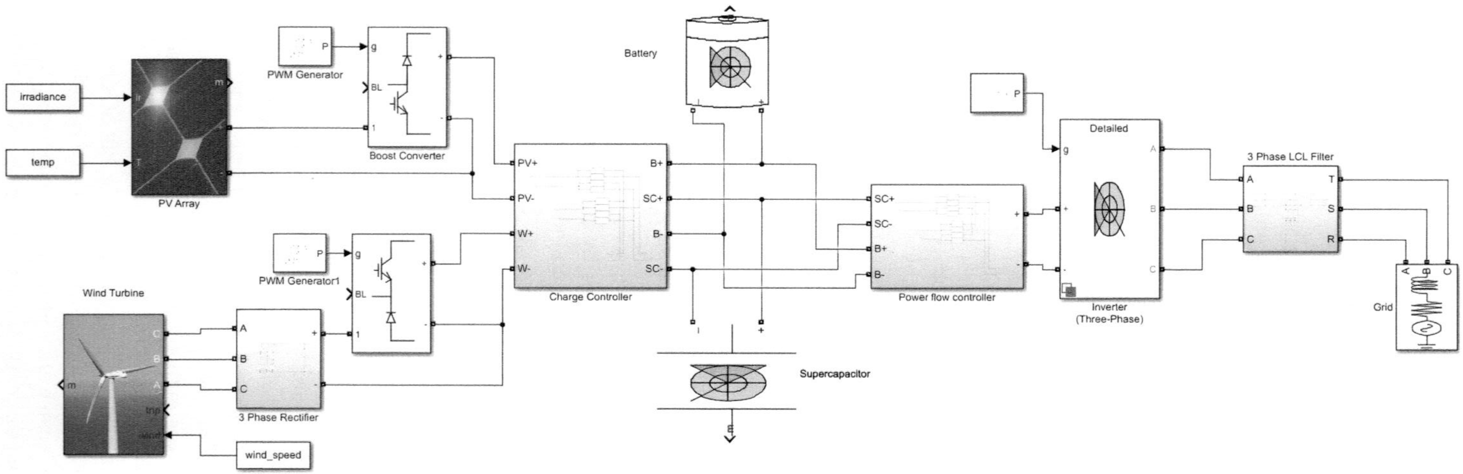

FIGURE 9.4 Configuration of a hybrid energy storage system using SCs for enhanced wind–solar hybrid power setup performance.

depicts a hybrid renewable energy system integrating a photovoltaic (PV) array and a wind turbine to supply power to the grid. The PV array, influenced by solar irradiance and temperature, is connected to a boost converter controlled by a pulse width modulation (PWM) generator. Similarly, the wind turbine's output, influenced by wind speed, is rectified using a three-phase rectifier and regulated by a PWM-controlled boost converter. Both sources feed a charge controller, which manages the energy flow to a battery and a SC for energy storage. A power flow controller coordinates the stored energy to ensure smooth power delivery. The inverter converts the DC energy into three-phase AC, passing it through an inductor-capacitor-inductor (LCL) filter to reduce harmonics before supplying the grid. This setup emphasizes efficient energy conversion and storage management for reliable grid integration.

9.4 SCs IN ELECTRIC VEHICLES AND TRANSPORTATION

The automotive and transportation industries are increasingly focused on sustainable solutions, particularly through the adoption of EVs, hybrid electric vehicles (HEVs), and fuel cell vehicles (FCVs). SCs play a pivotal role in these advancements, addressing specific performance challenges associated with high-power demands, regenerative braking, and battery lifespan. This section examines how SCs enhance EVs and transportation systems by providing rapid energy response, supporting energy recovery, and improving overall efficiency.

9.4.1 Role of SCs in Electric Vehicle Systems

In EVs, SCs complement batteries by delivering rapid power bursts, reducing strain on the primary battery, and improving overall efficiency. They support acceleration by providing instant energy and enhance braking by capturing and storing energy during deceleration [44–49]. By handling high-power demands, SCs minimize battery cycling and depth of discharge, extending battery lifespan [50–54]. Their fast charge–discharge capability boosts energy efficiency, particularly in stop-and-go urban driving, making them a valuable addition to EV power systems [55–58].

9.4.2 SCs for Regenerative Braking

One of the primary applications of SCs in EVs and HEVs is in regenerative braking. This technology allows vehicles to recover kinetic energy during braking, which can then be stored and used to power the vehicle later [59–61]. SCs are ideal for regenerative braking due to their high-power density, allowing rapid energy absorption and release to maximize captured energy. Their excellent cycle stability enables them to endure millions of charge–discharge cycles with minimal capacity loss, making them highly reliable. Additionally, by capturing braking energy, SCs reduce stress on batteries, preventing excessive power demands, minimizing wear, and improving efficiency in energy storage systems. SCs are commonly integrated into regenerative braking systems in hybrid buses, trams, and electric trains, where frequent stops and starts make efficient braking energy recovery critical for operational cost and energy savings.

9.4.3 Application in Hybrid Electric and Fuel Cell Vehicles

In HEVs and FCVs, SCs are often used to manage the high-power demands and rapid energy shifts associated with hybrid powertrains [59–61]. SCs enhance hybrid vehicle performance by delivering high power for initial acceleration, supporting quick starts, and overtaking maneuvers. They also smooth energy fluctuations by acting as a buffer between fuel cells or internal combustion engines, ensuring stable power output for dynamic driving conditions. Additionally, SCs improve efficiency in hybrid systems by optimizing power distribution, reducing fuel consumption, and balancing energy between the primary power source and storage systems, making vehicles more sustainable and responsive.

9.4.4 Integration with Power Electronics in Transportation

Integrating SCs in transportation systems requires advanced power electronics to manage energy flow between SCs, batteries, and electric motors [62–65]. SCs enhance hybrid vehicle performance by delivering high power for initial acceleration, supporting quick starts, and overtaking maneuvers. They also smooth energy fluctuations by acting as a buffer between fuel cells or internal combustion engines, ensuring stable power output for dynamic driving conditions. Additionally, SCs improve efficiency in hybrid systems by optimizing power distribution, reducing fuel consumption, and balancing energy between the primary power source and storage systems, making vehicles more sustainable and responsive. In modern EVs, these power electronics systems are typically embedded within battery management systems (BMS) and EMS to optimize energy flow and enhance overall vehicle performance [66, 67].

9.5 GRID AND POWER QUALITY APPLICATIONS

With the rapid transformation of modern power grids to incorporate renewable energy sources and meet increasing demand for reliability, grid stability and power quality have become paramount. SCs have emerged as effective tools in grid applications because they respond instantaneously to fluctuations, stabilize voltage and frequency, and support power quality. This section examines how SCs are used in grid applications, including UPS, voltage stabilization, frequency regulation, and their role in advanced EMS.

9.5.1 Voltage Stabilization in Power Grids

Voltage instability in power grids can lead to fluctuations that affect the performance of sensitive equipment and lead to potential outages. SCs are essential for stabilizing grid voltage by swiftly compensating for rapid voltage drops or spikes, ensuring a consistent power supply. They support reactive power needs by working alongside advanced power electronics to inject reactive power into the grid, maintaining voltage stability amid fluctuating demands. Additionally, SCs help reduce wear on traditional grid components, such as transformers and switchgear, by smoothing out voltage variations, potentially extending their lifespan and improving overall grid efficiency. Voltage stabilization is particularly beneficial in distributed energy grids with high levels of renewable integration, where wind or solar variability can cause frequent voltage swings. In such systems, SCs can help maintain the quality of service and reduce downtime [68–72].

9.5.2 Frequency Regulation and Grid Stabilization

Frequency stability is essential in maintaining synchronization across the grid. Variations in load or generation can lead to frequency fluctuations, impacting grid reliability and even triggering cascading failures. SCs play a vital role in frequency regulation by providing a fast response for primary frequency control, instantly injecting or absorbing power to counteract small frequency deviations and maintain system stability. They also balance short-term supply-demand mismatches by storing excess energy during low demand and discharging it when demand rises, ensuring grid reliability. SCs reduce reliance on traditional spinning reserves by responding swiftly to frequency variations, offering a more cost-effective and efficient solution for maintaining grid stability. In regions where renewable energy penetration is high, frequency regulation is crucial due to the unpredictable nature of renewable sources. SCs provide the speed and reliability needed to mitigate these frequency variations effectively [73, 74].

9.6 SCs IN UPS

Power interruptions are unacceptable in critical applications, such as hospitals, data centers, and telecommunications. SCs in UPS systems provide immediate power backup, ensuring continued operation during power outages [75–78]. SCs complement batteries by delivering instantaneous power during transition periods, ensuring seamless operation until backup generators or batteries take over. By handling high-power demands in the initial seconds of an outage, SCs reduce battery stress, extend lifespan, and minimize maintenance requirements. Their high cycle stability and rapid response make them ideal for backup systems, efficiently managing frequent power interruptions and providing reliable short-term energy storage. UPS systems with SCs are especially valuable in facilities with sensitive data or equipment, where even a brief power fluctuation could result in significant losses.

9.7 ENERGY MANAGEMENT IN MICROGRIDS

Microgrids, which are small-scale, decentralized power systems that can operate independently or in conjunction with the main grid, benefit significantly from SCs. In microgrids, SCs support in the following:

- *Smooth integration of renewable sources*: SCs absorb and release energy as renewable generation fluctuates, helping maintain consistent power quality.

- *Short-term energy storage and balancing*: SCs act as a local energy buffer for microgrids, absorbing excess generation and releasing it during demand peaks to stabilize the system.
- *Black-start capabilities*: SCs provide the rapid energy needed to restart isolated microgrid systems after an outage, particularly in off-grid or remote areas where power restoration is challenging.

Microgrids are increasingly popular in remote communities, campuses, and industrial facilities. By incorporating SCs, these microgrids can achieve greater reliability and resilience, even with high levels of intermittent renewable generation [79–81].

9.8 INDUSTRIAL APPLICATIONS

Reliable power, rapid energy discharge, and high cycle stability are critical in industrial settings. SCs have gained attention in various industrial applications due to their ability to handle power surges, sustain frequent charging and discharging, and provide reliable backup for high-demand operations. From heavy machinery and robotics to material handling and automated systems, SCs are essential in improving energy efficiency and operational stability. This section explores key industrial applications, highlighting the advantages of SCs in manufacturing, warehousing, and automation.

9.8.1 Power Backup for Industrial Equipment

Maintaining uninterrupted power in industries with sensitive or high-power equipment is crucial to avoid production downtime, equipment damage, and data loss. SCs provide an effective power backup solution by

- *Delivering instant energy during power interruptions*: SCs respond within milliseconds, providing the initial power needed to keep critical equipment running until an alternative power source is engaged.
- *Reducing reliance on batteries and generators*: SCs minimize the cycling demands on batteries and can serve as the first line of defense in short outages by supplying immediate backup power.
- *Improving reliability in UPS*: For high-demand machinery, SCs within UPS systems ensure smooth transitions during power fluctuations, maintaining the integrity of industrial processes and equipment.

Power backup SCs are commonly used in industries like semiconductor manufacturing, where even brief power disruptions can lead to costly defects in sensitive production processes.

9.8.2 SCs in Heavy Machinery

Industrial heavy machinery, such as cranes, excavators, and forklifts, frequently demands high power output, especially during peak operation cycles [82]. SCs enhance these applications by

- *Providing rapid bursts of energy for lifting and acceleration*: SCs deliver energy quickly without straining the primary power source when high power is needed for short durations.
- *Storing energy from regenerative braking*: Heavy machinery often includes regenerative braking systems that capture energy when the equipment slows down. SCs effectively store this energy for reuse, reducing fuel or electricity consumption.
- *Reducing wear on primary power sources*: By handling high-power demands, SCs reduce stress on batteries or engines, leading to less wear and extending the life of primary power sources.

These benefits have made SCs popular in ports and construction sites, where cranes, forklifts, and other heavy machinery require frequent high-power boosts and energy storage capabilities.

9.8.3 Automation and Robotics

In industries reliant on automation and robotics, SCs support operational efficiency and reliability by

- *Providing energy for high-speed operations*: Robotics systems often require immediate energy for rapid movement and operation. SCs respond quickly, ensuring that robots perform with high precision and speed.
- *Enhancing efficiency in pick-and-place applications*: Automated robotic arms in assembly lines benefit from SCs that provide quick energy for repetitive tasks, improving productivity and energy use.
- *Supporting power demands in autonomous mobile robots (AMRs)*: AMRs used in warehousing and manufacturing require rapid power bursts for navigation and material handling. SCs allow these robots to recharge quickly and operate efficiently, reducing downtime and maximizing productivity.

In automation, SCs enable smoother operations and minimize downtime due to their high-power density and quick energy response, which is essential for maintaining continuity in fast-paced manufacturing environments [83–86].

9.8.4 Material Handling and Warehousing

Material handling equipment, such as automated guided vehicles (AGVs) and conveyor systems, requires efficient power management for reliable operation in warehouses and logistics centers. SCs enhance these applications by

- *Providing rapid energy for loading and unloading tasks*: SCs offer the quick power bursts needed for lifting, loading, and moving heavy materials.
- *Reducing charging time for AGVs:* In applications where AGVs are frequently deployed, SCs enable fast charging, minimizing downtime and allowing for extended operational hours.

- *Improving efficiency in high-cycling operations*: Warehouses with high-frequency picking and sorting tasks benefit from the robust cycling ability of SCs, which handle continuous power demands without significant degradation.

SCs have proven particularly useful in automated distribution centers, where efficiency and uptime are critical for meeting e-commerce and retail logistics demand.

9.9 CHALLENGES AND FUTURE DIRECTIONS

Despite the growing applications and advantages of SCs in various fields, their limitations and cost challenges remain. Addressing these challenges is crucial to unlocking the full potential of SCs and expanding their use in areas where energy storage and power delivery are critical. This section explores some key challenges associated with SCs, including cost, energy density, materials, and integration complexities. It also discusses potential future directions, focusing on advancements in materials science, hybrid systems, and emerging applications.

9.9.1 Key Challenges

SCs, especially those utilizing advanced materials like graphene, face cost-related challenges that hinder widespread adoption in large-scale and budget-sensitive applications. The high price per kilowatt-hour of stored energy is primarily driven by expensive material costs, including graphene, carbon nanotubes (CNTs), and ionic liquids, which require sophisticated production methods. Additionally, precision manufacturing processes and strict quality control add to production expenses, making SCs less accessible for cost-conscious industries. To improve affordability, efforts are focused on developing scalable manufacturing techniques and optimizing material efficiency, ensuring that SC technology becomes economically viable for broader applications.

9.9.1.1 Energy Density Limitations

Despite their superior power density, SCs have lower energy density compared to batteries, limiting their effectiveness in long-duration applications. This poses challenges for systems requiring a sustained energy supply, such as EVs and large-scale storage, where batteries remain the preferred choice. Additionally, SCs often need to be physically larger to match battery capacity, which can be a disadvantage in space-constrained environments. To overcome these limitations, ongoing research focuses on enhancing SC energy density using advanced materials, including hybrid structures that integrate electrostatic and faradaic storage mechanisms, aiming to make SCs more viable for extended energy applications.

9.9.1.2 Material Stability and Durability

SCs are known for their long cycle life, ensuring material stability under varying environmental conditions remains a challenge. High temperatures can degrade the performance and lifespan of certain SC materials, particularly in industrial and automotive settings. Additionally, electrolyte stability is a concern, as prolonged exposure to high voltage can lead to breakdown, reducing energy capacity and cycle life. Enhancing material stability under extreme temperature and voltage conditions is crucial for maintaining reliable performance across diverse applications, driving the need for continued research and innovation in SC technology.

9.9.1.3 System Integration Complexities

Integrating SCs into hybrid energy systems presents challenges in managing energy flow and balancing power sources:

1. *Control systems for hybrid storage*: Coordinating SCs with batteries or fuel cells requires sophisticated algorithms for optimal management.
2. *Voltage balancing*: In systems with varying requirements, ensuring consistent voltage levels can be complex and may need extra electronics.

Developing advanced control and EMS is crucial for smooth integration and efficient operation in hybrid systems combining SCs with other storage technologies.

9.9.2 Future Directions

Advancements in material science are improving SC performance by enhancing energy density, durability, and cost-effectiveness. Materials like graphene, CNTs, and metal oxides (MOs) are increasing energy storage capabilities, making SCs more competitive with batteries. Hybrid materials combining electrochemical and electrostatic properties are being developed to enhance cycle stability and extend SC lifespan. Research also targets scalable production methods for advanced materials to make SCs economically viable for energy storage applications. These advancements may lead to new SCs, such as solid-state and flexible types, enabling applications in wearable electronics, IoT devices, and portable electronics.

9.9.2.1 Hybrid Energy Storage Systems

Hybrid energy storage systems, which integrate the high-power density of SCs with the high energy density of batteries or fuel cells, offer significant advantages for modern energy applications. They enable optimized energy management by dynamically balancing power and energy demands, utilizing SCs for rapid bursts and batteries for sustained output. Additionally, these systems enhance efficiency and longevity by reducing cycling stress on batteries, extending their operational life and minimizing maintenance. Increasingly adopted in EVs, grid stabilization, and industrial systems requiring both short-term and long-term energy storage, hybrid solutions are becoming essential for reliable energy management. Advancements in control algorithms and integrated energy systems will be key to fully harnessing their potential, driving greater sustainability and performance improvements.

9.9.2.2 Emerging Applications and Markets

As SC technology evolves, applications expand beyond traditional energy storage. Flexible, lightweight SCs suit wearable electronics and IoT devices, offering quick power bursts and rapid recharging. In smart grids, SCs support energy balancing, ensuring stable power quality amid fluctuating renewables. Autonomous systems like drones, robots, and self-driving vehicles leverage SCs' high-power delivery for maneuverability and efficient energy recovery. With ongoing advancements, SCs are set to become a standard power solution for diverse applications needing fast energy delivery and high cycle stability.

9.9.3 The Path Forward

To unlock SCs' potential, continuous research, development, and commercialization are necessary. Developing hybrid storage solutions by integrating SCs with batteries will create versatile energy systems to meet diverse power demands. Investing in sustainable materials is crucial for advancing environmentally friendly, cost-effective SC production and ensuring broader adoption. Additionally, establishing industry standards for performance, safety, and reliability is essential as SCs integrate into various applications, fostering their widespread use in modern energy systems. Ultimately, the development of SCs can significantly contribute to sustainable energy solutions. By addressing limitations and leveraging opportunities, SCs can transform power electronics and fields needing efficient energy storage and rapid power delivery.

9.10 CONCLUSION

SCs are reshaping the landscape of power electronics and energy storage, offering unique advantages in applications requiring high-power density, rapid charge and discharge capabilities, and long cycle life. As seen throughout this chapter, their versatility makes them valuable across a diverse range of fields, from transportation and grid stability to industrial automation and renewable energy systems. Their role as complementary energy storage devices, particularly when combined with batteries in hybrid systems, highlights their growing importance in addressing modern energy challenges.

The characteristics that distinguish SCs—namely, their fast response time, high efficiency, and durability—address specific demands that conventional batteries and other storage technologies struggle to meet. For instance, in EVs, SCs provide quick energy bursts for acceleration; in renewable energy systems, they help smooth out power fluctuations and stabilize grid frequency. Industrial automation supports high-frequency cycling operations with minimal degradation, while UPSs offer reliable backup power during short outages.

However, challenges remain, particularly regarding the cost, energy density, and integration complexities of SCs. These limitations currently restrict their use in long-duration and large-scale energy storage applications, where batteries or other forms of storage are more practical. As a result, ongoing research is focused on improving the energy density of SCs and developing cost-effective materials and manufacturing processes. Emerging materials such as graphene and CNTs are expected to drive advancements, potentially making SCs more competitive and accessible.

Future directions for SCs include the development of hybrid energy storage systems, integration into smart grids, and deployment in novel applications such as wearable electronics and IoT devices. With their ability to meet the demands of modern and evolving technologies, SCs are likely to become increasingly integrated into the infrastructure of sustainable energy systems. The combination of SCs with intelligent EMS and renewable resources also positions them as critical components in transitioning to a low-carbon economy.

In conclusion, SCs hold substantial potential to enhance power quality, increase efficiency, and contribute to the resilience of energy systems. As research and development efforts continue, SCs are expected to expand into new domains, supporting the shift toward smarter and more sustainable power electronics. By addressing current challenges and capitalizing on new materials and technologies, SCs can become essential enablers of innovation across various applications, paving the way for a more flexible, reliable, and sustainable energy future.

9.11 AUTHOR AI DECLARATION

The authors hereby declare the use of generative AI and AI-assisted technologies in the writing process. While preparing this work, the authors used the ChatGPT tool/service to rephrase several statements in the manuscript. All AI-generated content was reviewed and edited by the authors as needed. The authors take full responsibility for the content of the publication.

ACKNOWLEDGMENT

The Editors acknowledge the use of Grammarly as a tool to assist in editing this book chapter. The suggestions made by Grammarly were carefully considered and incorporated where appropriate, but the author is solely responsible for the final content.

REFERENCES

1. S.-M. Chen, R. Ramachandran, V. Mani, and R. Saraswathi, "Recent advancements in electrode materials for the high performance electrochemical SCs: A review," *International Journal of Electrochemical Science* 9 (2014): 4072–4085. doi: 10.1016/S1452-3981(23)08076-8
2. Y.-G. Wang, L. Cheng, and Y.-Y. Xia, "Electrochemical profile of nano-particle CoAl double hydroxide/active carbon supercapacitor using KOH electrolyte solution," *Journal of Power Sources* 153 (2006): 191–196. doi: 10.1016/j.jpowsour.2005.04.009
3. Y. Zhu *et al.*, "Ultra-long lifespan asymmetrical hybrid supercapacitor device based on hierarchical NiCoP@C@LDHs electrode," *Electrochimica Acta* 334 (2020): 135589. doi: 10.1016/j.electacta.2019.135589
4. V. Sedlakova *et al.*, "Supercapacitor equivalent electrical circuit model based on charges redistribution by diffusion," *Journal of Power Sources* 286 (2015): 58–65. doi: 10.1016/j.jpowsour.2015.03.122
5. T. Barić, H. Glavaš, and E. K. Nyarko, "Second-order supercapacitor model and its parameter extraction," *International Journal of Numerical Modelling: Electronic Networks, Devices and Fields* 33 (2020): e2727. doi: 10.1002/jnm.2727

6. L. E. Helseth, "Modelling SCs using a dynamic equivalent circuit with a distribution of relaxation times," *Journal of Energy Storage* 25 (2019): 100912. doi: 10.1016/j.est.2019.100912
7. M. Lewandowski, and M. Orzylowski, "Fractional-order models: The case study of the supercapacitor capacitance measurement," *Bulletin of the Polish Academy of Sciences-Technical Sciences* 65 (2017): 449–457. doi: 10.1515/bpasts-2017-0050
8. R. Prasad, K. Kothari, and U. Mehta, "Flexible fractional supercapacitor model analyzed in time domain," *IEEE Access* 7 (2019): 122626–122633. doi: 10.1109/ACCESS.2019.2938543
9. F. Rafik, H. Gualous, R. Gallay, A. Crausaz, and A. Berthon, "Frequency, thermal and voltage supercapacitor characterization and modeling," *Journal of Power Sources* 165 (2007): 928–934. https://doi.org/10.1016/j.jpowsour.2006.12.021
10. D. Torregrossa, M. Bahramipanah, E. Namor, R. Cherkaoui, and M. Paolone, "Improvement of dynamic modeling of supercapacitor by residual charge effect estimation," *IEEE Transactions on Industrial Electronics* 61 (2014): 1345–1354. doi: 10.1109/TIE.2013.2259780
11. N. S. Jayalakshmi, D. N. Gaonkar, R. P. Karthik, and P. Prasanna, "Intermittent power smoothing control for grid connected hybrid wind/PV system using battery-EDLC storage devices," *Archives of Electrical Engineering* 69 (2020): 433–453. doi: 10.24425/aee.2020.133036
12. S. D. G. Jayasinghe, D. M. Vilathgamuwa, and U. K. Madawala, "A dual inverter-based supercapacitor direct integration scheme for wind energy conversion systems," *IEEE Transactions on Industry Applications* 49 (2013): 1023–1030. doi: 10.1109/tia.2013.2252318
13. W. Lee, and H. Cha, "A supercapacitor remaining energy control method for smoothing a fluctuating renewable energy power," *Journal of Electrical Engineering & Technology* 10 (2015): 146–154. doi: 10.5370/jeet.2015.10.1.146
14. Y. Sahri *et al*, "Energy management system for hybrid PV/wind/battery/fuel cell in microgrid-based hydrogen and economical hybrid Battery/Super capacitor energy storage," *Energies* 14 (2021): 31. doi: 10.3390/en14185722
15. A. Berrueta, J. Sacristán, J. López, J. L. Rodríguez, A. Ursúa, and P. Sanchis, "Inclusion of a supercapacitor energy storage system in DFIG and full-converter PMSG wind turbines for inertia emulation," *IEEE Transactions on Industry Applications* 59 (2023): 3754–3763. doi: 10.1109/tia.2023.3249145
16. S. Cui, X. W. Yan, and A. Siddique, "Inertia and primary frequency modulation strategy for a doubly fed induction generator based on supercapacitor energy storage control," *International Journal of Green Energy* 19 (2022): 941–955. doi: 10.1080/15435075.2021.1975124
17. M. C. Sousounis, J. K. H. Shek, and B. G. Sellar, "The effect of SCs in a tidal current conversion system using a torque pulsation mitigation strategy," *Journal of Energy Storage* 21 (2019): 445–459. doi: 10.1016/j.est.2018.11.032
18. V. Timmers, A. Egea-Alvarez, A. Gkountaras, and L. Xu, "Control and power balancing of an off-grid wind turbine with co-located electrolyzer," *IEEE Transactions on Sustainable Energy* 15 (2024): 2349–2360. doi: 10.1109/tste.2024.3418043
19. M. Ghodbane-Cherif, S. Skander-Mustapha, and I. Slama-Belkhodja, "Start-up system design for small scale autonomous DFIG wind turbine," *Wind Engineering* 45 (2021): 753–768. doi: 10.1177/0309524x211028756
20. A. F. Obando-Montaño, C. Carrillo, J. Cidrás, and E. Díaz-Dorado, "A STATCOM with supercapacitors for low-voltage ride-through in fixed-speed wind turbines," *Energies* 7 (2014): 5922–5952. doi: 10.3390/en7095922
21. A. Yasin, "Energy management of a stand-alone DC microgrid based on PV/wind/battery/diesel gen. Combined with super-capacitor," *International Journal of Renewable Energy Research* 9 (2019): 1811–1826.

22. M. Bagheri-Sanjareh, M. H. Nazari, and S. H. Hosseinian, "Coordination of hybrid energy storage system, photovoltaic systems, smart lighting loads, and thermostatically controlled loads for microgrid frequency control," *International Transactions on Electrical Energy Systems* 31 (2021). doi: 10.1002/2050-7038.12976
23. K. Citalingam, and Y. I. Go, "Hybrid energy storage design and dispatch strategy evaluation with sensitivity analysis: Techno-economic-environmental assessment," *Energy Storage* 4 (2022). doi: 10.1002/est2.353
24. F. Díaz-González *et al.*, "A hybrid energy storage solution based on supercapacitors and batteries for the grid integration of utility scale photovoltaic plants," *Journal of Energy Storage* 51 (2022). doi: 10.1016/j.est.2022.104446
25. O. Elma, and U. S. Selamogullari, "A comparative sizing analysis of a renewable energy supplied stand-alone house considering both demand side and source side dynamics," *Applied Energy* 96 (2012): 400–408. doi: 10.1016/j.apenergy.2012.02.080
26. P. Thounthong, V. Chunkag, P. Sethakul, S. Sikkabut, S. Pierfederici, and B. Davat, "Energy management of fuel cell/solar cell/supercapacitor hybrid power source," *Journal of Power Sources* 196 (2011): 313–324. doi: 10.1016/j.jpowsour.2010.01.051
27. M. Moghadam, and N. Ghaffarzadeh, "Suppressing solar PV output fluctuations by designing an efficient hybrid energy storage system controller," *Unconventional Resources* 4 (2024). doi: 10.1016/j.uncres.2024.100077
28. P. Nambisan, and M. Khanra, "Optimal energy management of battery supercapacitor aided solar PV powered agricultural feed mill using Pontryagin's minimum principle," *IEEE Transactions on Power Electronics* 37 (2022): 2216–2225. doi: 10.1109/tpel.2021.3104347
29. B. Panda, and A. Ghoshal, "An ANN based switching network for optimally selected photovoltaic array with battery and supercapacitor to mitigate the effect of intermittent solar irradiance," *Energy Sources Part a-Recovery Utilization and Environmental Effects* 44 (2022): 5784–5811. doi: 10.1080/15567036.2022.2088897
30. M. Uno, and M. Yamamoto, "Family of transformerless pulse-width modulation converters integrating voltage equalisers for PV panels and energy storage modules," *IET Power Electronics* 12 (2019): 1487–1498. doi: 10.1049/iet-pel.2018.6158
31. H. H. H. Awaji, A. A. Alhussainy, A. H. Alobaidi, S. Alghamdi, S. Alghamdi, and M. Alruwaili, "Real-time energy management simulation for enhanced integration of renewable energy resources in DC microgrids," *Frontiers in Energy Research* 12 (2024). doi: 10.3389/fenrg.2024.1458115
32. A. Bharatee, P. K. Ray, A. Ghosh, and G. Panda, "Implementation of SOC-based power management algorithm in a grid-connected microgrid with hybrid energy storage devices," *International Journal of Emerging Electric Power Systems* (2024). doi: 10.1515/ijeeps-2024-0149
33. G. V. B. Kumar, K. Palanisamy, and E. De Tuglie, "Ramp-rate control for power quality improvement of renewable grid-integrated microgrid with hybrid energy storage system," *Frontiers in Energy Research* 12 (2024). doi: 10.3389/fenrg.2024.1387908
34. K. Shirinda, K. Kusakana, and M. Ostraszewski, "Combinatorial optimization of a fuzzy logic-controlled grid connected photovoltaic with hybrid energy storage systems using time of use tariff," *Journal of Energy Storage* 99 (2024). doi: 10.1016/j.est.2024.113251
35. B. Younes, H. A. Abbas, K. Laroussi, A. Bousbaine, O. Fergani, and A. Mazari, "Optimizing DC microgrid systems for efficient electric vehicle battery charging in Ain El Ibel, Algeria," *International Journal of Engineering* 37 (2024): 1891–1900. doi: 10.5829/ije.2024.37.10a.03
36. P. B. Nempu, J. N. Sabhahit, D. N. Gaonkar, and V. S. Rao, "Novel power smoothing technique for a hybrid AC-DC microgrid operating with multiple alternative energy sources," *Advances in Electrical and Computer Engineering* 21 (2021): 99–106. doi: 10.4316/aece.2021.02011

37. M. S. Alam, F. S. Al-Ismail, and M. A. Abido, "Power management and state of charge restoration of direct current microgrid with improved voltage-shifting controller," *Journal of Energy Storage* 44 (2021). doi: 10.1016/j.est.2021.103253
38. B. Allaoua, K. Asnoune, and B. Mebarki, "Energy management of PEM fuel cell/supercapacitor hybrid power sources for an electric vehicle," *International Journal of Hydrogen Energy* 42 (2017): 21158–21166. doi: 10.1016/j.ijhydene.2017.06.209
39. K. A. Mamun *et al.*, "Systematic modeling and analysis of on-board vehicle integrated novel hybrid renewable energy system with storage for electric vehicles," *Sustainability* 14 (2022). doi: 10.3390/su14052538
40. M. S. Masaki, L. J. Zhang, and X. H. Xia, "A hierarchical predictive control for supercapacitor-retrofitted grid-connected hybrid renewable systems," *Applied Energy* 242 (2019): 393–402. doi: 10.1016/j.apenergy.2019.03.049
41. D. Rekioua *et al.*, "Optimized power management approach for photovoltaic systems with hybrid battery-supercapacitor storage," *Sustainability* 15 (2023). doi: 10.3390/su151914066
42. I. Tégani, A. Aboubou, M. Y. Ayad, M. Becherif, M. Bahri, and IEEE, "Power Flow Management in WT/FC/SC Hybrid System Using Flatness Based Control," in *3rd International Symposium on Environmental Friendly Energies and Applications (EFEA)*, Paris, France, 2014, New York: IEEE, 2014.
43. P. Sinha, and A. Sharma, "The prospect of supercapacitors in integrated energy harvesting and storage systems," *Nanotechnology* 35 (2024). doi: 10.1088/1361-6528/ad5a7b
44. K. Alobeidli, and V. Khadkikar, "A new ultracapacitor state of charge control concept to enhance battery lifespan of dual storage electric vehicles," *IEEE Transactions on Vehicular Technology* 67 (2018): 10470–10481. doi: 10.1109/tvt.2018.2871038
45. N. Herath, P. Binduhewa, L. Samaranayake, J. Ekanayake, S. Longo, and IEEE, "Design of a Dual Energy Storage Power Converter for a Small Electric Vehicle," in *IEEE International Conference on Industrial and Information Systems (ICIIS)*, Peradeniya University, Faculty of Engineering, Peradeniya, Sri Lanka, 2017, pp. 350–355, 2017.
46. H. Marzougui, A. Kadri, M. Amari, F. Bacha, and IEEE, "Energy Management of Fuel Cell Vehicle with Hybrid Storage System: A Frequency Based Distribution," in *6th International Conference on Control, Decision and Information Technologies (CoDIT)*, Paris, France, 2019, pp. 1853–1858, 2019.
47. N. M. Tien, V. D. Thanh, B. H. Nguyên, M. C. Ta, J. P. F. Trovao, and IEEE, "Sizing of Battery/Supercapacitor Hybrid Energy Storage System for Electric Vehicles," in *IEEE Vehicle Power and Propulsion Conference (VPPC)*, Merced, CA, 2022, 2022.
48. Y. Wu *et al.*, "Driving style-aware energy management for battery/supercapacitor electric vehicles using deep reinforcement learning," *Journal of Energy Storage* 73 (2023). doi: 10.1016/j.est.2023.109199
49. Q. Xun, Y. J. Liu, J. Zhao, E. A. Grunditz, and IEEE, "Modelling and Simulation of Fuel Cell/Supercapacitor Passive Hybrid Vehicle System," in *11th Annual IEEE Energy Conversion Congress and Exposition (ECCE)*, Baltimore, MD, 2019, pp. 2690–2696, 2019.
50. S. Q. Li, C. H. Gu, P. F. Zhao, and S. Cheng, "Adaptive energy management for hybrid power system considering fuel economy and battery longevity," *Energy Conversion and Management* 235 (2021). doi: 10.1016/j.enconman.2021.114004
51. B. Noh, and IEEE, "Supercapacitor Assisted Hybrid Electric Vehicle Powertrain and Power Selection using Fuzzy Rule-Based Algorithm," in *AEIT International Conference of Electrical and Electronic Technologies for Automotive (AEIT AUTOMOTIVE)*, Turin, Italy, pp. 1–5, 2020.

52. P. V. Sidharthan, Y. Kashyap, and IEEE, "Intelligent Power Allocation Strategy of Hybrid Source System in Solar Electric Vehicle," in *6th International Conference on Green Energy and Applications (ICGEA)*, Singaore, pp. 93–98, 2022.
53. S. P. Vishnu, Y. Kashyap, and R. V. Castelino, "Adaptive intelligent hybrid energy management strategy for electric vehicles," *Energy Storage* 5 (2023). doi: 10.1002/est2.436
54. Y. M. Ye, H. C. Wang, B. Xu, and J. F. Zhang, "An imitation learning-based energy management strategy for electric vehicles considering battery aging," *Energy* 283 (2023). doi: 10.1016/j.energy.2023.128537
55. A. Benhammou, H. Tedjini, M. A. Hartani, R. M. Ghoniem, and A. Alahmer, "Accurate and efficient energy management system of fuel cell/battery/supercapacitor/AC and DC generators hybrid electric vehicles," *Sustainability* 15 (2023). doi: 10.3390/su151310102
56. V. S. Panaparambil, Y. Kashyap, and R. V. Castelino, "A review on hybrid source energy management strategies for electric vehicle," *International Journal of Energy Research* 45 (2021): 19819–19850. doi: 10.1002/er.7107
57. A. Upadhyaya, C. Mahanta, and IEEE, "Adaptive Energy Management System based on Pontryagin's Minimum Principle for Battery/Supercapacitor Electric Vehicle Considering Topographic Information," in *20th IEEE Vehicle Power and Propulsion Conference (IEEE VPPC)*, Milan, Italy, 2023, 2023.
58. W. Yaïci, L. Kouchachvili, E. Entchev, M. Longo, and IEEE, "Dynamic Simulation of Battery/Supercapacitor Hybrid Energy Storage System for the Electric Vehicles," in *International Conference on Renewable Energy Research and Applications*, Brasov, Romania, 2019, pp. 460–465, 2019.
59. M. Paredes, J. A. Pomilio, A. A. Santos, and IEEE, "Combined Regenerative and Mechanical Braking in Electric Vehicle," in *12th Brazilian Power Electronics Conference (COBEP)*, Gramado, Brazil, 2013, pp. 935–941, 2013.
60. J. Partridge, and D. I. Abouelamaimen, "The role of supercapacitors in regenerative braking systems," *Energies* 12 (2019). doi: 10.3390/en12142683
61. Z. Y. Zou, J. Y. Cao, B. G. Cao, and W. Chen, "Evaluation strategy of regenerative braking energy for supercapacitor vehicle," *ISA Transactions* 55 (2015): 234–240. doi: 10.1016/j.isatra.2014.09.011
62. P. N. Borza, M. Machedon-Pisu, and M. C. Carp, "Hybrid Electrical Storage Solutions for Developing Reliable Transport Systems," in *14th International Renewable Energy Storage Conference (IRES)*, Dusseldorf, Germany, 2020, vol. 6, pp. 206–212, 2021.
63. D. Iannuzzi, D. Lauria, and P. Tricoli, "Optimal design of stationary supercapacitors storage devices for light electrical transportation systems," *Optimization and Engineering* 13 (2012): 689–704. doi: 10.1007/s11081-011-9160-4
64. F. Ortenzi, M. Pasquali, P. P. Prosini, A. Lidozzi, and M. Di Benedetto, "Design and validation of ultra-fast charging infrastructures based on supercapacitors for urban public transportation applications," *Energies* 12 (2019). doi: 10.3390/en12122348
65. A. Rufer, P. Barrade, D. Hotellier, and S. Hauser, "Sequential supply for electrical transportation vehicles: Properties of the fast energy transfer between supercapacitive tanks," *Journal of Circuits Systems and Computers* 13 (2004): 941–955. doi: 10.1142/s0218126604001763
66. D. X. Xin, J. L. Li, and C. A. Liu, "Research on the application and control strategy of energy storage in rail transportation," *World Electric Vehicle Journal* 14 (2023). doi: 10.3390/wevj14010003
67. J. Y. Zhang, J. L. Tao, Y. M. Hu, and L. H. Ma, "An energy management strategy based on DDPG with improved exploration for battery/supercapacitor hybrid electric vehicle," *IEEE Transactions on Intelligent Transportation Systems* 25 (2024): 3999–4008. doi: 10.1109/tits.2023.3327105

68. A. Agrawal, and R. Gupta, "Coordinated control of hybrid DERs enabled grid-interactive residential PCM with hybrid bus layout," *IEEE Systems Journal* 16 (2022): 4607–4618. doi: 10.1109/jsyst.2021.3137096
69. E. Farrokhi, P. Safari, H. Ghoreishy, and IEEE, "A Rule-Based Energy Management Strategy with Current Estimation for Controlling Grid Connected Hybrid Energy Storage System," in *14th Power Electronics, Drive Systems, and Technologies Conference (PEDSTC)*, Babol, Iran, 2023.
70. M. M. Mahmoud, M. M. Aly, H. S. Salama, and A. M. M. Abdel-Rahim, "An internal parallel capacitor control strategy for DC-link voltage stabilization of PMSG-based wind turbine under various fault conditions," *Wind Engineering* 46 (2022): 983–992. doi: 10.1177/0309524x211060684
71. F. Naseri, E. Farjah, Z. Kazemi, E. Schaltz, T. Ghanbari, and J. L. Schanen, "Dynamic stabilization of DC traction systems using a supercapacitor-based active stabilizer with model predictive control," *IEEE Transactions on Transportation Electrification* 6 (2020): 228–240. doi: 10.1109/tte.2020.2964423
72. S. B. Siad, A. Iovine, G. Damm, L. Galai-Dol, and M. Netto, "Nonlinear hierarchical easy-to-implement control for DC MicroGrids," *Energies* 15 (2022). doi: 10.3390/en15030969
73. F. Asghar, M. Talha, and S. H. Kim, "Fuzzy logic-based intelligent frequency and voltage stability control system for standalone microgrid," *International Transactions on Electrical Energy Systems* 28 (2018). doi: 10.1002/etep.2510
74. X. K. Li, M. H. Wang, W. T. Jiang, C. Y. Dong, Z. Xu, and X. H. Wu, "An autonomous finite-time backstepping control for decentralized automatic power sharing of hybrid energy storage systems in DC microgrids," *IEEE Transactions on Smart Grid* 15 (2024): 3348–3360. doi: 10.1109/tsg.2024.3353733
75. W. Choi, J. W. Howze, and P. Enjeti, "Fuel-cell powered uninterruptible power supply systems: Design considerations," *Journal of Power Sources* 157 (2006): 311–317. doi: 10.1016/j.jpowsour.2005.07.058
76. A. Lahyani, P. Venet, A. Troudi, and IEEE, "Design of Power Sharing System between Supercapacitors and Battery in an Uninterruptible Power Supply," in *33rd IEEE International Telecommunications Energy Conference (INTELEC)*, Amsterdam, The Netherlands, 2011, 2011.
77. A. M. A. Malkawi, M. I. Alawneh, and A. Bashaireh, "A seamless start-up for a hybrid uninterruptible power supply based on a diesel generator and supercapacitor energy storage," *Results in Engineering* 24 (2024). doi: 10.1016/j.rineng.2024.103418
78. A. Stepanov, I. Galkin, L. Bisenieks, and IEEE, "Implementation of supercapacitors in uninterruptible power supplies," in *2007 European Conference on Power Electronics and Applications*, Aalborg, Denmark, 2007, pp. 4949–4955, 2007.
79. S. R. Subbanna, and M. Rajini, "Energy Management System with a Reduced Sensor for DC Microgrid," in *Innovations in Energy management and Renewable Resources (IEMRE) Conference*, Kolkata, India, 2021.
80. C. J. Yin, H. W. Wu, F. Locment, and M. Sechilariu, "Energy management of DC microgrid based on photovoltaic combined with diesel generator and supercapacitor," *Energy Conversion and Management* 132 (2017): 14–27. doi: 10.1016/j.enconman.2016.11.018
81. Y. R. Zhang, and Y. W. Li, "Energy management strategy for supercapacitor in droop-controlled DC microgrid using virtual impedance," *IEEE Transactions on Power Electronics* 32 (2017): 2704–2716. doi: 10.1109/tpel.2016.2571308
82. Z. Y. Xu, H. Y. Yi, D. Liu, R. Zhang, and X. Luo, "Design a hybrid energy-supply for the electrically driven heavy-duty hexapod vehicle," *Journal of Bionic Engineering* 20 (2023): 1434–1448. doi: 10.1007/s42235-023-00351-z

83. A. S. Kulkarni, S. M. Kulkarni, and IEEE, "Design of an Energy Regenerative System for a Robotic Leg Prosthesis," in *International Conference on Advanced Mechatronics, Intelligent Manufacture, and Industrial Automation (ICAMIMIA)*, Surabaya, Indonesia, pp. 97–103, 2017.
84. A. Senfelds, M. Vorobjovs, D. Meike, O. Bormanis, and IEEE, "Power Smoothing Approach within Industrial DC Microgrid with Supercapacitor Storage for Robotic Manufacturing Application," in *IEEE International Conference on Automation Science and Engineering (CASE)*, Gothenburg, Sweden, 2015, pp. 1333–1338, 2015.
85. M. F. R. Lee, and A. Nugroho, "Intelligent energy management system for Mobile robot," *Sustainability* 14 (2022). doi: 10.3390/su141610056
86. S. S. Nardekar, and S. J. Kim, "Untethered magnetic soft robot with ultra-flexible wirelessly rechargeable micro-supercapacitor as an onboard power source," *Advanced Science* 10 (2023). doi: 10.1002/advs.202303918

10 Challenges and Future Directions in Supercapacitor Research

Anusha Srinivas, Ramesh Chandran K, Sreeja P C, and Sapna Nehra

10.1 INTRODUCTION

Supercapacitors (SCs) have been widely evolved in energy storage technology, which is currently undergoing drastic changes. This is because of their distinct electrochemical properties, such as high power density, fast charge and discharge cycles, and excellent cycle stability [1–3]. In contrast to traditional battery cells, which are based on faradaic reactions for energy storage, SCs can store energy by electrostatic processes or surface redox reactions, enabling them to operate efficiently over hundreds of thousands of cycles with minimal degradation [4–7].

SCs are considered crucial components for the future shift to sustainable energy. Their capability to deliver significant power quickly makes them ideal for tasks like capturing kinetic energy in electric vehicles (EVs), managing energy distribution from renewable sources, and supplying backup power for electronic devices. Moreover, their ability to rapidly adjust output to satisfy power needs complements battery use in hybrid systems, where they support peak loads while potentially prolonging battery lifespan [8–10]. However, as SCs evolve for broader market acceptance and energy system integration, they encounter multiple technological and economic challenges [11, 12]. The main drawback is their comparatively low energy density relative to lithium-ion batteries, limiting their applicability in situations needing an extended energy supply. Additionally, the high expense of advanced electrode materials, issues with optimizing electrolytes, device design, and long-term stability pose significant obstacles. Environmental issues related to material sourcing and disposal further complicate matters, highlighting the necessity for sustainable and scalable manufacturing methods [13–15].

The fact that SC research is a varied activity requires a multifaceted approach to innovation. Moreover, breakthroughs in materials science—especially the development of nanostructured carbon, metal oxides (MOs), and conductive polymers—have generated new opportunities to enhance SC performance. Concurrently, researchers have strongly favored the hybrid configuration of SCs with other storage technologies to optimize the balance of energy and power density to the highest level [16].

The chapter describes some of the major problems preventing SCs from being widely available and provides a detailed analysis of the issues related to materials, manufacturing, device performance, and sustainability. In addition, it explains

 DOI: 10.1201/9781003535836-10

the latest technical achievements aimed at addressing these constraints and explores directions that appear more promising, such as a new type of electrode, intelligent material systems, and artificial intelligence (AI)-assisted design optimization. The chapter concludes by examining the new roles of SCs in key sectors such as electric mobility, renewable energy integration, consumer electronics, and aerospace, highlighting their potential to transform global energy paradigms.

10.2 MAJOR CHALLENGES IN CURRENT TECHNOLOGIES

Despite the numerous benefits of SCs, they also possess certain drawbacks. These challenges impede their widespread commercialization and practical integration into various high-demand applications. It is essential to have a clear understanding of these limitations, as this will enable informed decisions for future research and development. Table 10.1 provides a comprehensive overview of the main challenges in SC technologies.

10.2.1 Low Energy Density

One of SCs' most significant challenges is their inherently low energy density [17, 18]. Energy density refers to the energy stored per unit volume or mass. While SCs excel in power density—delivering large bursts of energy over short durations—their ability to store significant energy for extended periods remains limited compared to conventional batteries like lithium-ion.

This limitation arises primarily from the charge storage mechanism in SCs, which is either electrostatic (as seen in Electric Double Layer Capacitor (EDLCs)) or pseudocapacitive (involving reversible surface redox reactions). Unlike batteries, which store energy through bulk faradaic processes within electrode materials, SCs are restricted by surface phenomena. Consequently, their energy densities typically range from 5 to 10 Wh/kg, compared to over 100 Wh/kg for lithium-ion batteries [19].

TABLE 10.1
Summary of Major Challenges in Supercapacitor Technologies

Challenge	Description	Impact
Low energy density	Limited energy storage capability due to surface-based charge storage mechanisms.	Restricts use in applications needing prolonged energy delivery (e.g., EV propulsion systems).
High material and processing cost	Expensive nanomaterials (graphene, metal oxides) and complex fabrication methods hinder affordability and scale-up.	Limits commercial viability and widespread deployment.
Electrode and electrolyte limitations	Mismatch in conductivity, stability, and operating range; issues with redox stability and voltage window.	Affects performance, safety, and durability of supercapacitors.
Recycling and environmental issues	Limited infrastructure for recycling, hazardous precursors, and high life cycle footprint.	Raises sustainability concerns and contributes to electronic waste.

Low energy density restricts the use of SCs in applications requiring prolonged energy supply, such as EV propulsion systems or grid-scale energy storage. Addressing this issue requires innovative materials with higher specific capacitance and the development of hybrid systems that combine the high-power density of SCs with the high energy density of batteries [20].

10.2.2 High Cost of Materials and Processing

Another barrier to the widespread adoption of SCs is the high cost associated with advanced materials and fabrication processes. High-performance SCs often rely on nanostructured materials such as graphene, carbon nanotubes (CNTs), MOs, and conducting polymers (CPs), which are expensive to synthesize and process at scale. Moreover, the precision required in electrode fabrication, including pore size control, surface modification, and structural uniformity, increases manufacturing complexity and cost. The production of high-purity electrolytes and sophisticated separators further contributes to the overall system cost [21–23].

Scalability also poses economic concerns. While laboratory-scale methods yield promising materials with exceptional properties, translating these results to industrial-scale production while maintaining performance and cost-effectiveness remains a significant challenge. For commercial viability, it is essential to identify abundant, low-cost precursors and develop simple, green synthesis routes that reduce energy consumption and chemical waste [24, 25].

10.2.3 Electrode and Electrolyte Limitations

The performance of an SC largely depends on the nature and compatibility of its electrode materials and electrolytes. Each component introduces specific limitations affecting efficiency, stability, and safety.

Electrode Limitations: Existing carbon-based electrodes produced through traditional methods are widely used, but they often exhibit little to no capacitance because they only rely on surface area. The pseudocapacitive family of materials can achieve significantly higher capacitance, including CPs (such as PANI and PPy) and MOs (e.g., MnO_2 and RuO_2). However, these compounds greatly lack electrical conductivity, which impacts overall mechanical stability, and they have a very limited cycle life. The main challenge is thus how to prepare composite materials that can deliver high specific capacitance while also maintaining their mechanical and cyclic stability over the long term [24, 26–28].

Electrolyte Limitations: Electrolytes are essential in determining the operational voltage window, ionic conductivity, and temperature tolerance of SCs. Aqueous electrolytes are certainly non-hazardous and inexpensive, but they have a narrow voltage range (~1 V), which limits energy density [29]. Inorganic electrolytes and ionic liquids provide larger voltage windows (up to 3.5 V); however, they are more expensive, flammable, and often toxic [30]. Ensuring compatibility between the electrolytes and electrodes, while maintaining performance across temperature ranges and operating conditions without the need for replacements, is an area of intense research [31, 32].

10.2.4 Recycling and Environmental Issues

Sustainability in the environment is a growing concern in the energy storage world, and SCs are not being left out. Even though they are widely regarded as safer and greener than batteries due to their non-explosive and non-toxic nature, several challenges are still to be addressed [33–35].

10.2.4.1 Material Sourcing

Some high-performance products, such as ruthenium oxide or graphene, require a lot of energy or dangerous chemicals [36, 37].

All these activities lead to environmental destruction and harm workers' health.

10.2.4.2 Recycling Challenges

Unlike batteries, eco-friendly SC recycling practices are available to a much lesser extent. The complexity of the electrode designs, the use of compounds, and the lack of standardized patterns all contribute to making the recovery and reuse of the components difficult. Often, spent SCs are found in waste conditions, highlighting the electronic waste problem [38–40].

10.2.4.3 Life Cycle Impact

Their environmental footprint can only be understood and corrected using a complete life cycle assessment (LCA) of SCs, including the phases of raw material extraction, manufacturing, use, and end-of-life disposal. The environmental burden can be decreased by using biodegradable materials, green chemistry, and closed-loop manufacturing processes with innovative technology [41, 42].

10.2.5 Performance Degradation Over Long-Term Cycles

Even though SCs are praised mainly due to their long cycle life, in many cases even over 100,000 cycles, the loss of their performance with time still exists inevitably, especially while working under a high voltage or high temperature environment.

10.2.5.1 Electrode Aging

Over repeated charge–discharge cycles, electrode materials can suffer from structural breakdown, loss of active sites, or detachment from current collectors. In pseudocapacitive materials, swelling and shrinking during redox reactions can lead to cracking and mechanical failure, reducing capacitance and energy efficiency [43–46].

10.2.5.2 Electrolyte Decomposition

Electrolytes may degrade over time, especially at elevated voltages and temperatures. Decomposition products can block pores, increase internal resistance, and reduce ionic mobility, affecting overall performance [47, 48].

10.2.5.3 Packaging and Interface Stability

The interfaces between electrodes, electrolytes, and separators must remain chemically and mechanically stable during operation. Poor interface stability can lead to leakage, internal short-circuiting, and capacity fade [49].

Researchers are exploring advanced coating techniques, self-healing materials, and in situ monitoring systems to combat these issues, enhance reliability, and extend device lifetimes. Understanding the fundamental degradation mechanisms at the molecular level through advanced characterization techniques and modeling is essential for engineering more durable systems [50–52].

10.3 RECENT ADVANCES IN MATERIALS AND DESIGN

Continuous innovation in materials science and engineering has significantly enhanced the performance of SCs [53]. The integration of advanced materials and novel device architectures has led to improvements in energy density, cycle life, and operational flexibility, enabling SCs to overcome many of their traditional limitations [54, 55].

10.3.1 Innovations in Nanomaterials, Hybrid Electrodes, and Novel Device Architectures

The development of SC performance has always depended on advancements in materials and the evolution of design methodology. Nanoscience has indeed presented numerous possibilities for refining the electrochemical properties of materials at the nanometer scale. Consequently, this has had a direct influence on the energy storage, conductivity, and cycling stability of materials [56]. The most prominent among these are carbon nanomaterials, which enhance the previously mentioned factors through their extraordinary surface area, conductivity, and resistance to mechanical stress [57].

Materials such as graphene, activated carbon, CNTs, and carbon aerogels have shown great potential. For example, graphene, which has a two-dimensional structure and a high conductivity of over 10^6 S/m, provides extensive active areas and minimal resistance to power and energy density improvements [58].

Furthermore, by incorporating heteroatoms like nitrogen, sulfur, or phosphorus into these carbon materials, one can create enhanced wettability, and where in place, surface redox reactions bring about pseudocapacitive behavior. These changes, which are initially helpful in realizing the inherent EDLC defects, can lead to an added capacitance. The other benefit is that creating hierarchical structures using micropores for storage and meso-/macropores for ion transport aids in improving the rate performance since the passageways are wider and the ions can cover the distance more easily [59].

Figure 10.1 depicts the details of electrode nanomaterials and their applications in SCs [60].

The usage can lead to a significant improvement in the SC capability of hybrid electrodes. These integrate the advantage of both DLC and pseudocapacitive materials, that is, the TMOs (like MnO_2, $NiCo_2O_4$) and CPs (for instance, PANI, PPy). Although MOs are capable of high theoretical capacitance, their low electrical conductance and high mechanical stiffness are challenging [61]. In a different example, researchers created a combination of conducting carbon matrices

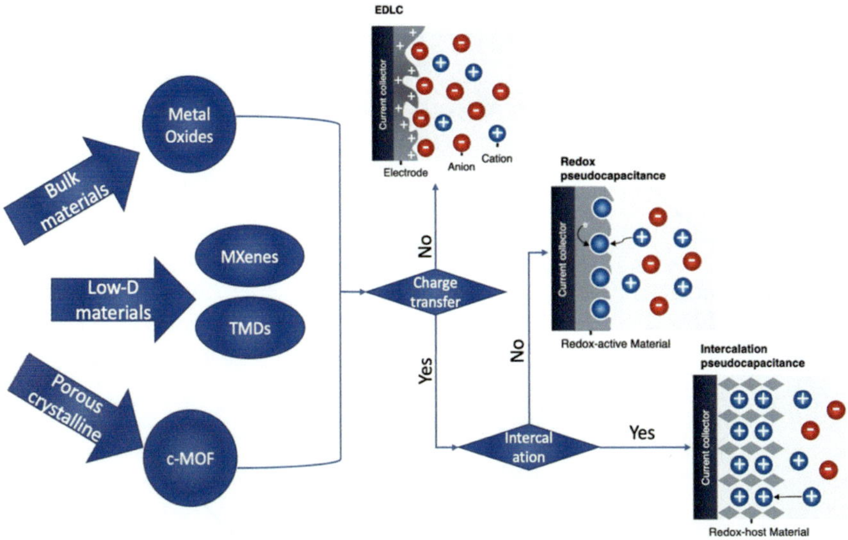

FIGURE 10.1 Electrode nanomaterials and their applications in supercapacitors. (Adapted from [63] under a Creative Commons CC-BY-NC-ND 4.0 License.)

to form viable composite electrodes, as they offer a balanced high capacitance, good conductivity, and long cycle life. Take, for example, the case of graphene—MnO_2 hybrids, which not only have a high energy density but also a good rate capability [62].

Most recently, new materials such as metal organic frame works (MOFs), covalent organic frame works (COFs), and MXenes have become quite popular. This is because MOFs and COFs have porous and tunable structures and are high-surface-area materials, resulting in ion diffusion and redox activity enhancements [63]. MXenes, which refer to a type of hydrophilic surface of 2D transition metal carbides and nitrides such as Ti_3C_2Tx, combine metallic conductivity with hydrophilic surfaces and are powerful contenders for quick electric charge and energy storage in water-friendly electrolytes. As a result, the tradition plus this type of material equals the appearance of the latest and versatile, high-performance SCs [64].

Figure 10.2 illustrates that combining machine learning (ML), quantum mechanics, molecular mechanics, and hybrid techniques for modeling nanomaterial-based SCs could result in breakthroughs in energy storage. ML can expedite material discovery by predicting optimal nanostructures, while quantum mechanics aids in modeling electron interactions at small scales. Molecular mechanics guarantees a precise understanding of material properties, and hybrid methods merge multiple approaches for more accurate simulations.

Furthermore, device architecture plays an equally important role in realizing the potential of advanced materials. Asymetric super capacitors (ASCs) connect dissimilar electrode materials at the cell's anode and cathode sides, which exploit the expanded operating voltage window, thereby making higher energy densities

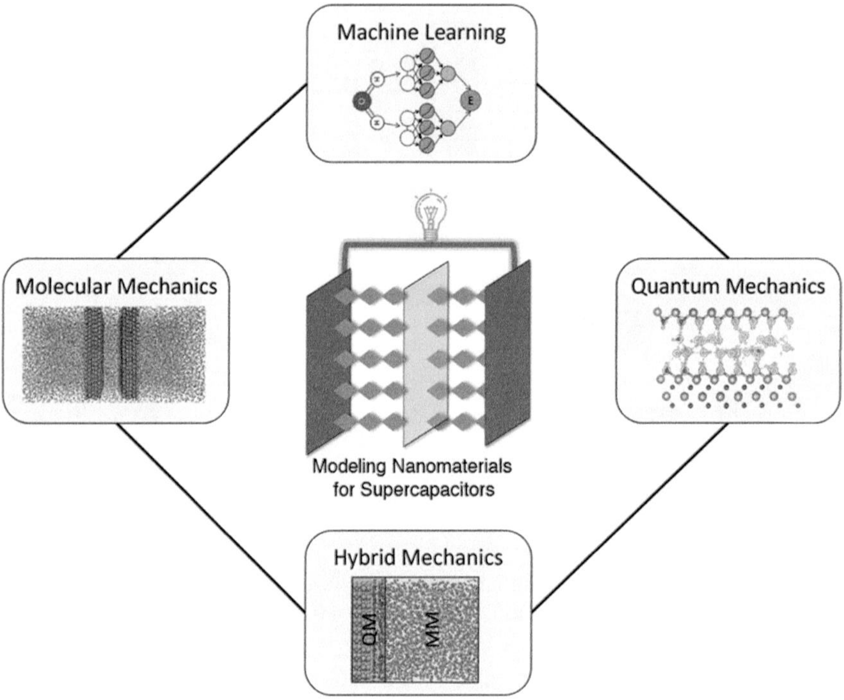

FIGURE 10.2 Modeling nanomaterials supercapacitors involving ML, quantum, molecular, and hybrid mechanics. (Adapted from under a Creative Commons CC BY 4.0 License.)

available [65]. Another trend gaining popularity is flexible and wearable devices that include stretchable substrates and printable electrode inks, which, in some cases, provide mechanical flexibility and are also used in smart textiles, medical devices, and portable electronics. Solid-state SCs with gel or polymer electrolytes not only enhance safety but also allow for miniaturization, thereby maintaining the same performance of the device.

Micro-supercapacitors (MSCs) are made by lithographic or printing methodologies, making them advantageous for the on-chip integration of energy storage and their suitability for the Internet of Things (IoT) collection. Moreover, using 3D printing and laser-scribing techniques to manufacture electrodes has made it possible to produce the electrodes with made-to-order geometries and the maximum porosity. These architectural creations, in combination with advanced materials, are a comprehensive method of defeating the performance limitations that the conventional SC systems faced.

In summary, the combination of novel nanomaterials, hybrid electrode systems, and groundbreaking device architectures has significantly advanced the capabilities of SCs [66]. This progress brings us closer to realizing the practical applications of SCs in the real world, where they demand high energy and power densities, long lifetimes, and mechanical flexibility [67].

10.4 SUSTAINABLE MANUFACTURING AND COST REDUCTION

The commercialization of SCs on a large scale relies heavily on the invention of non-polluting manufacturing technologies, along with efficient production methods [68]. The history of synthesizing electrode materials through traditional means has always been focused on the cost of materials, energy consumption, and toxic substances that have raised economic and environmental concerns. Recently, the search for biosynthesis has become a priority on researchers' agendas, utilizing principles such as eco-friendly solvents, low-temperature processing, and making the most of renewable resources. For example, biomass-derived carbons from agrowaste, like coconut shells, sugar cane bagasse, and sawdust, may provide a sustainable and economical alternative to traditional sources of activated carbon [69].

Transforming from laboratory-scale experiments to industrial applications is still a very difficult thing to do. Some techniques, for example, roll-to-roll coating, spray deposition, and screen printing, are being researched to produce large-area, even electrodes [70]. Besides, changing the electrolytes used from water-rich to expensive and toxic decreases the risks but lowers production costs [71]. To ensure SCs remain maintainable and ecological and can still give quick and efficient results, we must combine these green and scalable methods into practice. As consumer electronics and EVs demand increases, the manufacturing industry must rethink and adopt sustainable design and production methods to minimize its environmental footprint [72].

10.5 HYBRID ENERGY STORAGE AND INTEGRATION

Hybrid energy storage systems (HESS) that combine SCs with batteries are becoming increasingly vital to address the diverse demands of modern energy applications [73]. While batteries excel in delivering high energy density over longer durations, they are limited in power output and cycle life. Conversely, SCs offer high power density and rapid charge–discharge capabilities but suffer from limited energy storage [74]. Combined, these technologies can deliver synergistic benefits, providing immediate power boosts and sustained energy supply [75].

One notable example of HESS application is in EVs, where SCs handle regenerative braking and acceleration surges, while batteries manage long-distance travel [59]. This inclusion benefits the battery and enhances the overall system's efficiency [76]. Similarly, in renewable energy systems such as solar and wind, SCs can absorb energy fluctuations, making grid connections smoother and reducing reliance on oil-based backup systems [77]. As the need for energy storage continues to expand, mixed designs will become more robust and will significantly aid in making energy infrastructures more flexible, resilient, and efficient [78].

10.6 EMERGING AND FUTURE APPLICATIONS

Shortly after the development of new materials and design technologies, SCs have flooded the market due to their wide range of applications. This new class of consumer electronics provides fast-charging features for devices like smartphones, tablets, and laptops [79].

Their lightweight and compact design makes them particularly suitable for wearables like fitness trackers and smartwatches, where rapid energy bursts and robust construction are essential. In addition to this, the flexible and stretchable SCs, embedded in the fabric or affixed to the skin, are advancing the next generation of health monitoring devices [56].

SCs form the backbone of aerospace and defense applications by powering crucial operations requiring quick energy transfer and high resilience to extreme weather conditions [75]. They maintain functionality in harsh environments, such as high vibrations and elevated temperatures, making them ideal for satellites, unmanned aircraft, and missile systems [59]. Moreover, the current rise of the IoT has created a demand for storage solutions that are compact, secure, and capable of fast charging [76, 77].

SCs have been increasingly integrated into sensor networks, smart houses, and industrial automation systems, thus setting the path for cleaner and more efficient technologies in everyday life [78].

10.7 FUTURE RESEARCH DIRECTIONS

The future of SC research focuses on developing intelligent, multifunctional devices capable of adapting to real-time energy demands. Smart materials, such as stimuli-responsive polymers and phase-change materials, are being explored for their potential to dynamically adjust capacitance in response to environmental changes. Additionally, self-healing materials are also being investigated to extend device lifetimes by repairing internal structural damage caused by mechanical stress or cycling fatigue.

AI is set to revolutionize how SCs are designed, tested, and optimized. ML algorithms can analyze large datasets to predict material performance, identify optimal electrode combinations, and forecast failure modes. This data-driven approach speeds up discovery cycles and minimizes the need for extensive experimental trials. Additionally, integrating AI into SC-powered systems enables real-time monitoring, predictive maintenance, and adaptive energy management, representing a significant advancement in intelligent energy storage.

10.8 CONCLUSION

SCs have emerged as promising energy storage devices, offering rapid charge–discharge rates, long cycle life, and high-power density. However, several challenges must be addressed to realize their full potential in modern and future energy systems.

These include improving energy density, lowering material and production costs, enhancing sustainability, and minimizing environmental impact. Advancements in nanomaterials, hybrid electrode systems, and innovative architectures are helping to overcome these limitations.

The integration of SCs into hybrid energy systems and their expanding role in innovative technologies, wearable electronics, and renewable energy grids underscore their growing significance. Future research focused on intelligent materials, AI-driven optimization, and multifunctional capabilities will further enhance the role of SCs in next-generation energy solutions. As the global demand for reliable, efficient, and sustainable energy storage continues to increase, SCs are poised to become indispensable components of the energy landscape.

10.9 AUTHOR AI DECLARATION

The authors hereby declare the use of generative AI and AI-assisted technologies in the writing process. While preparing this work, the authors used the ChatGPT tool/service to rephrase several statements in the manuscript. All AI-generated content was reviewed and edited by the authors as needed. The authors take full responsibility for the content of the publication.

ACKNOWLEDGMENT

The Editors acknowledge the use of Grammarly as a tool to assist in editing this book chapter. The suggestions made by Grammarly were carefully considered and incorporated where appropriate, but the author is solely responsible for the final content.

REFERENCES

1. Olabi, Abdul Ghani, Qaisar Abbas, Ahmed Al Makky, and Mohammad Ali Abdelkareem. "Supercapacitors as next generation energy storage devices: Properties and applications." *Energy* 248 (2022): 123617. https://doi.org/10.1016/j.energy.2022.123617
2. Karthikeyan, Sangapillai, Balasubramaniam, Narenthiran, Aruncachalam Sivanantham, Laxmi Deepak Bhatlu, and Thirupathy Maridurai. "Supercapacitor: Evolution and review." *Materials Today: Proceedings* 46 (2021): 3984–3988. https://doi.org/10.1016/j.matpr.2021.02.526
3. Yadlapalli, Ravindranath Tagore, RamaKoteswara Rao Alla, Rajani Kandipati, and Anuradha Kotapati. "Super capacitors for energy storage: Progress, applications and challenges." *Journal of Energy Storage* 49 (2022): 104194. https://doi.org/10.1016/j.est.2022.104194
4. Shukla, Ashok, Srinivasan Sampath, and Kunjukrishna Pillai Vijayamohanan. "Electrochemical supercapacitors: Energy storage beyond batteries." *Current Science* 79, 12 (2000): 1656–1661. http://www.jstor.org/stable/24104124
5. Salanne, Mathieu, Benjamin Rotenberg, Katsuhiko Naoi, Katsumi Kaneko, Pierre-Louis Taberna, Clare P. Grey, Bruce Dunn, and Patrice Simon. "Efficient storage mechanisms for building better supercapacitors." *Nature Energy* 1, 6 (2016): 1–10. https://doi.org/10.1038/nenergy.2016.70

6. Dutta, Arindam, Shirsendu Mitra, Mitali Basak, and Tamal Banerjee. "A comprehensive review on batteries and supercapacitors: Development and challenges since their inception." *Energy Storage* 5, 1 (2023): e339. https://doi.org/10.1002/est2.339
7. Zhang, Jiawei, and Minghua Chen. "Supercapacitors." *Towards Next Generation Energy Storage Technologies: From Fundamentals to Commercial Applications* (2024): 331–368. https://doi.org/10.1002/9783527845316.ch9
8. Kumar, Yedluri Anil, Nipa Roy, Tholkappiyan Ramachandran, Mohamed Hussien, Md Moniruzzaman, and Sang Woo Joo. "Shaping the future of energy: The rise of supercapacitors progress in the last five years." *Journal of Energy Storage* 98 (2024): 113040. https://doi.org/10.1016/j.est.2024.113040
9. Lakshmi, K. C. Seetha, and Balaraman Vedhanarayanan. "High-performance supercapacitors: A comprehensive review on paradigm shift of conventional energy storage devices." *Batteries* 9, 4 (2023): 202. https://doi.org/10.3390/batteries9040202
10. Whittingham, M. Stanley. "History, evolution, and future status of energy storage." *Proceedings of the IEEE* 100, Special Centennial Issue (2012): 1518–1534. https://doi.org/10.1109/JPROC.2012.2190170
11. Ghosh, Sourav, Sarita Yadav, Ambika Devi, and Tiju Thomas. "Techno-economic understanding of Indian energy-storage market: A perspective on green materials-based supercapacitor technologies." *Renewable and Sustainable Energy Reviews* 161 (2022): 112412. https://doi.org/10.1016/j.rser.2022.112412
12. Abdel, Maksoud, Muhammad Ibrahim Ahmed. Ramy Amer Fahim, Ahmed Esmail Shalan, M. Abd Elkodous, S. O. Olojede, Ahmed I. Osman, Charlie Farrell,Ala'a H. Al-Muhtaseb, A. S. Awed, A. H. Ashour, and David W. Rooney. "Advanced materials and technologies for supercapacitors used in energy conversion and storage: A review." *Environmental Chemistry Letters* 19 (2021): 375–439. https://doi.org/10.1007/s10311-020-01075-w
13. Sharma, Rahul, Harish Kumar, Gaman Kumar, Saloni Sharma, Ranjan Aneja, Ashok K. Sharma, Ramesh Kumar, and Parvin Kumar. "Progress and challenges in electrochemical energy storage devices: Fabrication, electrode material, and economic aspects." *Chemical Engineering Journal* 468 (2023): 143706. https://doi.org/10.1016/j.cej.2023.143706
14. Khan, Humaira Rashid, and Abdul Latif Ahmad. "Supercapacitors: Overcoming current limitations and charting the course for next-generation energy storage." *Journal of Industrial and Engineering Chemistry* (2024). https://doi.org/10.1016/j.jiec.2024.07.014
15. Gür, Turgut M. "Review of electrical energy storage technologies, materials and systems: Challenges and prospects for large-scale grid storage." *Energy & Environmental Science* 11, 10 (2018): 2696–2767. https://doi.org/10.1039/C8EE01419A
16. Jampani, Prashanth H., Ayakkannu Manivannan, and Prashant N. Kumta. "Advancing the supercapacitor materials and technology frontier for improving power quality." *The Electrochemical Society Interface* 19, no. 3 (2010): 57. https://doi.org/10.1149/2.F07103if
17. Yan, Jun, Qian Wang, Tong Wei, and Zhuangjun Fan. "Recent advances in design and fabrication of electrochemical supercapacitors with high energy densities." *Advanced Energy Materials* 4, 4 (2014): 1300816. https://doi.org/10.1002/aenm.201300816
18. Guo, Wei, Chang Yu, Shaofeng Li, and Jieshan Qiu. "Toward commercial-level mass-loading electrodes for supercapacitors: Opportunities, challenges and perspectives." *Energy & Environmental Science* 14, 2 (2021): 576–601. https://doi.org/10.1039/D0EE02649B
19. Bhojane, Prateek. "Recent advances and fundamentals of pseudocapacitors: Materials, mechanism, and its understanding." *Journal of Energy Storage* 45 (2022): 103654. https://doi.org/10.1016/j.est.2021.103654

20. Adebayo, Damola Habeeb, Joshua Adenrele Ajiboye, Ugochukwu Daniel Okwor, Aminu Labaran Muhammad, Chikadibia Daniel Ugwuijem, Emmanuel Kenechukwu Agbo, and Victor Ikechukwu Stephen. "Optimizing energy storage for electric grids: Advances in hybrid technologies." *Management* 10 (2025): 11. https://doi.org/10.30574/wjaets.2025.14.2.0053
21. Dissanayake, Kavishka, and Dulsha Kularatna-Abeywardana. "A review of supercapacitors: Materials, technology, challenges, and renewable energy applications." *Journal of Energy Storage* 96 (2024): 112563. https://doi.org/10.1016/j.est.2024.112563
22. Kandasamy, Senthil Kumar, and Kannan Kandasamy. "Recent advances in electrochemical performances of graphene composite (graphene-polyaniline/polypyrrole/activated carbon/carbon nanotube) electrode materials for supercapacitor: A review." *Journal of Inorganic and Organometallic Polymers and Materials* 28 (2018): 559–584. https://doi.org/10.1007/s10904-018-0779-x
23. Samtham, Manopriya, Diwakar Singh, Krishnaiah Hareesh, and Rupesh S. Devan. "Perspectives of conducting polymer nanostructures for high-performance electrochemical capacitors." *Journal of Energy Storage* 51 (2022): 104418. https://doi.org/10.1016/j.est.2022.104418
24. Conway, Brian. E., and Wendy. G. Pell. "Power limitations of supercapacitor operation associated with resistance and capacitance distribution in porous electrode devices." *Journal of Power Sources* 105, 2 (2002): 169–181. https://doi.org/10.1016/S0378-7753(01)00936-3
25. Baptista, Joana Monteiro, Jagdeep S. Sagu, Upul Wijayantha Kg, and Killian Lobato. "State-of-the-art materials for high power and high energy supercapacitors: Performance metrics and obstacles for the transition from lab to industrial scale–A critical approach." *Chemical Engineering Journal* 374 (2019): 1153–1179. https://doi.org/10.1016/j.cej.2019.05.207
26. Rajagopal, Sivakumar, Rameez Pulapparambil Vallikkattil, Mohamed Ibrahim Mohamed, and Dimiter Georgiev Velev. "Electrode materials for supercapacitors in hybrid electric vehicles: Challenges and current progress." *Condensed Matter* 7, 1 (2022): 6. https://doi.org/10.3390/condmat7010006
27. Da Silva, Leonardo M., Reinaldo Cesar, Cássio M.R. Moreira, Jéferson H.M. Santos, Lindomar G. De Souza, Bruno Morandi Pires, Rafael Vicentini, Willian Nunes, and Hudson Zanin. "Reviewing the fundamentals of supercapacitors and the difficulties involving the analysis of the electrochemical findings obtained for porous electrode materials." *Energy Storage Materials* 27 (2020): 555–590. https://doi.org/10.1016/j.ensm.2019.12.015
28. Ali, Basant A., and Nageh K. Allam. "A first-principles roadmap and limits to design efficient supercapacitor electrode materials." *Physical Chemistry Chemical Physics* 21, 32 (2019): 17494–17511. https://doi.org/10.1039/C9CP02614B
29. Fic, Krzysztof, Mikolaj Meller, Jakub Menzel, and Elzbieta Frackowiak. "Around the thermodynamic limitations of supercapacitors operating in aqueous electrolytes." *Electrochimica Acta* 206 (2016): 496–503. https://doi.org/10.1016/j.electacta.2016.02.077
30. Saha, Moumita, Ambrish Kumar, Rahul Kanaoujiya, Kamalakanta Behera, and Shruti Trivedi. "A comprehensive review of novel emerging electrolytes for supercapacitors: Aqueous and organic electrolytes versus ionic liquid-based electrolytes." *Energy & Fuels* 38, 10 (2024): 8528–8552. https://doi.org/10.1021/acs.energyfuels.4c00685

31. Johns, Phil A., Matthew R. Roberts, Yasuaki Wakizaka, James H. Sanders, and John R. Owen. "How the electrolyte limits fast discharge in nanostructured batteries and supercapacitors." *Electrochemistry Communications* 11, 11 (2009): 2089–2092. https://doi.org/10.1016/j.elecom.2009.09.001
32. Pal, Bhupender, Shengyuan Yang, Subramaniam Ramesh, Venkataraman Thangadurai, and Rajan Jose. "Electrolyte selection for supercapacitive devices: A critical review." *Nanoscale Advances* 1, 10 (2019): 3807–3835. https://doi.org/10.1039/C9NA00374F
33. Muzaffar, Aqib, Ahamed M. Basheer, and Chaudhery Mustansar Hussain. "Green supercapacitors: Latest developments and perspectives in the pursuit of sustainability." *Renewable and Sustainable Energy Reviews* 195 (2024): 114324. https://doi.org/10.1016/j.rser.2024.114324
34. Ahmed, Shakeel, Priynka Sharma, Satyaranjan Bairagi, Nelson Pynadathu Rumjit, Shivani Garg, Akbar Ali, Chin Wei Lai, Seyyed Mojtaba Mousavi, Seyyed Alireza Hashemi, and Chaudhery Mustansar Hussain. "Nature-derived polymers and their composites for energy depository applications in batteries and supercapacitors: Advances, prospects and sustainability." *Journal of Energy Storage* 66 (2023): 107391. https://doi.org/10.1016/j.est.2023.107391
35. Luanwuthi, Santamon, Thanyapak Akkharaamnuay, Arisa Phukhrongthung, and Channarong Puchongkawarin. "Comparative environmental impact assessment of activated carbon electrodes for supercapacitors." *Green Chemistry* 26, 16 (2024): 9209–9219. https://doi.org/10.1039/D4GC02700K
36. Banik, Uchhwas, MD Tanvir Uddin Malik, SM Sultan Mahmud Rahat, and Abul K. Mallik. "Performance enhancement of ruthenium-based supercapacitors: A review." *Energy Storage* 5, 3 (2023): e380. https://doi.org/10.1002/est2.380
37. Ates, Murat, and Carlos Fernandez. "Ruthenium oxide–carbon-based nanofiller-reinforced conducting polymer nanocomposites and their supercapacitor applications." *Polymer Bulletin* 76, 5 (2019): 2601–2619. https://doi.org/10.1007/s00289-018-2492-x
38. Padha, Bhavya, Sonali Verma, Aamir Ahmed, Madhav P. Chavhan, Prerna Mahajan, and Sandeep Arya. "From trash to treasure: Crafting electrochemical supercapacitors with recycled waste materials." *Progress in Energy* 6, 1 (2024): 012005. https://doi.org/10.1088/2516-1083/ad139c
39. Chakraborty, Abhishek, Judith Gomes, Sunil Kumar, Kanchan Upadhyay, and Raunak K. Tamrakar. "Recycling, sustainability, ecological, economic, and safety aspects of supercapacitors." *Nanostructured Materials for Energy Storage* 4 (2024): 1607–1619. https://doi.org/10.1002/9783527838851.ch43
40. Thakur, Anukul K., Mandira Majumder, Archana S. Patole, and Shashikant P. Patole. "Commercial aspects, safety regulations, environmental and health impacts, and recycling strategies of supercapacitors" (2023). https://doi.org/10.1039/BK9781837672479-00477
41. Cossutta, Matteo, Viliam Vretenar, Teresa A. Centeno, Peter Kotrusz, Jon McKechnie, and Stephen J. Pickering. "A comparative life cycle assessment of graphene and activated carbon in a supercapacitor application." *Journal of Cleaner Production* 242 (2020): 118468. https://doi.org/10.1016/j.jclepro.2019.118468
42. Kamali, A. Kamal, Edis Glogic, Nilanka M. Keppetipola, Guido Sonnemann, Thierry Toupance, and Ludmila Cojocaru. "Prospective life cycle assessment of two supercapacitor architectures." *ACS Sustainable Chemistry & Engineering* 11, 44 (2023): 15898–15909. https://doi.org/10.1021/acssuschemeng.3c04007
43. Liu, Tianyu, and Yat Li. "Addressing the Achilles' heel of pseudocapacitive materials: Long-term stability." *InfoMat* 2, 5 (2020): 807–842. https://doi.org/10.1002/inf2.12105

44. Mahala, Siddharth, Kavian Khosravinia, and Amirkianoosh Kiani. "Unwanted degradation in pseudocapacitors: Challenges and opportunities." *Journal of Energy Storage* 67 (2023): 107558. https://doi.org/10.1016/j.est.2023.107558
45. Khosravinia, Kavian. "Addressing electrode-specific degradation in the production and performance of electrochemical energy storage systems" (2023). https://hdl.handle.net/10155/1663
46. Pameté, Emmanuel, Lukas Köps, Fabian Alexander Kreth, Sebastian Pohlmann, Alberto Varzi, Thierry Brousse, Andrea Balducci, and Volker Presser. "The many deaths of supercapacitors: Degradation, aging, and performance fading." *Advanced Energy Materials* 13, 29 (2023): 2301008. https://doi.org/10.1002/aenm.202301008
47. Chen, Xuecheng, Yuping Wu, and Rudolf Holze. "Ag(e)ing and degradation of supercapacitors: Causes, mechanisms, models and countermeasures." *Molecules* 28, 13 (2023): 5028. https://doi.org/10.3390/molecules28135028
48. Liu, Yinghui, Bénédicte Réty, Camélia Matei Ghimbeu, Benoît Soucaze-Guillous, Pierre-Louis Taberna, and Patrice Simon. "Understanding ageing mechanisms of porous carbons in non-aqueous electrolytes for supercapacitors applications." *Journal of Power Sources* 434 (2019): 226734. https://doi.org/10.1016/j.jpowsour.2019.226734
49. Lungoci, Carmen M., and Ioan D. Oltean. "Comparative analysis for the supercapacitors packaging characteristics." In *2010 IEEE 16th International Symposium for Design and Technology in Electronic Packaging (SIITME)*, 93–98. IEEE (2010). https://doi.org/10.1109/SIITME.2010.5653578
50. Cheng, Yan, Xiao, Kunming Pan, and Huan Pang. "Development and application of self-healing materials in smart batteries and supercapacitors." *Chemical Engineering Journal* 380 (2020): 122565. https://doi.org/10.1016/j.cej.2019.122565
51. Zhou, Shaopeng, Deping Chen, Baoyu Du, Pan Wang, Xiucai Wang, Wenbo Zhu, Si Liu, Peng Xiao, and Jianwen Chen. "Study of the in-situ test setup and analysis methods for self-healing properties of metallized film capacitors." *Review of Scientific Instruments* 95, 4 (2024). https://doi.org/10.1063/5.0194057
52. Sanyal, Simpy, SeonJu Park, Ramachandran Chelliah, Su-Jung Yeon, Kaliyan Barathikannan, Selvakumar Vijayalakshmi, Ye-Jin Jeong, Momna Rubab, and Deog Hawn Oh. "Emerging trends in smart self-healing coatings: A focus on micro/nanocontainer technologies for enhanced corrosion protection." *Coatings* 14, 3 (2024): 324. https://doi.org/10.3390/coatings14030324
53. Dissanayake, Kavishka, and Dulsha Kularatna-Abeywardana. "A review of supercapacitors: Materials, technology, challenges, and renewable energy applications." *Journal of Energy Storage* 96 (2024) 112563. https://doi.org/10.1016/j.est.2024.112563
54. Patel, Ashutosh, Suresh Kumar Patel, Raghvendra S. Singh, and R. P. Patel. "Review on recent advancements in the role of electrolytes and electrode materials on supercapacitors." *Discover Nano* 19 (1) (2024): 188. https://doi.org/10.1186/s11671-024-04053-1
55. Iqbal, Muhammad Faisal, Farooq Nasir, Fiza Shabbir, Zaheer Ud Din Babar, Muhammad Farooq Saleem, Kaleem Ullah, Nana Sun, and Faizan Ali. "Supercapacitors: An emerging energy storage system." *Advanced Energy and Sustainability Research* (2025): 2400412. https://doi.org/10.1002/aesr.202400412
56. Simon, Patrice, and Yury Gogotsi. "Capacitors." *Nature Materials* 7, 11 (2008): 845–854. https://doi.org/10.1038/nmat2297
57. Zhu, Yanfeng, et al. "Supercapacitors produced by activation of graphene." *Science* 332, 6037 (2011): 1537–1541. https://doi.org/10.1126/science.1200770
58. Wang, Xue, et al. "Composite films for high-performance supercapacitors." *Advanced Materials* 24, 24 (2012): 3451–3456. https://doi.org/10.1002/adma.201200455

59. Zhai, Yu, et al. "Carbon materials for chemical capacitive energy storage." *Advanced Materials* 23, 42 (2011): 4828–4850. https://doi.org/10.1002/adma.201100984
60. Bi, Sheng, Lisanne Knijff, Xiliang Lian, Alicia van Hees, Chao Zhang, and Mathieu Salanne. "Modeling of nanomaterials for supercapacitors: Beyond carbon electrodes." *ACS Nano* 18, 31 (2024): 19931–19949. https://doi.org/10.1021/acsnano.4c01787
61. Naguib, Michael, et al. "Two-dimensional nanocrystals produced by exfoliation of Ti_3AlC_2." *Advanced Materials* 23, 37 (2011): 4248–4253. https://doi.org/10.1002/adma.201102306
62. Zhang, Limin, et al. "Flexible supercapacitors based on paper substrates: A review." *Journal of Materials Chemistry A* 1, 21 (2013): 7031–7039. https://doi.org/10.1039/C5CS00174A.
63. Gao, Wei, et al. "Direct laser writing of micro-supercapacitors on hydrated graphene oxide films." *Nature Nanotechnology* 6, 8 (2011): 496–500. https://doi.org/10.1038/nnano.2011.110
64. Beidaghi, Majid, and Yury Gogotsi. "Capacitive energy storage in micro-supercapacitors: Recent advances and future challenges." *Energy & Environmental Science* 7, 3 (2014): 867–884. https://doi.org/10.1039/C3EE43526A
65. Conway, Brian E.*Electrochemical Supercapacitors: Scientific Fundamentals and Technological Applications*. Springer Science & Business Media (1999). https://doi.org/10.1007/978-1-4757-3058-6
66. Wang, Kecheng, et al. "Recent advances in solid-state flexible supercapacitors for portable electronics." *Energy Storage Materials* 24 (2020): 44–71. https://doi.org/10.1016/j.ensm.2019.07.024
67. Zhang, Li, and Xiu Song Zhao. "Carbon-based materials as supercapacitor electrodes." *Chemical Society Reviews* 38, 9 (2009): 2520–2531. https://doi.org/10.1039/B813846J
68. Sevilla, Marta, and Antonio B. Fuertes. "Sustainable porous carbons with a superior performance for CO_2 capture." *Energy & Environmental Science* 5, 3 (2012): 5636–5646. https://doi.org/10.1039/C0EE00784F
69. Pang, Shujie, et al. "Low-cost and environmentally Friendly electrode materials for supercapacitors: Recent progress and future perspectives." *Energy Storage Materials* 35 (2021): 191–211. https://doi.org/10.1016/j.ensm.2020.11.003
70. El-Kady, Maher F., and Richard B. Kaner. "Scalable fabrication of high-power graphene micro-supercapacitors for flexible and on-chip energy storage." *Nature Communications* 4 (2013): 1475. https://doi.org/10.1038/ncomms2446
71. Zhang, Kuan, et al. "Advances in scalable fabrication of electrodes for flexible supercapacitors: roll-to-roll, spray, and printing technologies." *Nano Energy* 92 (2022): 106761. https://doi.org/10.1016/j.nanoen.2021.106761
72. Li, Yue, et al. "Green electrolytes and their applications in supercapacitors: A review." *Journal of Power Sources* 476 (2020): 228870. https://doi.org/10.1016/j.jpowsour.2020.228870
73. Burke, Andrew. "Ultracapacitors: Why, how, and where is the technology." *Journal of Power Sources* 91, 1 (2000): 37–50. https://doi.org/10.1016/S0378-7753(00)00485-7
74. Lu, Xiaochuan, et al. "Hybrid electrochemical energy storage: Principles, progress, and future directions." *Chemical Reviews* 113, 8 (2013): 6557–6593. https://doi.org/10.1021/acs.chemmater.4c02203
75. Miller, John R., and Patrice Simon. "Electrochemical capacitors for energy management." *Science* 321, 5889 (2008): 651–652. https://doi.org/10.1126/science.1158736
76. Khalid, Waqas, et al. "A review on flexible and wearable supercapacitors based on carbonaceous nanomaterials." *Journal of Energy Storage* 51 (2022): 104401. https://doi.org/10.1016/j.est.2022.104401
77. Zhang, Xin, et al. "Supercapacitors for aerospace applications: A review." *Journal of Power Sources* 482 (2021): 228926. https://doi.org/10.1016/j.jpowsour.2020.228926

78. Salgado, Elkin A., et al. "Supercapacitor-based energy storage systems for IoT applications: Recent advances and challenges." *Sensors* 22, 9 (2022): 3486. https://doi.org/10.3390/s22093486
79. Zhang, Lu, Xiaohui Zhang, and Xiaobo Ji. "Flexible and stretchable supercapacitors for wearable electronics." *Advanced Materials* 32, 15 (2020): 1904752. https://doi.org/10.1002/adma.201904752

11 Synthesis of Metal–Organic Frameworks (MOFs)-Based Supercapacitors in Energy Storage

P. Chinnappan Santhosh, Suresh Jayakumar, and A. V. Radhamani

11.1 INTRODUCTION

The global energy and environmental crises have accelerated efforts to develop sustainable, high-performance energy storage systems. In 2015, the United Nations launched the 2030 Agenda for Sustainable Development, which features 17 goals to promote inclusive growth and address major global issues. Goals 7: Affordable and Clean Energy and 13: Climate Action are particularly linked to advancing energy solutions like batteries and supercapacitors, which facilitate renewable power use and help reduce greenhouse gas emissions. The Intergovernmental Panel on Climate Change (IPCC) reports that the Earth's average surface temperature has increased by about 1.1°C since pre-industrial times due to fossil fuel use, resulting in severe climate disruptions such as heat waves, flooding, and rising sea levels. Thus, clean, safe, and efficient energy storage systems mitigate these impacts and support a sustainable energy transition [1]. Supercapacitors are characterized by internal resistances typically ranging from 5 to 10 mΩ, allowing for high current flow and faster charging while generating minimal heat, an issue often associated with conventional batteries. Notably, their operational stability under extreme temperatures, particularly in sub-zero environments, provides a substantial advantage for applications in harsh climates [2, 3]. Supercapacitors are categorized into three types based on charge storage behavior: Electric Double-Layer Capacitors (EDLC), pseudocapacitors, and hybrid capacitors. EDLCs use a physical mechanism with electrostatic ion accumulation at the electrode–electrolyte interface, without electron transfer or redox reactions. Carbon materials like Activated carbon (AC), Carbon Nanotube (CNT), graphene, and reduced graphene oxide (rGO) are common in EDLCs due to their conductivity and surface area. Although EDLCs offer cyclic stability and rapid charge–discharge rates, their energy density is limited by a lack of faradaic processes.

In contrast, pseudocapacitors utilize fast, reversible redox reactions at or near electroactive surfaces, including transition metal oxides (e.g., MnO_2, NiO, RuO_2) and conducting polymers (e.g., PANI, PPy). These faradaic processes provide higher

DOI: 10.1201/9781003535836-11

capacitance and energy storage than EDLCs but often result in reduced cycle life and rate capability from structural changes during cycling. Depending on the electrode composition, hybrid systems can be further classified into battery-type hybrids, composite-type hybrids, and asymmetric configurations, each optimized for specific performance criteria [4]. Among the various materials studied, porous materials have demonstrated increased electrochemical performance. This is due to their structural advantages, such as an enhanced surface area, which increases reaction kinetics. Metal–organic frameworks (MOFs), known as porous coordination polymers (PCPs), are crystalline structures with two-dimensional or three-dimensional porous architecture [5]. It was discovered in the 1990s and utilized in various fields due to its high surface area (1000–10,000 m^2/g), surpassing other porous materials, such as zeolites and carbon-based materials. It is formed by self-assembling metal nodes (ions or clusters) with negatively charged organic linkers, which act as a bridge [6]. This chapter explores various synthesis methods of MOF and MOF-based electrodes, primarily focusing on recent supercapacitor advancements. In addition, it discusses future perspectives and challenges related to the MOF-based electrode in supercapacitor applications.

11.2 SYNTHESIS OF MOF

Over the past two decades, the synthesis of MOFs has gained significant attention due to their ability to form various nanostructures with tunable morphology. MOFs are generally synthesized using two methods: first, the metal salt and ligand are prepared separately and then combined, or the metal salt and ligand solution is dissolved in the solvent. The key objective of MOF synthesis is to create specific inorganic building units without degrading the integrity of the organic ligand [7]. The MOF can be synthesized by various techniques, such as hydrothermal or solvothermal synthesis, microwave-assisted synthesis, electrochemical synthesis, and sonochemical synthesis, as shown in Figure 11.1.

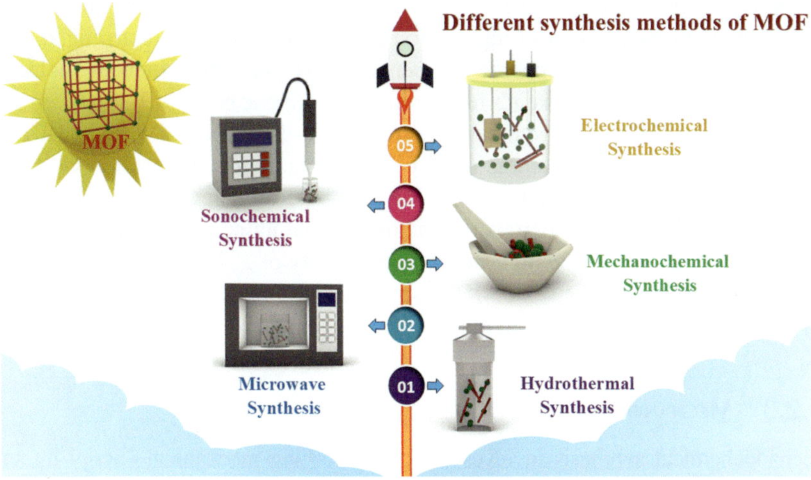

FIGURE 11.1 Illustration of the various synthesis methods of MOF.

11.2.1 Solvothermal or Hydrothermal Synthesis

One of the effective ways to synthesize MOFs is by using the hydrothermal or solvothermal approach, due to its repeatability, scalability, and simplicity. Fundamentally, both hydrothermal and solvothermal synthesis involve heating the precursor molecules with the solvents in a PTFE-linked autoclave. When water serves as a solvent in the synthesis process, the method is referred to as the hydrothermal method. A solvent other than water, such as ethanol, N,N-dimethylformamide, or toluene, is known as solvothermal synthesis. The main advantages of these techniques are their ability to produce MOF with various morphologies by varying reaction time, temperature, and solvent volume.

Additionally, these processes can be easily scaled up for large-scale productions. Recently, Feng et al. used the one-pot hydrothermal method to synthesize hollow carbon nanospheres (HCNS)-embedded Ni nanoparticles (Ni NPs) anchored on rGO conductive matrix. The precursors are dissolved in DMF/ethanol solvent, transferred into Teflon, treated at 160°C for 16 h, and calcinated at 1000°C under an argon atmosphere. The MOF-derived HCNS@Ni NP/rGO achieved a superior performance of 1589 F/g. It also delivers a capacity retention of 46.2% at a high current density of 50 A/g [8].

11.2.2 Microwave Synthesis

Microwave synthesis is emerging as one of the most effective approaches to synthesizing MOFs, offering significant advantages such as reducing reaction times from hours to seconds and decreasing power consumption. It also provides precise control over the surface morphology, size, and phase of the MOF during synthesis. Microwave radiation lies between radio and infrared frequencies and relies on two mechanisms for operation: dipole rotation and ionic conduction. In this synthesis process, the solution is prepared with a suitable solvent, then transferred to Teflon and placed inside the microwave for the required temperature and duration. The microwave reactor converts the electromagnetic wave into thermal energy by interacting with the permanent dipole moment, as these dipoles are arranged in the electromagnetic fields. At specific frequencies, this interaction induces molecules collisions, increasing the system's temperature and kinetic energy [9, 10]. Chen et al. utilized microwave-assisted synthesis to create zinc-doped Ni-MOF using HCl as a structural modulator. The authors mixed the precursors for the reaction with N,N-dimethylformamide (DMF) and added HCl drop by drop. They then transferred this mixture into a household microwave reactor and maintained it for 6 min. The synthesized MOF exhibited the highest specific capacity of 237.4 mAh/g at 1 A/g. The group also synthesized the same MOF using the hydrothermal method, demonstrating a lower particular capacity than the microwave-assisted MOF [11].

11.2.3 Mechanochemical Synthesis

Mechanochemical synthesis involves directly absorbing mechanical energy by solid precursors during grinding or milling processes. In the ball milling process, the energy required to initiate the chemical reaction is generated through friction between the milling balls and the chemical precursors. High-impact forces are necessary to break bonds and drive chemical reactions, while low-impact forces may lead to elastic deformation without triggering a chemical reaction. Early mechanochemical synthesis relied on manual mortar and pestle techniques; however, modern advancements have introduced highly precise ball mills that enable consistent and controlled synthesis. The main advantage of this technique is the absence of harmful solvents. There are three approaches: neat grinding (NG), which uses only solid precursors without solvents; to enhance the efficiency of NG, hydrated reactants are added, producing water as a byproduct and consequently increasing ion mobility—a condition known as liquid-assisted grinding (LAG). Advancing LAG, ion and LAG techniques have been developed, wherein both a solvent and a salt are added during the milling process. This dissolves solid precursors and promotes a homogeneous reaction environment, improving overall reaction efficiency [12]. Recently, Hameed et al. prepared MOF-derived $NiCo_2O_4$ using three different organic linkers. Among these, the 1,3,5-benzene tricarboxylic acid (H_3BTC) linker used in the MOF achieved the highest specific capacitance of 913.5 F/g. Interestingly, the prepared ASC exhibits the highest energy density of 73.83 Wh/kg, and after 10,000 charge–discharge cycles, the device regained 98.6% of its initial capacitance [13].

11.2.4 Electrochemical Synthesis

Compared to other synthesis processes, electrochemical synthesis holds a significant position due to its short duration and mild reaction conditions during the synthesis process [14]. MOF can be deposited over the substrate using three approaches: (1) Electrode superficial nucleation (ESN), (2) Indirect bipolar electrodeposition (IBED), and (3) Electrophoretic deposition (EPD). The ESN operates by applying an external voltage in the circuit, causing ions in the solution to move toward the electrode and form MOF nucleation. This method is divided into two types: anode dissolution and cathode deposition. In IBED, the particles become polarized when a voltage is applied between the conductive particles released into the electrolyte solution. As a result, oxidation occurs on the positively polarized side, releasing metal ions that react with negatively charged ligands to form MOF. Unlike the other two approaches, EPD consists of two steps: pre-synthesis of MOF and deposition on the substrate using an electric field [15]. Recently, Khan et al. reported that NiMn-terephthalic acid MOF (NMT MOF) and NMT MOF/MnS were synthesized using the electrodeposition method. The SEM of NMT MOF shows a less dense, flake-like structure, and as the deposition time of MnS increases, the density of the flakes also increases [16].

11.2.5 Sonochemical Synthesis

This synthesis technique used mechanical vibration frequencies (20 kHz to 10 MHz) to synthesize MOF. These ultrasound waves interact with a solution containing precursors to synthesize MOF, generating high- and low-pressure zones. The low-pressure zones fall below the solution's vapor pressure, forming small bubbles, also known as cavities. As these cavities grow and eventually collapse, that is a phenomenon known as the cavitation process. This process, which involves bubble formation, expansion, and implosion, releases significant amounts of energy. The temperature can reach approximately 5000 K, and the pressure can exceed 1000 bar. The key aim of this synthesis method is to develop a fast, energy-efficient process at ambient temperature while synthesizing [17]. In this context, Shakeel et al. synthesized Cu-MOF using hydrothermal and sonochemical methods, comparing their electrochemical performance. In the hydrothermal method, methanol is used as a solvent, pyridine-4-carboxylic acid serves as an organic linker, and copper chloride ($CuCl_2$) acts as a precursor, which are blended to create a homogeneous solution. This mixture was then placed in an autoclave for 12 h at 180°C. For the sonochemical method, $CuCl_2$ was dissolved in deionized water, and pyridine-4-carboxylic acid was added dropwise during sonication. The sonication was conducted at 40% amplitude and a power of 600 watts, lasting 45 min at 30°C, with a cyclic interval of 2 s of sonication followed by 1 s of rest. The sonochemically synthesized Cu-MOF displayed thin flakes compared to those produced by the hydrothermal method. Consequently, it achieved the highest specific capacity of 594.2 C/g, outperforming the hydrothermal counterpart, which showed 462.0 C/g. The ASC device prepared using the sonochemically synthesized Cu-MOF yields an energy density of 74.9 Wh/kg and a power density of 13765 W/kg [18].

11.3 MOF-BASED ELECTRODES FOR SUPERCAPACITORS

In supercapacitors, MOF-based electrodes are considered promising materials because they possess a large surface area, tunable porosity, and redox-active metal centers. These distinctive characteristics facilitate efficient ion transport and charge storage, making MOFs well-suited for high-performance energy storage applications. Among the various MOFs, those incorporating Cu, Ni, Co, Zn, and Mn have been extensively researched, thanks to their multiple oxidation states, exceptional electrochemical activity, and strong metal–ligand interactions. The metal ions in these frameworks participate in faradaic redox reactions, significantly enhancing the charge storage capacity of MOF electrodes. Copper, nickel, and cobalt demonstrate vigorous redox activity, which promotes high capacitance, while zinc and manganese contribute to structural stability and environmental compatibility. The amalgamation of these properties makes them ideal candidates for supercapacitor applications.

11.3.1 Ni-MOF-based Electrode for Supercapacitors

Nickel-based MOFs are attractive in supercapacitor research due to their properties, including impressive specific capacitance, stability, and efficiency. Nickel is not only widely accessible and cost-effective but also readily forms highly coordinated structures, making it an excellent choice for MOF-based electrodes. Their robust redox activity sets Ni-MOFs apart, which enhances charge storage and energy conversion efficiency. Additionally, their thermal and chemical stability ensures long-term durability, a crucial characteristic for practical applications. However, one of the primary challenges associated with MOFs is their limited electrical conductivity, which can hinder overall performance. Fortunately, researchers are addressing this issue by modifying their structures and combining them with conductive materials.

A. Gowdhaman et al. successfully synthesized NiO/Ni–rGO composites through hydrothermal processes followed by calcination at 400°C in an argon atmosphere. To optimize performance, the materials were prepared with different ratios of Ni-MOF to GO. The 1:0.5 ratio demonstrated the highest specific capacity of 649.22 C/g at a current density of 3 A/g, significantly surpassing that of pure NiO (122.55 C/g at a current density of 4 A/g). This enhancement is attributed to the synergistic effect between NiO/Ni nanoparticles and rGO, which enhances electron transport and ion diffusion, leading to superior charge storage. Additionally, after 5,000 cycles, the composite material retained 81.1% of its original capacity at 20 A/g, which is notably higher than the retention value of 63% observed for pure NiO [19].

Further, X. Zhang, S. Yang, W. Lu, and colleagues synthesized $Ti_3C_2T_x$/Ni-MOF microbelts using an MXene-induced method, where the MXenes served both as structure-directing agents and conductive substrates. The $Ti_3C_2T_x$ was prepared through HF etching, followed by the refluxing of $Ni(NO_3)_2{\cdot}6H_2O$ and terephthalic acid in a mixed solvent under argon protection, with the addition of $Ti_3C_2T_x$ for composite formation. The Ni-MOF microbelts show the performance of 786 F/g. In comparison, the $Ti_3C_2T_x$/Ni-MOF composites achieved an impressive 1124 F/g at a current density of 1 A/g, showcasing a significant enhancement due to the synergistic effects of the MXenes and Ni-MOFs. This composite exhibited better rate capability, retaining 62% at 20 A/g, compared to 41% for the Ni-MOFs alone. Electrochemical impedance spectroscopy revealed a reduced charge-transfer resistance of 9.35 Ω for the $Ti_3C_2T_x$/Ni-MOFs, enhancing redox kinetics. Moreover, the composite maintained 83.2% capacitance retention after 4000 cycles, significantly outperforming the Ni-MOFs, which retained only 50.4%, as shown in Figure 11.2. The improved performance is attributed to the high conductivity, plentiful redox sites, and optimized ion transport pathways afforded by the 2D microbelt structure and the integration of MXenes [20]. Another promising approach to enhance the performance of Ni-MOF involves elemental doping. For example, Li et al. synthesized Mo-doped Ni-MOF materials using a solvothermal method, meticulously adjusting the Ni:Mo ratio (1:0, 1:1, 1:2, 1:3) to investigate the impact of Mo on electrochemical performance. Among the synthesized materials, the M-NMN-1 electrode (Ni:Mo = 1:1) showed excellent electrochemical performance, delivering a specific capacity of 802 C/g, which represents a significant 84% enhancement over the undoped Ni-MOF (435.6 C/g). Even at higher current densities, M-NMN-1 maintained its superior performance, reaching 416 C/g at 20 A/g, while Ni-MOF showed a lower capacity of 193.2 C/g, and M-NMN-2 reached only 341.8 C/g. However, increasing the Mo ratio beyond 1:1 resulted in diminishing returns, as excessive Mo led to material agglomeration, which obstructed active sites and degraded performance. What makes M-NMN-1 particularly effective? The presence of Mo clusters contributes to stabilizing the Ni-MOF structure, while the self-assembled nanosheets and nanorods facilitate efficient charge transport pathways. When assessed in a battery–supercapacitor hybrid device, M-NMN-1 achieved an impressive energy density of 59 Wh/kg at 802 W/kg, along with outstanding cyclic stability of 93% over 20,000 cycles, significantly outperforming the other compositions [21].

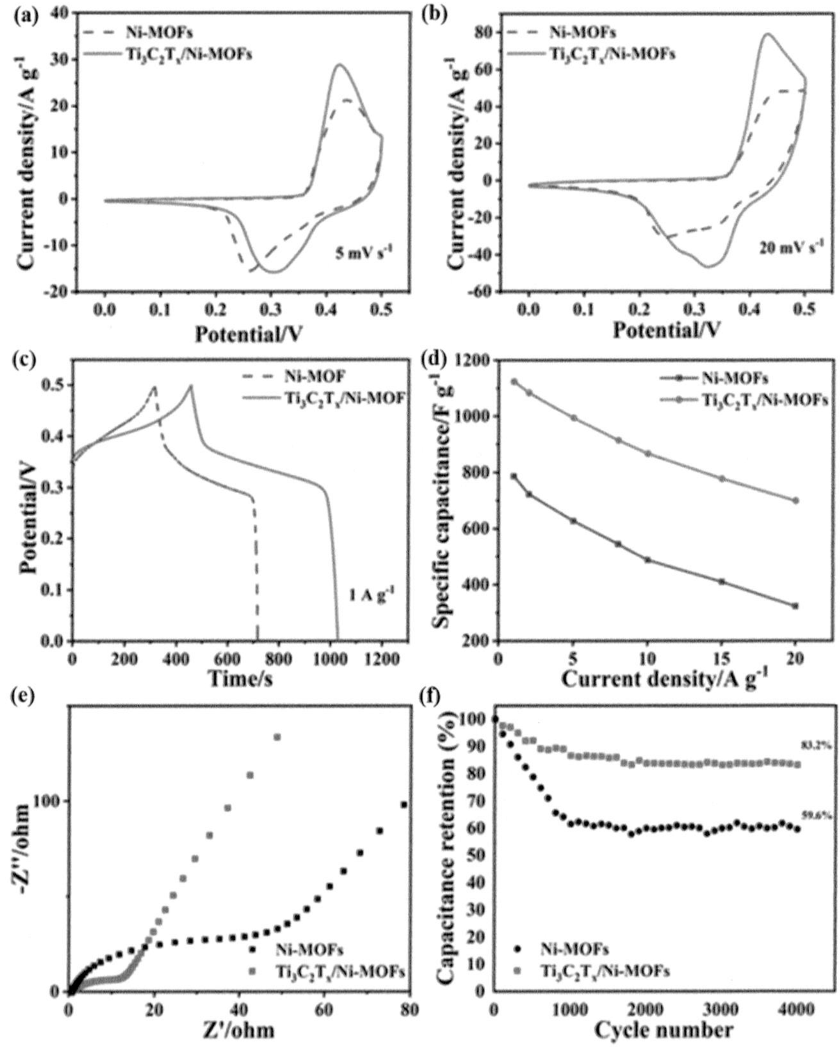

FIGURE 11.2 Presentation of the electrochemical performance of the synthesized Ni-MOFs and $Ti_3C_2T_x$/Ni-MOFs: (a and b) CV curves at scan rates of 5 and 20 mV/s; (c) GCD profiles at a current density of 1 A/g; (d) rate capability at various current densities; (e) Nyquist plots; and (f) stability test conducted at a current density of 10 A/g. (Reprinted with permission from [20]. © Elsevier 2021.)

11.3.2 Mn-MOF-based Electrode for Supercapacitors

Manganese oxides (MnO, Mn_3O_4, Mn_2O_3, MnO_2) are abundant, cost-effective, and environmentally friendly, positioning them as promising candidates for supercapacitor electrodes. Manganese-based MOFs demonstrate outstanding charge storage capabilities; however, their development is limited by challenges such as restricted ligand compatibility with Mn^{2+} and instability in aqueous electrolytes. Nevertheless, pursuing higher energy density has sparked increasing interest in Mn-MOFs as advanced electrode materials. Rong et al. developed an innovative 3D Mn-based MOF, Mn-BGPD, by integrating Mn^{2+} with isoindolediacetic acid and various isomers of H_2L_1 as organic linkers. This material demonstrated exceptional electrochemical properties, achieving a good performance of 832.6 F/g at a current density of 1 A/g and maintaining 102.2 F/g at 8 A/g within a three-electrode system. When combined with rGO in an ASC, the material reached an energy density of 35.5 Wh/kg, and after 4000 cycles, the cyclic stability is measured to be 46.7% [22]. Similarly, P.-P. Sun et al. successfully synthesized a robust 3D Mn-MOF using hydrothermal. This framework boasts an orthorhombic structure (Pbcn space group) and has the formula $C_{10}H_{10}Mn_3N_4O_{12}$. A small amount of copper powder was incorporated to enhance its electrochemical performance, forming Mn-MOF-Cu. This modified material demonstrated significantly improved conductivity, specific capacitance, and cycling stability. The electrode achieved a remarkable electrochemical performance of 1606 F/g at a current density of 0.5 A/g, surpassing that of the pristine Mn-MOF, which had a capacitance of 1106 F/g, representing a 45% increase. Furthermore, Mn-MOF-Cu maintained the cyclic retention of 83.73% for 10,000 cycles, underscoring its exceptional, prolonged durability. The enhanced charge transport is primarily attributed to copper's role in optimizing the conductivity of the 3D Mn-MOF, thereby establishing it as a highly effective electrode material for supercapacitors [23].

11.3.3 Co-MOF-based Electrode for Supercapacitors

Due to their porosity and tunable architectural design, cobalt-based MOFs have gained prominence as key candidates for supercapacitor applications. Since their initial utilization by Lee et al. in 2012, extensive research has been conducted to evaluate their electrochemical performance, particularly in lithium hydroxide (LiOH) electrolytes. Co-MOF films demonstrate impressive pseudocapacitive behavior, achieving the electrochemical performance of 206.76 F/g in a 1 M LiOH solution. Furthermore, these materials exhibit an energy density of 7.18 Wh/kg with remarkable cyclic retention, with only a 1.5% loss in their initial capacitance observed for 1000 cycles [24]. Recently, Kshetri et al. developed a flexible and highly conductive Co-MOF electrode by integrating it onto an MXene-carbon nanofiber mat (MX-CNF). This innovative architecture was then transformed into Co-PC@MX-CNF. This capacitive material demonstrated excellent performance, with a specific capacitance of 426.7 F/g at 1 A/g, owing to its high surface area, wettability, and conductivity. To fully harness its potential, Co-PC@MX-CNF was combined with MnO_2@Co_3O_4-PC@MX-CNF to create a flexible wearable hybrid supercapacitor (FW-HSC). This device achieved an impressive energy density of 72.5 Wh/kg at 832.4 W/kg while maintaining 90.36% capacitance retention over extended cycling periods. Notably, two FW-HSC cells effectively operate a digital clock and light up an LED, as shown in Figure 11.3 [25].

Further, H. Zhang, J. Wang, Y. Sun et al. successfully synthesized a nanowire microsphere Co-MOF through a one-step solvothermal method, eliminating the need for calcination and significantly enhancing the electrochemical performance compared to traditional Co-MOFs. Unlike the conventional prismatic cobalt–benzene-1,3,5-tricarboxylate

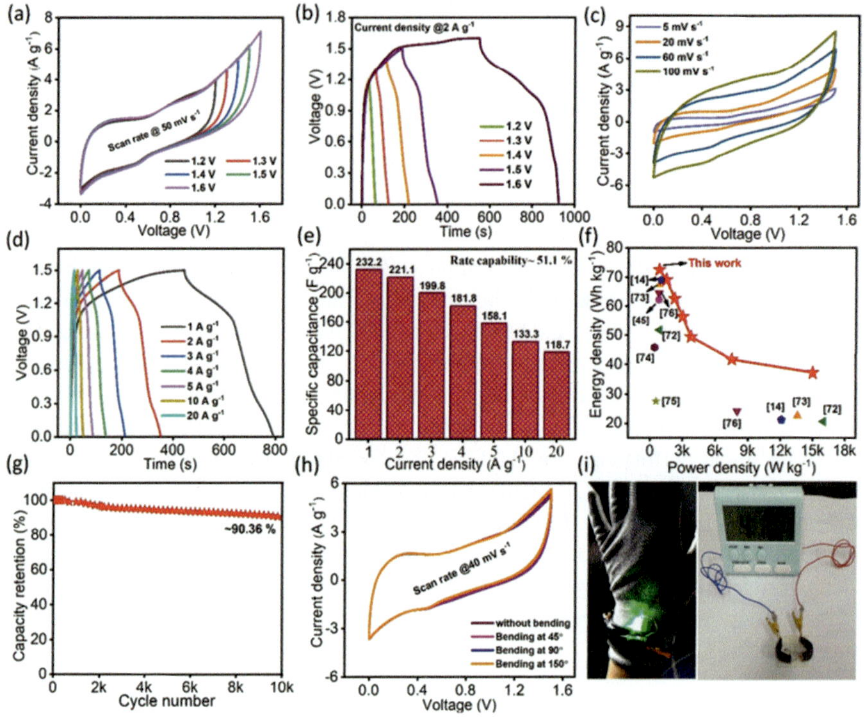

FIGURE 11.3 (a and b) The figure displays the CV and GCD curves for the flexible solid-state supercapacitor, (c and d) illustrating the CV and GCD plots at different scan rates and current densities. Part (e) shows specific capacitance with different current densities. Part (f) shows the Ragone plot. Part (g) depicts the stability curve. Part (h) illustrates CV curves at a scan rate of 40 mV/s, both before and during bending. Part (i) demonstrates the device powering a green LED (left) and a digital clock (right). (Reprinted with permission from [25]. © Elsevier 2022.)

(Co–BTC), the newly developed material exhibits a unique wire-spherical morphology, offering increased redox-active sites and improved ion transport, which results in superior capacitance and stability. The transformation of MOFs into a nanowire (1D) morphology effectively elevates reaction sites and electron mobility; when these nanowires entangle to form microspheres, the mechanical and chemical stability of the material is further enhanced. This innovative product is created by modifying the Co–BTC method, incorporating a saturated urea solution, and raising the reaction temperature to 180°C. The final product is a purple–red powder, obtained after filtration, washing, and drying. SEM confirms that, in contrast to the prismatic Co–BTC, this novel material consists of microspheres composed of intertwined nanowires, as shown in Figure 11.4. Electrochemical tests reveal that the specific capacity of the nanowire microsphere-shaped Co–BTC (CBNWM) reached 1020 F/g at 5 mV/s, and cyclic stability of 81.4%. Moreover, a flexible ASC utilizing CBNWM and AC as electrodes exhibits an energy density of 34.4 Wh/kg at a power density of 375.3 W/kg, demonstrating its capability to power a thermometer and hygrometer for 26 min after just 30 sec of charging [26].

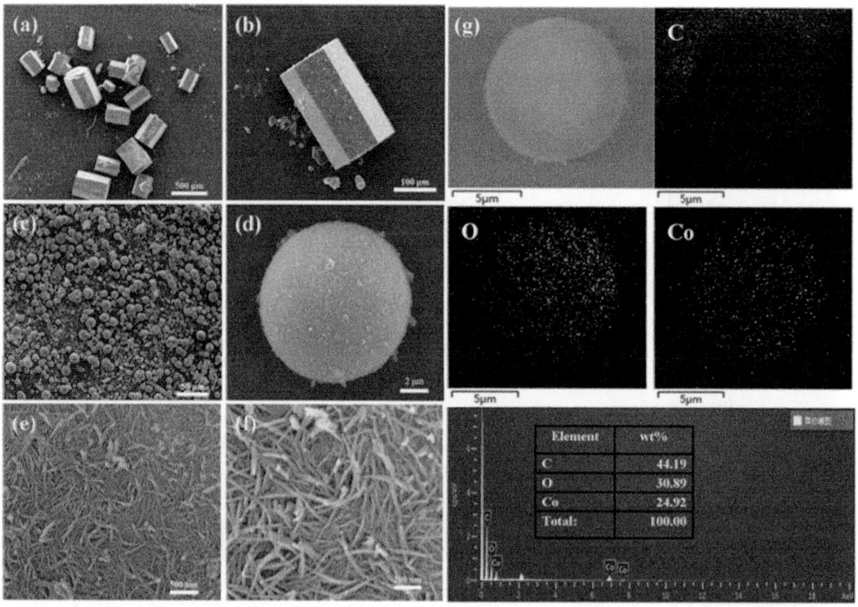

FIGURE 11.4 The SEM images of cobalt benzene tricarboxylate (CBC) are shown in (a) and (b). In contrast, the images of nanowire microsphere-shaped cobalt benzene tricarboxylate (CBNWM) are depicted in (c), (d), (e), and (f). The energy-dispersive X-ray spectroscopy (EDS) mapping of the nanowire microsphere-shaped cobalt benzene tricarboxylate is illustrated in (g). (Reprinted with permission from [26]. © Elsevier 2021.)

11.3.4 Cu-MOF-based Electrode for Supercapacitors

M.K. Singh et al. prepared a novel Cu-based MOF known as HMRL-1, which was created using a solvothermal method. This method utilized a tetracarboxylic linker, H_4L (5,5′-MDIP), and Cu^{2+} salt, as illustrated in Figure 11.5. The resulting 3D porous framework features redox-active Cu^{2+} centers and ion transport channels. The framework was composited with rGO through ultrasonication for enhanced performance in supercapacitor applications. The HMRL-1/rGO composite exhibited hybrid capacitive behavior, combining the EDLC provided by rGO and the pseudocapacitance from the Cu-MOF. This synergistic interaction resulted in an 82% increase in specific capacitance compared to rGO alone, achieving 366 F/g at 1 A/g while maintaining good cyclic stability. Furthermore, a solid-state supercapacitor device fabricated with the HMRL-1/rGO composite exhibited a maximum energy density of 57.2 Wh/kg, retaining 46.9% of its capacitance at a current density of 10 A/g [27].

Recently, J. Khan et al. investigated the potential of isonicotinic acid-based Cu-MOF as an innovative electrode material for hybrid supercapacitors, particularly emphasizing its charge storage characteristics and electrochemical performance. As verified by structural analyses, the authors successfully achieved a nanoflower-like morphology through hydrothermal synthesis. Further, Electrochemical studies conducted using a three-electrode system revealed a predominant redox-active nature, contributing to 93.7% Faradic charge storage, alongside a remarkable specific capacitance of 781.87 F/g.

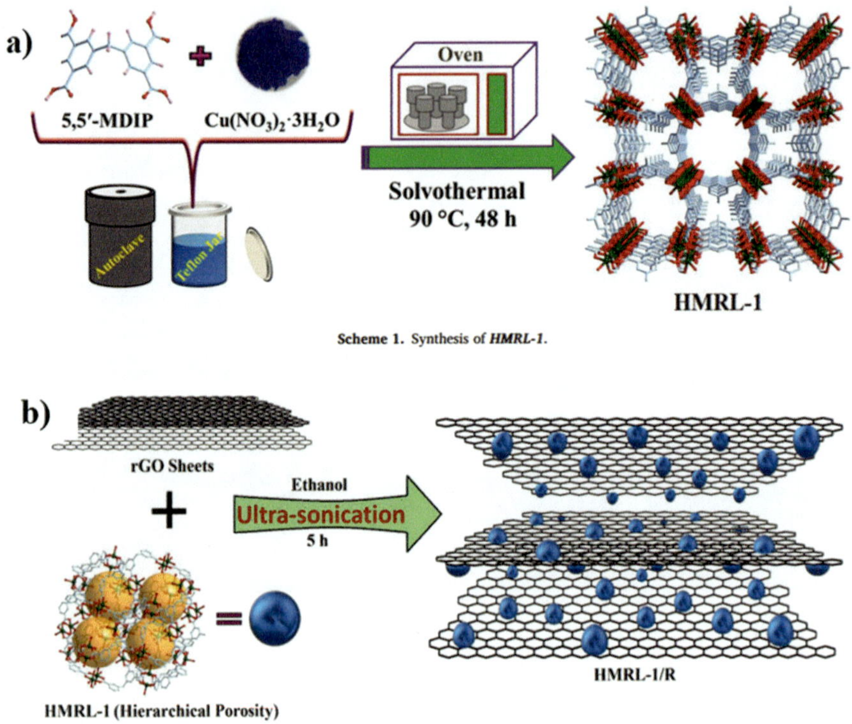

FIGURE 11.5 Illustration of the synthesis process of HMRL-1/rGO. (Reprinted with permission from [27]. © Elsevier 2021.)

Furthermore, ASC devices were assembled using Cu-MOF and activated carbon as electrodes, demonstrating impressive energy storage capabilities. It delivered a maximum energy density of 48.50 Wh/kg at a low current density of 1.3 A/g; even at 10 A/g; it exhibited an energy density of 33.32 Wh/kg. Moreover, this device offered a notable power density of 8667 W/kg and strong durability, retaining 93.47% of its capacity at 8 A/g after 5000 cycles (Table 11.1). Analysis of the simulated charge storage mechanism revealed a predominantly diffusion-controlled process, accounting for 89.69%. This study highlights the importance of the distinctive coordination of copper centers with isonicotinic acid, which facilitates enhanced charge transfer and structural stability, thereby positioning Cu-MOF as a promising candidate for high-performance supercapacitors. These findings pave the way for advancements in optimizing MOF-based electrodes, addressing critical energy density and conductivity challenges, and herald the development of next-generation energy storage solutions [28].

11.3.5 Zn-MOF-based Electrode for Supercapacitors

Due to their large surface area, tunable porosity, and exceptional structural stability, Zn-MOFs have become standout materials for supercapacitor electrodes. Initially introduced by Yaghi et al. in 1995, Zn-MOFs are composed of Zn^{2+} ions coordinated with

TABLE 11.1
The Electrochemical Performances of Various MOFs

MOF	Ligand	Specific Capacitance	Stability	Ref.
Ni-MOF@PPy	1,4-Benzenedicarboxylic acid (PTA)	1815.4 F/g @ 1 A/g	90.2% for 3000 cycles @ 10 A/g	[31]
$NiCo_{2-x}Mn_xO_4$@ Ni-MOF	Benzenedicarboxylic acid (PTA)	3543.5 F/g @ 6 mA/g	93.7% for 15,000 cycles @10 mA/g	[32]
MoS_2/Mn-MOF/ MWCNTs	Terephthalic acid	862.7 F/g @ 1 A/g	71.4% for 5000 cycles @ 8 A/g	[33]
Graphene/ZIF-9/ (CoCu)	Benzoimidazole	2668 F/g @ 1 A/g	100% for 1200 cycles @ 20 A/g	[34]
e-Fe-doped Co MOF	2-Methylimidazole	700 C/g @ 1 A/g	88% for 10,000 cycles @ 20 A/g	[35]
Cu(I) MOF	2-(1*H*-1,2,4-triazol-1-yl) pyridine (2TzPy)	508 C/g @ 1 A/g	96.5% for 10,000 cycles @ 10 A/g	[36]
rGO/Cu-BDC MOF	1,4-Benzene dicarboxylate (H_2BDC)	656.4 mAh/g @ 2 A/g	91.7% for 10,000 cycles @ 6 A/g	[37]
Zn-MOF	1,3-5 Benzenetricarboxylic acid	683 F/g @ 0.5 A/g	111% for 1000 cycles @ 0.5 A/g	[30]

various organic linkers, such as trimesic acid (H_3BTC), 2-methylimidazole (H-meIM), and terephthalic acid (BDC). These frameworks' robust coordination, bonding, and hierarchical pore structure significantly enhance ion transport and charge storage capabilities. When compared to alternative MOFs, Zn-MOFs exhibit superior chemical stability and adaptability, positioning them as ideal candidates for the development of metal oxide and carbon-based electrodes, which ultimately contribute to high-performance supercapacitor applications [29]. Recent studies, including H. Esfandian and M. M. Sadegh et al., have investigated the potential of Zn-MOFs through mixed-linker strategies to enhance their electrochemical properties. By varying the ratios of two ligands during synthesis, researchers have been able to tailor the porosity and surface area of these frameworks, which significantly influences their electrochemical performance. A total of five different variants of Zn-MOF (designated as MOF-0 to MOF-4) were synthesized and assessed using electrochemical measurements in a 6 M KOH electrolyte. Among these, MOF-1, with an optimal ligand composition, demonstrated outstanding performance, achieving a specific capacitance of 683 F/g at a current density of 0.5 A/g, surpassing the specific capacitance of the other variants in terms of charge storage capability and cycling stability [30].

11.3.5.1 Challenges and Future Perspective

Despite their numerous advantages, MOFs remain hindered by some challenges to their large-scale production use as a supercapacitor and electrode material.

- Many pristine MOFs show low electronic conductivity, which significantly limits their use as electrodes, despite their advantages like porous structure and surface area.

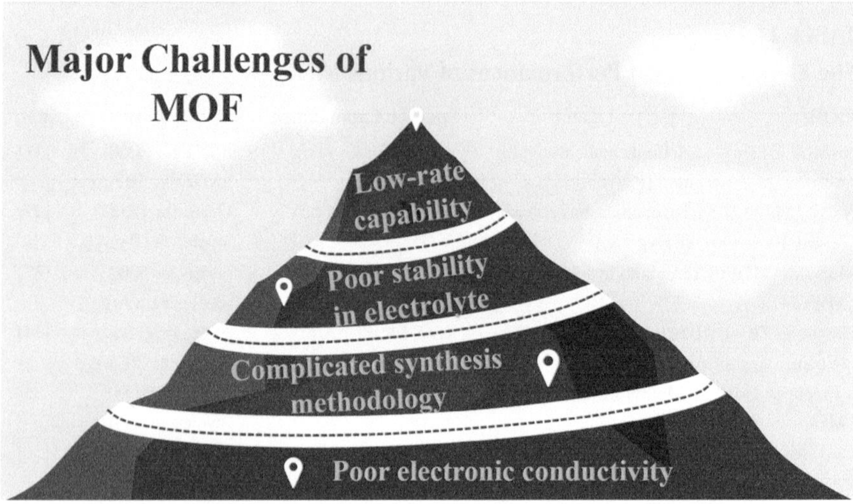

FIGURE 11.6 Illustration of the significant challenges associated with MOF.

- The pristine MOF and their composites are not stable in electrolytes and require enhancement for long-term electrochemical stability.
- Numerous MOFs are not engineered for supercapacitor application, which lacks the important electrochemical properties required for energy storage.
- Several MOF-derived materials show low-rate capability; it must be improved to meet the high energy density and power density demands.
- The synthesis of many MOFs relies on complex procedures and the use of costly ligands, which can hinder large-scale production (Figure 11.6).

However, MOF remains one of the promising classes of electrodes for energy storage applications. Therefore, the research community has proposed many solutions to address these limitations. The following section explains some of the important approaches.

- One practical approach to overcoming poor conductivity is to combine it with high-conductivity materials, such as rGO, CNTs, and other conductive additives.
- Introducing heteroatoms like N, S, and P to enhance the electrochemical performance of MOF-based materials. The enhancement is due to the defects created by the dopants and MOF.
- The appropriate electrolyte is essential for achieving high specific capacitance and long-term stability.
- In-situ characterization studies and theoretical modeling are essential to better understand the electrochemical mechanism of MOF. These provide valuable insight into MOF-based supercapacitors.
- The fabrication of 2D films using MOF increases the active sites for electrochemical reactions.

- For commercial applications, a large quantity of electrode materials is required, and a straightforward and cost-effective synthesis technique is necessary, such as one-pot synthesis.

11.4 AUTHOR AI DECLARATION

The authors hereby declare the use of generative AI and AI-assisted technologies in the writing process. While preparing this work, the authors used the ChatGPT tool/service to rephrase several statements in the manuscript. All AI-generated content was reviewed and edited by the authors as needed. The authors take full responsibility for the content of the publication.

ACKNOWLEDGMENT

The Editors acknowledge the use of Grammarly as a tool to assist in editing this book chapter. The suggestions made by Grammarly were carefully considered and incorporated where appropriate, but the author is solely responsible for the final content.

REFERENCES

1. https://www.wri.org/insights/2023-ipcc-ar6-synthesis-report-climate-change-findings, Last modified March 20, 2023.
2. S. Jayakumar, P.C. Santhosh, M.M. Mohideen, A.V. Radhamani, A comprehensive review of metal oxides (RuO_2, Co_3O_4, MnO_2 and NiO) for supercapacitor applications and global market trends, J Alloys Compd 976 (2024) 173170. https://doi.org/10.1016/j.jallcom.2023.173170.
3. P. Chinnappan Santhosh, S. Jayakumar, M.M. Mohideen, A.V. Radhamani, Advancements in supercapacitor technology through the utilization of NiCo2X4 (X = O, S, Se) based nanocomposites as electrodes: A comprehensive review, Journal of Industrial and Engineering Chemistry (2024). https://doi.org/10.1016/j.jiec.2024.05.021.
4. S. Jayakumar, P.C. Santhosh, S. Ramakrishna, A.V. Radhamani, 2D (Ti3C2Tx) MXene: A comprehensive review of advancements in synthesis protocols, applications in supercapacitors, sustainability targets and future prospects, J Energy Storage 97 (2024). https://doi.org/10.1016/j.est.2024.112741.
5. Q. Wang, D. Astruc, State of the art and prospects in metal–organic framework (MOF)-based and MOF-derived nanocatalysis, Chem Rev 120 (2019) 1438–1511.
6. C.W. Jones, Metal–organic frameworks and covalent organic frameworks: Emerging advances and applications, JACS Au 2 (2022) 1504–1505. https://doi.org/10.1021/jacsau.2c00376.
7. S. Khokhar, H. Anand, P. Chand, Current advances of nickel based metal organic framework and their nanocomposites for high performance supercapacitor applications: A critical review, J Energy Storage 56 (2022) 105897. https://doi.org/10.1016/j.est.2022.105897.
8. H. Wu, S. Li, Y. Liu, M. Liu, Y. Shi, Metal-organic framework-derived hollow carbon nanosphere@Ni/reduced graphene oxide composites for supercapacitor electrodes with enhanced performance, ACS Appl Nano Mater 6 (2023) 1582–1591. https://doi.org/10.1021/acsanm.2c04327.
9. V.F. Yusuf, N.I. Malek, S.K. Kailasa, Review on metal–organic framework classification, synthetic approaches, and influencing factors: Applications in energy, drug delivery, and wastewater treatment, ACS Omega 7 (2022) 44507–44531. https://doi.org/10.1021/acsomega.2c05310.

10. N.A. Khan, S.H. Jhung, Synthesis of metal-organic frameworks (MOFs) with microwave or ultrasound: Rapid reaction, phase-selectivity, and size reduction, Coord Chem Rev 285 (2015) 11–23.
11. Y. Chen, D. Ni, X. Yang, C. Liu, J. Yin, K. Cai, Microwave-assisted synthesis of honeycomblike hierarchical spherical Zn-doped Ni-MOF as a high-performance battery-type supercapacitor electrode material, Electrochim Acta 278 (2018) 114–123. https://doi.org/10.1016/j.electacta.2018.05.024.
12. S. Głowniak, B. Szczęśniak, J. Choma, M. Jaroniec, Mechanochemistry: Toward green synthesis of metal–organic frameworks, Materials Today 46 (2021) 109–124. https://doi.org/10.1016/j.mattod.2021.01.008.
13. A.M. Shahul Hameed, G. Shanmugam, E. Thirugnanasambandam, S. Kamaraj, Tweaking the efficiency of asymmetric hybrid supercapacitors with mechanochemically synthesized mixed-metal–organic framework-derived metal oxides, Energy Technol 12 (2024). https://doi.org/10.1002/ente.202301633.
14. V. M. V. G. Nageswaran, Review—direct electrochemical synthesis of metal organic frameworks, J Electrochem Soc 167 (2020) 155527. https://doi.org/10.1149/1945-7111/abc6c6.
15. H. Ren, T. Wei, Electrochemical synthesis methods of metal-organic frameworks and their environmental analysis applications: A review, ChemElectroChem 9 (2022) e202200196.
16. M.W. Khan, A.S. Khan, Rohma, S.H.S. Pai, G. Mohapatra, A. Kumar, R.B. Ali, Q.A. Sial, H. Seo, Electrodeposition of redox active nickel-manganese metal organic framework and manganese sulfide composite for supercapacitors, Electrochim Acta 513 (2025) 145554. https://doi.org/10.1016/j.electacta.2024.145554.
17. N. Stock, S. Biswas, Synthesis of metal-organic frameworks (MOFs): Routes to various MOF topologies, morphologies, and composites, Chem Rev 112 (2012) 933–969.
18. N. Shakeel, J. Khan, A.A. Al-Kahtani, Morphology-driven electrochemical attributes of Cu-MOF: A high-performance anodic material for battery supercapacitor hybrids, RSC Adv 14 (2024) 33941–33951.
19. A. Gowdhaman, S.A. Kumar, D. Elumalai, C. Balaji, M. Sabarinathan, R. Ramesh, M. Navaneethan, Ni-MOF derived NiO/Ni/r-GO nanocomposite as a novel electrode material for high-performance asymmetric supercapacitor, J Energy Storage 61 (2023) 106769. https://doi.org/10.1016/j.est.2023.106769.
20. X. Zhang, S. Yang, W. Lu, D. Lei, Y. Tian, M. Guo, P. Mi, N. Qu, Y. Zhao, MXenes induced formation of Ni-MOF microbelts for high-performance supercapacitors, J Colloid Interface Sci 592 (2021) 95–102. https://doi.org/10.1016/j.jcis.2021.02.042.
21. Q. Li, H. Guo, R. Xue, M. Wang, M. Xu, W. Yang, J. Zhang, W. Yang, Self-assembled Mo doped Ni-MOF nanosheets based electrode material for high performance battery-supercapacitor hybrid device, Int J Hydrogen Energy 45 (2020) 20820–20831. https://doi.org/10.1016/j.ijhydene.2020.05.143.
22. H. Rong, P. Song, G. Gao, Q. Jiang, X. Chen, L. Su, W.-L. Liu, Q. Liu, A three-dimensional Mn-based MOF as a high-performance supercapacitor electrode, Dalton Trans 52 (2023) 1962–1969. https://doi.org/10.1039/D2DT02857C.
23. P.-P. Sun, Y.-H. Zhang, X. Yu, Q. Shi, B. Tian, J. Gao, F.-N. Shi, Cu powder decorated 3D Mn-MOF with excellent electrochemical properties for supercapacitors, Inorganica Chim Acta 508 (2020) 119629. https://doi.org/10.1016/j.ica.2020.119629.
24. D.Y. Lee, S.J. Yoon, N.K. Shrestha, S.-H. Lee, H. Ahn, S.-H. Han, Unusual energy storage and charge retention in Co-based metal–organic-frameworks, Microporous Mesoporous Mater 153 (2012) 163–165. https://doi.org/10.1016/j.micromeso.2011.12.040.
25. T. Kshetri, D.D. Khumujam, T.I. Singh, Y.S. Lee, N.H. Kim, J.H. Lee, Co-MOF@MXene-carbon nanofiber-based freestanding electrodes for a flexible and wearable quasi-solid-state supercapacitor, Chem Eng J 437 (2022) 135338. https://doi.org/10.1016/j.cej.2022.135338.

26. H. Zhang, J. Wang, Y. Sun, X. Zhang, H. Yang, B. Lin, Wire spherical-shaped Co-MOF electrode materials for high-performance all-solid-state flexible asymmetric supercapacitor device, J Alloys Compd 879 (2021) 160423. https://doi.org/10.1016/j.jallcom.2021.160423.
27. M.K. Singh, A.K. Gupta, S. Krishnan, N. Guha, S. Marimuthu, D.K. Rai, A new hierarchically porous Cu-MOF composited with rGO as an efficient hybrid supercapacitor electrode material, J Energy Storage 43 (2021) 103301. https://doi.org/10.1016/j.est.2021.103301.
28. J. Khan, M.Z. Iqbal, B. Rubab, F. Jamshaid, A. Khan, N. Shakeel, A. Ahmed, A. A. Al-Kahtani, Isonicotinic acid-based copper-MOF: An exotic redox propertied electrode material for high energy asymmetric supercapacitor, J Energy Storage 72 (2023) 108655. https://doi.org/10.1016/j.est.2023.108655.
29. S. Sharma, P. Chand, S. Kaushik, A critical review of recent advancements in zinc based metal organic framework nanocomposites and their derivatives for supercapacitor applications with future perspectives and challenges, Sustain Mater Technol 41 (2024) e01045. https://doi.org/10.1016/j.susmat.2024.e01045.
30. H. Esfandian, M.M. Sadeghi, Preparation and application of mixed ligand Zn-MOF as electrode materials for supercapacitors applications, Electrochim Acta 509 (2025) 145352. https://doi.org/10.1016/j.electacta.2024.145352.
31. Z. Qin, Y. Xu, L. Liu, M. Liu, H. Zhou, L. Xiao, Y. Cao, C. Chen, Ni-MOF composite polypyrrole applied to supercapacitor energy storage, RSC Adv 12 (2022) 29177–29186. https://doi.org/10.1039/D2RA04939B.
32. S.D. Dhas, A.C. Mendhe, P.N. Thonge, A.M. Patil, Y. Kim, D. Kim, A high-performance battery–supercapacitor hybrid device and electrocatalytic oxygen evolution reaction based on $NiCo_{2-x}$ Mn_x O_4 @Ni-MOF ternary metal oxide core–shell structures, J Mater Chem A Mater 12 (2024) 21956–21970. https://doi.org/10.1039/D4TA02978J.
33. W. Peng, N. Song, Z. Su, J. Wang, K. Chen, S. Li, B. Wei, S. Luo, A. Xie, Two-dimensional MoS2/Mn-MOF/multi-walled carbon nanotubes composite material for high-performance supercapacitors, Microchem J 179 (2022) 107506. https://doi.org/10.1016/j.microc.2022.107506.
34. F. Mousavi, S. Ghasemi, M. Shamsipur, 2D/2D Nanocomposite of graphene-supported Cu-doped Co-MOF for high-performance supercapacitor applications, Energy Fuels 38 (2024) 23101–23113. https://doi.org/10.1021/acs.energyfuels.4c04124.
35. Y. Yuan, C. Han, Y. Fu, Z. Ye, Q. Shen, W. Feng, Y. Zhao, Structural energy storage composites based on etching engineering Fe-doped Co MOF electrode toward high energy density supercapacitors, J Power Sources 609 (2024) 234688. https://doi.org/10.1016/j.jpowsour.2024.234688.
36. S. Mishra, M.K. Singh, D. Pandey, D.K. Rai, A. Raghuvanshi, A two-dimensional semiconducting Cu(I)-MOF for binder and conductive additive-free supercapattery, J Mater Chem A Mater 12 (2024) 4534–4543. https://doi.org/10.1039/D3TA04708C.
37. A.A. Yadav, Y.M. Hunge, S. Majumder, A.-H.I. Mourad, M.M. Islam, T. Sakurai, S.-W. Kang, Multiplicative rGO/Cu-BDC MOF for 4-nitrophenol reduction and supercapacitor applications, J Colloid Interface Sci 677 (2025) 161–170. https://doi.org/10.1016/j.jcis.2024.08.060.

12 High-Performance MXene Supercapacitors

Ramesh Chandran K, Sapna Nehra, and Dinesh Kumar

12.1 INTRODUCTION

The race for adequate storage and efficient release of higher energy levels to absorb voltage distortions in electronics started when Howard I. Becker and his team, on behalf of General Electric Co., filed a patent in 1954 for an electrochemical capacitor using activated carbon [1]. Significant progress in materials research, especially in supercapacitors, has been fueled by the rising demand for environmentally friendly energy storage options [2]. These energy storage technologies have become well-known because of remarkable power density, extended cycle life, and quick charge–discharge speeds, which make them suitable for large-scale energy networks and handheld electronics [3]. MXenes are a novel class of two-dimensional (2D) materials that have shown unmatched promise among the materials investigated for supercapacitor electrodes [4].

Since their discovery in 2011, MXenes, a class of 2D transition metal (TM) carbides, nitrides, and carbonitrides, have fascinated scientists [5]. Their layered structure and surface terminations, like –OH, –O, and –F, provide MXenes with special qualities like excellent electrical conductivity, adjustable surface chemistry, and remarkable mechanical strength. These characteristics make MXenes ideal for supercapacitor applications, where power density and energy efficiency are crucial [6].

Because of their vast surface area and adjustable interlayer spacing, MXenes have a significant advantage as they can attain high charge storage capacity and efficient ion transport [7]. Their hydrophilic nature enables effective interaction with organic and aqueous electrolytes, further enhancing electrochemical performance [8]. The higher volumetric capacitances produced by MXenes than traditional materials have made MXenes a leader in next-generation energy storage technologies [9].

This chapter provides a detailed overview of the physics and engineering of high-performance MXene-based supercapacitors, including device fabrication, structural optimization, and material production [10]. Significant advancements in modifying the chemical and physical properties of MXenes to enhance electrochemical performance are thoroughly explored. The challenges of scalability, stability, and integration into practical devices are also examined in detail. MXene supercapacitors represent a groundbreaking advancement in the global transition toward cleaner and more efficient energy systems [11]. This chapter highlights recent successes and the potential for MXenes to transform energy storage technology fundamentally in the future.

DOI: 10.1201/9781003535836-12

12.2 OVERVIEW OF MXenes

MXenes are derived from the parent MAX phases, which consist of a TM, an A-group element, and carbon or nitrogen [12]. A selective chemical etching process is applied to the parent material, removing the A-group element and thereby creating a 2D layered structure [13]. The van der Waals forces are responsible for the interlayer interactions between these layers, and it is through these forces that the interlayer separation can be adjusted [14]. This property is essential for ion transport and intercalation in high-performance supercapacitors, given the structure's flexibility [15]. Random particle adsorption during the etching process leads to the formation of metal-terminated fluorine (–F), oxygen (–O), and hydroxyl (–OH) surfaces on MXenes, and these surface groups influence the results [16]. Intriguingly, the presence of these functional groups enhances the electrolyte-ion-dipole interaction, making the material more hydrophilic, providing it with a surface charge, and potentially boosting its electrochemical activity [17].

The chemicals in MXenes can always be manipulated in practical ways, such as replacing the TM and altering the etching process, which allows for the build-up of functionality based on the user's requirements [18]. Titanium, molybdenum, and vanadium are the most commonly chosen metals in the fabrication of MXenes, each offering different electrochemical and physical properties [19]. For example, MXenes made of titanium excel in electrical conductivity and stability; however, those made from molybdenum are predominantly pseudocapacitive [20]. The crystal chemical formula of a compound is modified by its surface chemistry, and thus, the supercapacitor's properties, energy, and power density can be better controlled [21]. Thanks to their multifunctional and extremely adaptable characteristics, MXenes are currently game-changing materials in energy storage technology [22].

The MXene synthesis method employs an etching approach that selectively interacts with the A-component, leaving the X-layer intact [23]. On average, fluoride-contaminated etchants, such as HF, HCl, and LiF, are used to etch the MXenes [24, 25]. Intercalation or ultrasonication technologies are often used, as the delamination of prepared aerogels into "single" or "few" layers provides customized activity [26].

MXenes, with their unique structure, have various applications in supercapacitors and possess several key properties. Few obstacles are encountered in electrical conduction because the TMs display metallic characteristics [27, 28]. The highest internal conductivity, which can be achieved with minimal losses in charge transport, can be realized by merging supercapacitors and batteries [29]. Improvements in inter-electrode spacing and edge configurations through intercalation in MXene porosity facilitate the efficient transfer of electrons with high-rate kinetics and a substantial number of ions through the matrix, respectively [30]. Furthermore, the additional adsorption of electrolyte ions on MXenes may arise from edge defects (vacancies), which is beneficial in this context.

The extraordinary hydrophilic properties of MXenes enhance electrochemical performance and flexibility, which is essential. In this context, compatibility with organic and aqueous electrolytes is extremely important [31]. MXenes not only offer superior structural stability and mechanical flexibility but also ensure a longer lifespan during capacitive battery operation [32]. The ability to perform under extreme environmental conditions while supporting rapid charge–discharge rates provides the most compelling evidence of their potential as supercapacitors. In principle, studies have shown that supercapacitors can employ this material as an energy source [33]. Therefore, it is confirmed that they demonstrate higher energy and power densities than conventional electrode materials, representing a more competitive option in advanced energy storage devices [34, 35].

12.3 MECHANISMS OF ENERGY STORAGE IN MXene-BASED SUPERCAPACITORS

MXene-based supercapacitors predominantly utilize two mechanisms: electric double layer capacitor (EDLC) and, to a lesser extent, pseudocapacitance [36]. EDLC is the process by which the ions of the electrolyte layer attach to the MXene surface, forming a charge separation region that occurs rapidly without any chemical reactions [37]. Thus, this guarantees rapid energy release and storage. Additionally, the surfaces of oxide and hydroxylated MXenes are significantly involved in pseudocapacitance and in the catalysis of redox reactions, which enhances this property [38, 39].

12.3.1 EDLC

Rapid energy storage and high-power density are made possible by the physical mechanism known as EDLC, which is essential to the functioning of MXene-based supercapacitors [40]. The formation of a charge separation layer at the interface between the electrode and the electrolyte depends on the adsorption of electrolyte ions onto the surface of MXene materials [41]. Due to their wide surface area and 2D nature, MXenes provide many active sites for ion adsorption, which is essential for optimizing charge storage.

When an external voltage is applied to the supercapacitor, positive and negative ions from the electrolyte move toward the oppositely charged MXene electrode surfaces [42]. Without penetrating the substance, these ions create the Helmholtz layer, a tightly bonded structure. The storage mechanism is entirely electrostatic because neither chemical nor phase changes are involved [43]. Due to the incredibly rapid charging and discharging cycles, MXene-based supercapacitors can produce high power outputs with superior cycle stability [44–46].

The EDLC mechanism is further improved by the special structure of MXenes, which are distinguished by high conductivity and adjustable surface chemistry. MXenes' hydrophilic surface terminations, such oxygen or hydroxyl groups, encourage ion accessibility and wetting in aqueous or organic electrolytes and conductivity guarantees effective charge transfer [47]. Furthermore, MXenes' layered structure offers interlayer gaps for ion accommodation, enhancing the material's capacitance even more.

Utilizing the materials' wide surface area, high conductivity, and advantageous surface chemistry, the electrostatic adsorption of ions powers the EDLC mechanism in MXene supercapacitors. The remarkable performance of MXene-based supercapacitors is supported by this mechanism [48].

12.3.2 Pseudocapacitance

Pseudocapacitance is an electrochemical energy storage mechanism typically inspired by the faradaic charge transfer processes via EDLC [49]. Figure 12.1 illustrates the timeline of key developments in pseudocapacitance, including new materials, mechanisms, and theoretical concepts, from 1962 to 2020. This energy

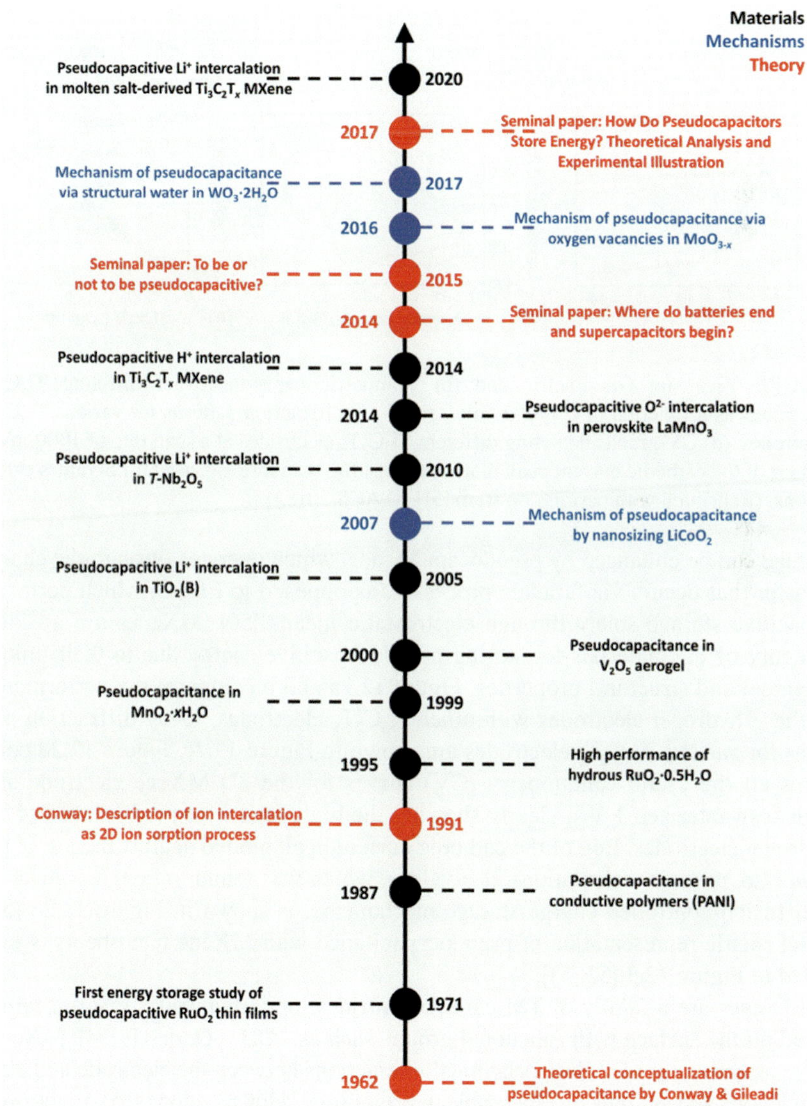

FIGURE 12.1 Timeline of significant advancements in pseudocapacitance: new pseudocapacitive materials, mechanisms, and theoretical concepts. (Reprinted with permission from [50]. © ACS 2020.)

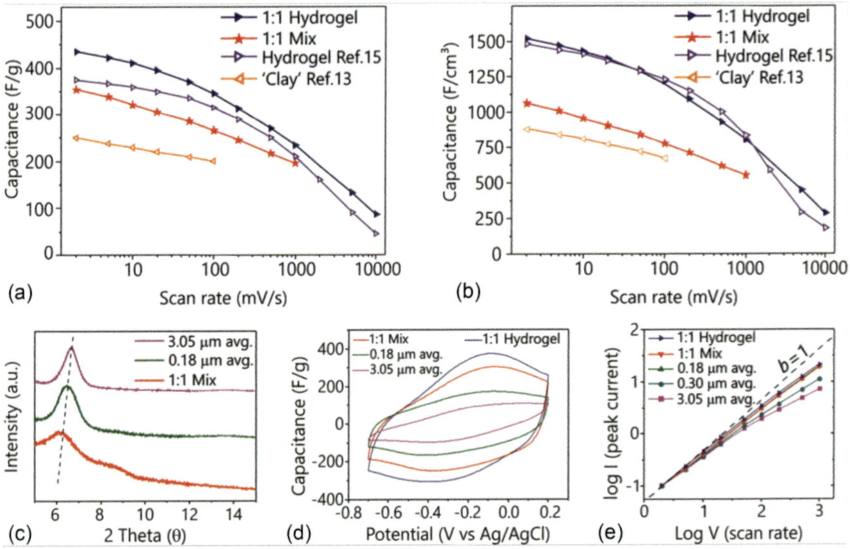

FIGURE 12.2 (a) Gravimetric and (b) volumetric capacitances of different $Ti_3C_2T_x$ electrodes as a function of CV scan rates. (c) X-ray diffraction patterns for various $Ti_3C_2T_x$ electrodes. (d) CV graphs depicting different $Ti_3C_2T_x$ electrodes at a scan rate of 1000 mV/s. (e) Log of the cathodic current peak plotted against the scan rate log, used for b-values calculations. (Reprinted with permission from [51]. © ACS 2018.)

storage can be enhanced by pseudocapacitance, which operates through the charge transfer that occurs via faradaic processes, as opposed to EDLC, which performs capacitive storage solely through electrostatic means [50]. MXenes are an ideal category of 2D materials for storing pseudocapacitive energy due to their unique chemical and structural properties. Figure 12.2a and b compares the performance of these hydrogel electrodes with other $Ti_3C_2T_x$ electrodes. X-ray diffraction patterns for various $Ti_3C_2T_x$ electrodes are shown in Figure 12.2c. Figure 12.2d compares all the cyclic voltammetry (CV) curves for the 2D MXene electrode at a high scan rate, i.e., 1 V/s, clearly showing the high capacitive performance of 1:1 hydrogel electrodes. Log of the cathodic current peak plotted against the log of the scan rate, used for calculations of b-values, which was found to be 0.5, indicating a diffusion-controlled charge storage mechanism, as shown in Figure 12.2e [51]. A schematic representation of pseudocapacitance with MXene nanosheets is provided in Figure 12.3 [52, 53].

MXenes are a family of TM carbides, nitrides, or carbonitrides that are terminated on the surface with functional groups such as –OH, –O, or –F [54]. MXenes serve as active sites for electrochemical interactions between the electrode and electrolyte ions and are thus indispensable in promoting redox reactions [55]. Electrolyte ions are released from the MXene surface and undergo redox processes when a potential difference is applied to the anode [56]. Through the electrolyte, charge storage occurs by the absorption of ions and the transfer of electrons into the material's near-surface layers [57].

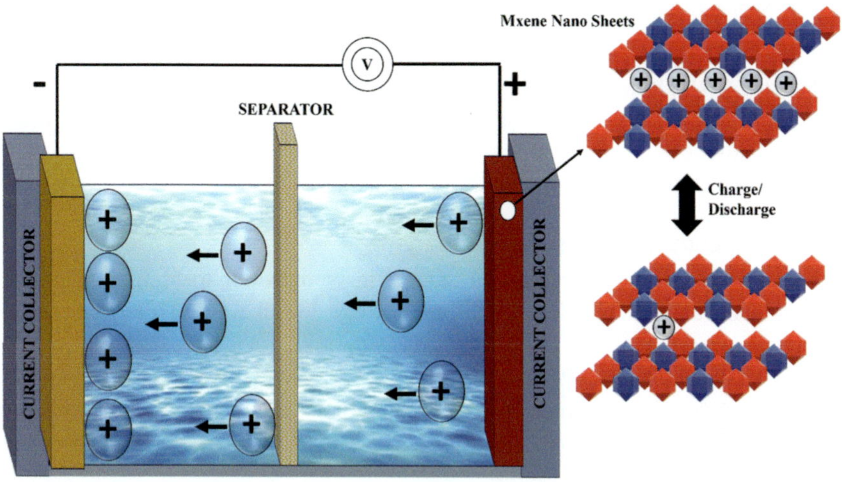

FIGURE 12.3 Schematic representation of pseudocapacitance with MXene nanosheets.

MXenes exhibit high conductivity due to their metallic-like character and enable fast charging and discharging by allowing an efficient electron transport pathway during redox reactions [58]. The layered structure of MXenes creates interlayer spaces that accommodate electrolyte ions, making the material more suitable for pseudocapacitive activity [59]. A rich redox chemistry is characterized by the numerous TM sites found in MXenes, leading to various faradaic reactions that contribute to energy storage [60, 61].

Pseudocapacitance is a mechanism responsible for both capacitive and battery-like behaviors, making the combination of MXene and supercapacitors a perfect choice for industries that require rapid charging and discharging cycles, as well as uninterrupted energy delivery [62]. By combining pseudocapacitance with MXenes and their inherent qualities—such as high conductivity, stability, and customizable surface chemistry—these materials are positioned as viable options for next-generation energy storage systems in portable devices and electric vehicles (EVs) [63].

12.4 APPLICATIONS OF HIGH-PERFORMANCE MXene SUPERCAPACITORS

The water-attracting nature of MXenes enhances their electrochemical behavior and applicability in both aqueous and organic electrolyte media. MXenes are admired for their exceptional mechanical strength, extensive structural reliability, and noted for their long-life efficiency [64]. The capability to operate under very demanding conditions, such as adverse weather or high charge–discharge rates, makes using these materials in commercial applications very feasible [65]. Utilizing these features, MXenes have outperformed some traditional electrode materials in terms of micro-energy and micro-power composites, thus becoming a viable option for the next generation of energy storage systems [66].

12.4.1 In EV

Batteries for EVs must be efficient, durable, and quickly cycle between charging and discharging [67]. Despite their large market share, traditional lithium-ion batteries suffer from low energy density, slow charging, and limited lifespan [68]. MXene supercapacitors offer a promising alternative or complement due to their superfast energy storage and operational range from 22°F to 86°F [69].

12.4.1.1 Speedy Charging and Discharging

MXene supercapacitors charge and discharge in seconds due to high conductivity and rapid ion transport, avoiding traditional battery slow charging [70].

12.4.1.2 High Power Density

MXene supercapacitors provide high power outputs, allowing EVs to accelerate quickly and perform better [71].

12.4.1.3 Longer Lifespan

MXene supercapacitors experience negligible capacity loss after millions of charge–discharge cycles, unlike lithium-ion batteries, which deteriorate over time.

12.4.1.4 Superior Energy Density

Combining MXenes with substances like metal oxides or conductive polymers achieves energy densities comparable to batteries, bridging the gap between conventional storage and supercapacitors.

12.4.1.5 Environmentally Friendly

MXenes' recyclability aligns with EV manufacturers' sustainability goals, and their production involves simple, scalable processes. MXene supercapacitors boast extended cycle life, high power density, and quick charging, making them promising for EV use. They can complement batteries for energy storage during regenerative braking and acceleration, where rapid energy capture and release are crucial [72]. Additionally, EV auxiliary systems like infotainment, heating, and cooling can run on MXene supercapacitors without straining the main battery. Their ability to deliver power bursts and recharge in seconds improves vehicle sustainability and efficiency, making them ideal for hybrid EVs and practical energy management applications [73].

12.4.2 Portable Electronics

The domain of storing power in portable equipment is innovating with the growing demand for the highly efficient MXene supercapacitors. These capacitors offer exceptional qualities such as fast charging speed, along with other attributes like power efficiency, wear resistance, and a long lifespan [74]. These advantages position MXene supercapacitors as ideal candidates for various upcoming devices, including wearables, laptops, smartphones, Internet of Things (IoT) devices, and medical equipment [75].

12.4.2.1 In Smartphones and Laptops

MXene supercapacitors can enhance energy storage. They improve performance by supplying immediate power bursts during high energy demand, like multitasking or gaming [76]. Their quick recharge minimizes downtime [77]. The mechanical flexibility allows easy integration into curved designs for wearable devices, such as fitness trackers or smartwatches, providing effective charging solutions for rapid technological advancements [78].

12.4.2.2 In Medical Devices

Pacemakers, glucose monitors, and diagnostic instruments are significant medical devices using MXene supercapacitors. Based on precise health measurements, these supercapacitors perform flawlessly. They also serve as a viable option for IoT devices needing sudden energy bursts for connectivity and data transmission. This project operates seamlessly and continuously on a compact scale [79–81].

12.4.2.3 In Lightweight Devices

Powering very light devices represents a major application of MXene supercapacitors. When connected to energy harvesting devices such as solar panels or kinetic energy harvesters, MXene supercapacitors ensure that electrical gadgets are charged with sustainable energy sources, enabling them to function smoothly [82]. With outstanding electrochemical performance and compatibility with various electrolytes, these supercapacitors can be used in nearly all types of instruments, including those for general and specialized purposes [83]. Due to their energy storage capabilities and their ability to meet the demand for faster charging, higher power delivery, and advanced durability, top-performance MXene supercapacitors advance developments in mobile electronics and serve as the foundation for the next generation of portable electronics, thanks to their unmatched flexibility and performance [84–86].

12.4.3 In Grid Storage

High-performance MXene supercapacitors are gaining popularity as a breakthrough technology for grid energy storage due to their superior electrochemical properties, such as rapid charging and discharging, high power density, long cycle life, and environmental versatility. These features make MXenes suitable for various grid-focused tasks, beneficial for new power systems facing sustainability, stability, and energy management challenges [87, 88].

MXene supercapacitors are common in grid storage applications. Continuous power transmission is vital for maintaining constant system frequencies. They can quickly regulate frequency changes [89]. The power grid stabilizes rapidly during high demand or low supply by absorbing or releasing energy as necessary [90].

MXene supercapacitors effectively connect renewable energy sources and regulate grid voltage. When wind and solar are the sole energy producers, they can store excess energy generated during peak production and release it during lower demand [91]. Their long cycle life reduces maintenance needs and enhances the reliability of renewable energy systems over time [92].

Grid performance improves with MXene supercapacitors that peak shave and level loads [93]. They store energy when the grid is not congested and release it during high demand, reducing the need for new grid development. This enhanced power capacity leads to better efficiency, stressing lithium-ion batteries and other storage systems less and extending their lifetimes [94–96].

In addition to that, the development of advanced power networks (also referred to as distributed energy systems) represents a third key MXene technology utilized. Most of the time, the flexibility of the grids, in conjunction with the operation of the new MXene supercapacitors, offers smooth and consistent performance during the switching or change of the sources [97–99]. Both energy-intensive industries and small domestic setups can also benefit from the provided scalability and flexibility.

Major features like smooth transitions to renewable energy, fast storage, and improved grid stability are achieved by high-performance MXene supercapacitors. Their excellent lifecycle, adaptability, and high power make them essential for a robust green energy system.

12.5 STRATEGIES TO ENHANCE PERFORMANCE OF MXene-BASED SUPERCAPACITORS

Achieving higher energy and power densities, longer cycle life, and improved stability for applications ranging from portable electronics to grid storage presents just a few of the contemporary energy storage challenges that necessitate enhancing the performance of MXene-based supercapacitors. The surface chemistry of MXenes can be altered to enhance ionic accessibility and redox activity, and molecules or ions can be intercalated to increase interlayer spacing for better ion transport. The benefits of each material are combined when MXenes are hybridized with graphene, metal oxides, or carbon nanotubes (CNTs), which boost conductivity and energy density while preventing layer restacking. Utilizing organic electrolytes with wide voltage windows or ionic liquids exemplifies electrolyte optimization that can further enhance energy storage [100–102].

12.5.1 Structural Optimization

Maximizing MXene supercapacitors involves reducing defects via structural optimization for mechanical stability, charge storage, and ion accessibility [103]. However, restacking may limit active surface sites and impede ion transport, despite the planar configuration promoting conductivity. Unlocking MXene supercapacitors' full potential hinges on this structural optimization [104].

Using the interlayer spacing trick directly improves MXene-based supercapacitors. Scientists create larger channels for efficient ion transport and increased capacitance by intercalating small molecules, ions, or polymers into the MXene layers. For example, inserting metal ions or organic compounds enhances electrochemical stability and ion transport, resulting in higher energy and power densities [105–107].

Creating porous MXene structures is crucial. These micro- or mesoporous structures enhance electrolyte ion access to electrode surfaces. Techniques like chemical etching, templating, and forming three-dimensional (3D) porous frameworks are employed. These methods significantly improve rate performance and surface area [108, 109].

An additional effective structural optimization method involves the hybridization of MXenes with other materials. To prevent stacking of non-conducting layers, one can create alternative pathways and interconnected networks of more conductive states, such as CNTs and graphenes with MXenes. Similarly, doped manganese oxide, which enhances the specific capacitance and, conversely, the specific area, enables the creation of pseudocapacitance in MXenes when conducting polymers or manganese oxides are added [110–112].

Integrating MXene electrodes enhances the mechanical flexibility and structural integrity of flexible substrates or polymer matrices, allowing operation under mechanical deformation, making them ideal for flexible energy storage applications.

Structural optimization of MXene-based supercapacitors can enhance stability, active surface area, and ion transport through interlayer spacing expansion, porosity, material hybridization, and mechanical reinforcement. Applying these methods together creates strong, high-performance MXene supercapacitors for various energy storage applications [113, 114].

12.5.2 Composites and Hybrids

Enhancing MXene-based supercapacitors may involve developing composites and hybrids to improve mechanical stability, energy density, and layer restacking issues [115]. These systems combine MXenes with complementary materials to optimize structural and electrochemical performance.

1. *MXene–Carbon Hybrids*: Carbon materials like graphene, CNTs, and activated carbon are integrated to produce MXene hybrids [116]. Spacers create physical barriers for ion transportation, maintaining separation between MXene sheets and preventing restacking [117]. Enhanced charge transfer increases sites for energy storage, as MXenes have high conductivity and large surface areas. An example with graphene shows that the flexibility of these materials and the conductivity of MXenes together create high performance through a large surface area, resulting in excellent synergistic effects [118, 119].
2. *MXene–Metal Oxide Hybrids*: Through faradaic redox processes, MXenes can be integrated with metal oxides, such as nickel oxide (NiO), manganese dioxide (MnO_2), or titanium dioxide (TiO_2), to induce pseudocapacitance. While preserving the quick charge–discharge properties of MXenes, these hybrids enhance energy density. Additionally, robust interactions between MXenes and metal oxides improve structural stability and overall electrochemical performance [120–122].
3. *MXene-Conducting Polymer Hybrids*: To capitalize on the high pseudocapacitive behavior of MXene-based systems, conducting polymers like polyaniline (PANI) and polypyrrole (PPy) are added. These polymers

significantly increase energy density by storing more charge through redox processes. The flexibility of conducting polymers further enhances the mechanical adaptability of the hybrids, making them suitable for wearable and flexible energy storage applications [123, 124].

4. *Three-Dimensional (3D) MXene Hybrids*: Combining MXenes with other materials to form 3D structures results in porous, networked structures that enhance ion accessibility and reduce internal resistance [125]. By enabling effective ion transport channels and maintaining strong structural integrity, these 3D hybrids—often created using methods like hydrothermal synthesis or freeze-drying—enhance electrochemical performance [126, 127].
5. *MXene-Biomaterial Hybrids*: Supercapacitor development can become more environmentally friendly and sustainable through the combination of MXenes with biomaterials like chitosan or cellulose. These hybrids are suitable for green energy storage systems due to their enhanced flexibility and biodegradability while preserving the superior electrochemical properties of MXenes [128].

The energy density, power density, and structural stability of MXenes improve with integration with conducting polymers, metal oxides, carbon materials, and other complementary elements. To enhance the performance of MXene-based supercapacitors and meet the requirements of contemporary energy storage applications, these composite and hybrid techniques are essential.

12.5.3 Electrolyte Engineering

Electrolyte engineering is essential for improving the energy density, power density, and cycle life of MXene-based supercapacitors, as the selection and design of electrolytes directly impact these parameters [129]. High-performance energy storage requires optimized electrolytes to promote effective ion transport, increase the operating voltage window, and ensure compatibility with the MXene electrode.

1. *Aqueous Electrolytes*: The high ionic conductivity and low cost of aqueous electrolytes, such as acidic (like H_2SO_4), alkaline (like KOH), or neutral (like Na_2SO_4) solutions, ensure economic viability for use. The small voltage window of MXenes, usually confined to ~1.2 V, limits energy density. However, performance and stability can be enhanced by utilizing mixed electrolytes or by adjusting the pH and ionic strength of these electrolytes. The hydrophilic surface and surface terminations of MXenes promote excellent wettability and compatibility with aqueous electrolytes [44, 130].
2. *Organic Electrolytes*: The energy density of MXene supercapacitors is significantly enhanced by the broader voltage window provided by organic electrolytes, such as those based on acetonitrile or propylene carbonate, which can reach up to 2.7–3.5 V. Applications requiring greater energy storage are particularly well-suited for these electrolytes.

The ionic conductivity and compatibility of organic electrolytes with MXenes can be optimized by customizing the solvent and salt combinations [131].

3. *Ionic Liquids*: Ionic liquids are sophisticated electrolytes characterized by minimal volatility, good thermal stability, and a broad voltage window. Engineering for compatibility with MXene electrodes is facilitated by variable viscosity and ion size. Ionic liquids ensure stable and effective functioning, essential for high-temperature or high-voltage applications [132].
4. *Gel and Solid-State Electrolytes*: Solid-state and gel electrolytes are essential for flexible and wearable supercapacitors. These electrolytes provide mechanical flexibility while maintaining ionic conductivity and are often based on polymers such as poly(vinyl alcohol) (PVA). The functionality of gels can be further enhanced by incorporating ionic liquids or salts. Solid-state electrolytes are ideal for wearable and portable electronics due to their reduced risk of leakage, which further improves device safety [133].
5. *Redox-Active Electrolytes*: Redox-active electrolytes, which contain species such as quinones or vanadium ions, impart pseudocapacitive properties to the electrolyte. By facilitating redox processes within the electrolyte, this approach significantly increases the energy density and enhances the functionality of MXene electrodes [134].
6. *Additives and Hybrid Electrolytes*: Stabilizers and surfactants are examples of additives that can enhance electrode–electrolyte interactions and ion transport. By balancing conductivity, voltage range, and thermal stability, hybrid electrolytes—which integrate ionic liquids or mix the characteristics of aqueous and organic systems—provide tailored solutions for specific applications [135].

Electrolyte engineering is a potent technique for improving the efficiency of MXene-based supercapacitors. Improved electrolytes that optimize ionic conductivity, voltage window, and compatibility with MXene electrodes make next-generation supercapacitors possible in a variety of applications.

12.6 CHALLENGES OF MXene-BASED SUPERCAPACITORS

MXene-based supercapacitors are more robust than other hydrogen energy storage options but face problems limiting their application. The primary issue is the shifting of MXene layers during electrode production, which reduces usable surface area and ion transport. Improving MXene-based supercapacitors in real-world applications can be achieved through innovative structural design, optimal electrolyte use, and cost-effective synthesis techniques [136, 137].

12.6.1 Challenges with Scalability

Issues in the synthesis and incorporation of MXene materials limit the scalability of MXene-based supercapacitors. A major problem is that MXenes are produced by etching parent MAX phases, often using hazardous chemicals like hydrofluoric acid, which pose safety, environmental challenges and hinder commercialization. Alternative fluoride-free etching methods are still in research and not ready for commercial use. Moreover, the cost and availability of high-quality MAX phases, which are precursors to MXenes, complicate scalability. Additionally, synthesis conditions can slightly affect MXenes' electrochemical properties, making industrial-scale production of defect-free MXenes challenging [138, 139].

Scalability is a key concern for MXene electrodes in supercapacitors. The van der Waals force causes MXene layers to compress, absorbing rare earth ions and limiting movement and surface area. Intercalation using hybrid materials to prevent restacking complicates fabrication and raises costs. Additionally, creating flexible supercapacitor devices with MXene electrodes is challenging due to the need for advanced processing equipment unavailable for mass production. Currently, we are developing scalable printing techniques, such as roll-to-roll methods, for MXene-based capacitors [140, 141].

12.6.2 Challenges with Stability

Long-term stability is crucial for optimal electrochemical and mechanical performance of MXene-based supercapacitors. In aqueous solutions, MXene surfaces can be compromised by oxygen or moisture due to their high surface energy and hydrophilic properties, leading to oxidation or degradation [142–144]. This can impede electron movement across the anode and negatively affect the electrochemical stability of supercapacitors, potentially reducing their cycle life as capacitive performance deteriorates over time. Although methods for functionalization and surface modification are being developed to improve MXene stability, the challenge of maintaining stability in practical applications remains unresolved.

Mechanical integrity of MXene electrodes, especially during cycling, leads to stability issues. Separated MXene layers may lose material or misalign due to structural changes from repeated charging and discharging (e.g., volume expansion). In flexible MXene-based supercapacitors, such as thin films, mechanical stress-induced deterioration is more pronounced [145–149]. The restacking of MXene layers after several cycles results in the loss of active sites, causing decreased energy and power density.

12.6.3 Challenges with Electrode Design

Incorporating these issues into MXene-based supercapacitor electrode design significantly enhances energy density, power density, and cycling stability [150]. A primary difficulty is the MXene sheet's tendency to restack due to van der Waals forces, especially during electrode production [151]. This restacking slows ion diffusion, reducing the surface area for ion adsorption, thereby decreasing charge storage capacity and efficiency. Strategies being explored include interlayer spacers, intercalating molecules, and composite materials to address this issue [152–154]. However, these solutions can introduce complexity and increase costs.

Propagation of mechanical damage in MXene electrodes post-cycling is challenging. Despite their excellent specific surface area for energy storage, MXenes face mechanical issues like electrode expansion and contraction [155]. This can lead to decreased performance from reduced electrical conductivity and potential delamination. Additionally, mechanical pressure from bending and stretching complicates matters for flexible MXene supercapacitors, making it harder to enhance the mechanical stability of MXene-based electrodes. Creating MXene-based supercapacitors with a balance of high conductivity, high capacitance, and long-term structural stability remains difficult [156–158].

12.7 FUTURE DIRECTIONS FOR MXene-BASED SUPERCAPACITORS

Future developments for MXene-based supercapacitors aim to enhance performance across various applications while tackling challenges like electrode design, scalability, and stability. Developing economical synthesis techniques, including fluoride-free etching and advanced large-scale manufacturing methods, is a promising direction. Research continues on surface modifications, such as protective coatings and hybrid materials, to improve durability and prevent oxidation for better stability. Additionally, advancements in electrolyte engineering, using ionic liquids and redox-active electrolytes, will expand operational voltage and increase energy density [159, 160].

12.7.1 Green Synthesis

MXenes show promise in energy storage, especially in supercapacitors. However, traditional synthesis methods pose safety and environmental risks due to harsh chemicals like hydrofluoric acid. More eco-friendly synthesis routes utilize safer chemicals and reduce energy and by-product waste, supporting global sustainability goals. This method also improves the safety and scalability of MXene production, increasing its potential for industrial applications.

Additionally, green synthesis could enhance the electrochemical properties of MXene-based supercapacitors. For example, ion-liquid functionalized MXene and bio-based reducing agents can be utilized effectively, increasing ion accessibility and capacitance. Green synthesis techniques may also enable coupling MXenes with biocarbon or biomass polymers to create hybrid structures with greater energy density and stability. It represents a promising path for MXene energy storage, providing an eco-friendly approach that addresses both technological and environmental challenges [161, 162].

12.7.2 Advanced Characterization

To reach its full potential, MXene must decrypt complex structure–property interactions that influence its electrochemical performance [163]. Advanced characterization methods provide insights into MXene surface chemistry, ion dynamics, and structural changes during charge–discharge cycles, including synchrotron-based X-ray studies, high-resolution electron microscopy, and in situ and operando spectroscopy. By tracking interactions between MXene surfaces and electrolyte ions in real time, these technologies will help researchers uncover the mechanisms behind energy density, capacitance, and stability.

Sophisticated characterization reveals flaws, phase changes, and degradation pathways in MXene-based electrodes, essential for effective material optimization. Atomic force microscopy (AFM) determines nanoscale surface characteristics vital for ion adsorption, while nuclear magnetic resonance (NMR) and Raman spectroscopy uncover ion transport channels. Combined with computer modeling and machine learning, these methods allow researchers to predict MXene properties for specific supercapacitor applications. For advanced MXene-based supercapacitors with enhanced performance and longevity, these complex techniques require further development.

12.7.3 Integration with Emerging Technologies

MXene-based supercapacitors store energy efficiently and adapt to modern application demands. They can be embedded in wearable and flexible electronics, offering lightweight, compact, and high-performance energy storage for portable devices and smart textiles. Advances in wireless sensor networks and IoT necessitate storage with high power delivery and rapid charging, making MXene ideal due to its high capacitance and improved conductivity.

The incorporation of advanced technologies like artificial intelligence (AI) and machine learning with MXene-based supercapacitors can revolutionize their design and use. MXene compounds can be optimized for specific applications such as grid storage, hybrid energy systems, and EVs through predictive modeling. Additionally, integrating MXenes with renewable energy technologies like solar and wind can develop hybrid solutions for eco-friendly power generation and storage. This approach not only expands applications for MXene-based supercapacitors but also fosters innovation in energy storage, making it crucial for ongoing research and development.

12.8 CONCLUSION

Developed from MXene, supercapacitors represent a breakthrough in high-performance energy storage. Their flexibility and customizable surfaces, along with excellent electrical conductivity, make MXenes adaptable for various applications, from grid-scale systems to handheld devices. Significant advancements in MXene synthesis, surface modification, and hybridization methods have enhanced electrochemical performance, improving capacitance, energy density, and stability, as discussed in this chapter. These advancements highlight the essential role of MXenes in meeting the growing demand for efficient, environmentally friendly, and adaptable energy storage systems.

Greater use of MXene for supercapacitors depends on overcoming major challenges, including sustainable synthesis, long-term structural stability during cycling, and integration with emerging technologies. Through characterization techniques, computational modeling, and interdisciplinary research, scientists can explore MXene's potential and ultimately manufacture next-generation supercapacitors that merge sustainability and high performance. While still in early stages, MXene supercapacitors could revolutionize energy storage technology for a sustainable and interconnected future.

12.9 AUTHOR AI DECLARATION

The authors hereby declare the use of generative AI and AI-assisted technologies in the writing process. While preparing this work, the authors used the ChatGPT tool/service to rephrase several statements in the manuscript. All AI-generated content was reviewed and edited by the authors as needed. The authors take full responsibility for the content of the publication.

ACKNOWLEDGMENT

The Editors acknowledge the use of Grammarly as a tool to assist in editing this book chapter. The suggestions made by Grammarly were carefully considered and incorporated where appropriate, but the author is solely responsible for the final content.

REFERENCES

1. Abdel, Maksoud, Ramy Amer Fahim, Ahmed Esmail Shalan, Mohamed Abd Elkodous, Samuel Oluwaseun Olojede, Ahmed I. Osman, Colin Farrell, Ahmed H. Al-Muhtaseb, Ahmed S. Awed, Ahmed H. Ashour, and David W. Rooney. "Advanced Materials and Technologies for Supercapacitors Used in Energy Conversion and Storage: A Review." *Environmental Chemistry Letters* 19, 1 (2021): 375–439. https://doi.org/10.1007/s10311-020-01075-w
2. Rahman, Md. Arafat, Jin-Hyuk Kim, and Shakhawat Hossain. "Recent Advances of Energy Storage Technologies for Grid: A Comprehensive Review." *Energy Storage* 4, 6 (2022): e322. https://doi.org/10.1002/est2.322
3. Radhakrishnan, Sithara, Minu, Mathew, and Chandra Sekhar Rout. "Microfluidic Sensors Based on Two-Dimensional Materials for Chemical and Biological Assessments." *Materials Advances* 3, 4 (2022): 1874–1904. https://doi.org/10.1039/D1MA00929J
4. Ulbricht, Mathias. "Advanced Functional Polymer Membranes." *Polymer* 47, 7 (2006): 2217–2262. https://doi.org/10.1016/j.polymer.2006.01.084
5. Nasrin, Kabeer, Vasudevan Sudharshan, Kaipannan Subramani, and Marappan Sathish. "Insights into 2D/2D MXene Heterostructures for Improved Synergy in Structure toward Next-Generation Supercapacitors: A Review." *Advanced Functional Materials* 32, 18 (2022): 2110267. https://doi.org/10.1002/adfm.202110267
6. Baig, Mutawara Mahmood, Ifthikar Hussain Gul, Sherjeel Mahmood Baig, and Faisal Shahzad. "2D MXenes: Synthesis, Properties, and Electrochemical Energy Storage for Supercapacitors – A Review." *Journal of Electroanalytical Chemistry* 904 (2022): 115920. https://doi.org/10.1016/j.jelechem.2021.115920

7. Zhong, Cheng, Yida Deng, Wenbin Hu, Jinli Qiao, Lei Zhang, and Jiujun Zhang. "A Review of Electrolyte Materials and Compositions for Electrochemical Supercapacitors." *Chemical Society Reviews* 44, 21 (2015): 7484–7539. https://doi.org/10.1039/C5CS00303B
8. Xiong, Dongbin, Xifei Li, Zhimin Bai, and Shigang Lu. "Recent Advances in Layered $Ti_3C_2T_x$ MXene for Electrochemical Energy Storage." *Small* 14, 17 (2018): 1703419. https://doi.org/10.1002/smll.201703419
9. Hui, Xiaobin, Xiaoli Ge, Ruizheng Zhao, Zhaoqiang Li, and Longwei Yin. "Interface Chemistry on MXene-Based Materials for Enhanced Energy Storage and Conversion Performance." *Advanced Functional Materials* 30, 50 (2020): 2005190. https://doi.org/10.1002/adfm.202005190
10. Naguib, Michael, Michel W. Barsoum, and Yury. Gogotsi. "Ten Years of Progress in the Synthesis and Development of MXenes." *Advanced Materials* 33, 39 (2021): 2103393. https://doi.org/10.1002/adma.202103393
11. Das, Saptarshi, Amritanand Sebastian, Eric Pop, Connor J. McClellan, Aaron D. Franklin, Tibor Grasser, Theresia Knobloch, Yury Illarionov, Ashish V. Penumatcha, Joerg Appenzeller, Zhihong Chen, Wenjuan Zhu, Inge Asselberghs, Lain-Jong Li, Uygar E. Avci, Navakanta Bhat, Thomas D. Anthopoulos, and Rajendra Singh. "Transistors Based on Two-Dimensional Materials for Future Integrated Circuits." *Nature Electronics* 4, 11 (2021): 786–799. https://doi.org/10.1038/s41928-021-00670-1
12. Prasankumar, Thibeorchews, Kaaviah Manoharan, Nadia K. Farhana, Shahid Bashir, K Ramesh, S. Ramesh, and Vigna K. Ramachandaramurthy. "Advancements and Approaches in Developing MXene-Based Hybrid Composites for Improved Supercapacitor Electrodes." *Materials Today Sustainability* 28 (2024): 100963. https://doi.org/10.1016/j.mtsust.2024.100963
13. Salim, Osama, Khaled A. Mahmoud, Kamal Kishore Pant, and Rakesh K. Joshi. "Introduction to MXenes: Synthesis and Characteristics." *Materials Today Chemistry* 14 (2019): 100191. https://doi.org/10.1016/j.mtchem.2019.08.010
14. Pham, Phuong V, Huy-Binh Do, M. Vasundhara, Van-Huy Nguyen, Trieu Nguyen, Hao Van Bui, Van-Duong Dao, Ram K. Gupta, Vinoth Kumar Ponnusamy, and Jin-Hong Park. "Layer-by-Layer Thinning of Two-Dimensional Materials." *Chemical Society Reviews* 53, 10 (2024): 37. https://doi.org/10.1039/D3CS00817G
15. Kadam, Sujit Anil, Komal Prakash Kadam, and Nihar R. Pradhan. "Advancements in 2D MXene-Based Supercapacitor Electrodes: Synthesis, Mechanisms, Electronic Structure Engineering, Flexible Wearable Energy Storage for Real-World Applications, and Future Prospects." *Journal of Materials Chemistry A* 12, 4 (2024): 17992–18046. https://doi.org/10.1039/D4TA00328D
16. Seredych, Mykola, Dana Hulicova-Jurcakova, Gao Qing (Max) Lu, and Teresa J. Bandosz. "Surface Functional Groups of Carbons and the Effects of Their Chemical Character, Density and Accessibility to Ions on Electrochemical Performance." *Carbon* 46, 11 (2008): 1475–1488. https://doi.org/10.1016/j.carbon.2008.06.027
17. Hantanasirisakul, Kanit, and Yury Gogotsi. "Electronic and Optical Properties of 2D Transition Metal Carbides and Nitrides (MXenes)." In *MXenes: Fundamentals and Applications*, 135–205. Boca Raton: CRC Press, 2023. https://doi.org/10.1201/9781003306511-10
18. Qureshi, Ahmadyar, Abdelhay Ahmed Hazem, Shabi Abbas Zaidi, Hafiz Tariq, Kishor Kumar Sadasivuni, Hanan Sheikh, and Maimoona Mohamed. "Emerging Trends in Niobium, Vanadium, and Molybdenum Based MXenes Applications." *Critical Reviews in Solid State and Materials Sciences* 49, 1 (2024): 141–162. https://doi.org/10.1080/10408436.2022.2150124

19. Hussain, Iftikhar, Umay Amara, Faiza Bibi, Abdul Hanan, Muhammad Nazim Lakhan, Irfan Ali Soomro, Amjad Khan, Irum Shaheen, Uzair Sajjad, Gokana Mohana Rani, Muhammad Sufyan Javed, Karim Khan, Muhammad Bilal Hanif, Mohammed A. Assiri, Sumanta Sahoo, Wail Al Zoubi, Debananda Mohapatra, and Kaili Zhang. "Mo-Based MXenes: Synthesis, Properties, and Applications." *Advances in Colloid and Interface Science* 324 (2024): 103077. https://doi.org/10.1016/j.cis.2023.103077
20. Li, Xinliang, Zhaodong Huang, Christopher E. Shuck, Guojin Liang, Yury Gogotsi, and Chunyi Zhi. "MXene Chemistry, Electrochemistry and Energy Storage Applications." *Nature Reviews Chemistry* 6, 6 (2022): 389–404. https://doi.org/10.1038/s41570-022-00384-8
21. Das, Pratteek, and Zong-Shuai Wu. "MXene for Energy Storage: Present Status and Future Perspectives." *Journal of Physics: Energy* 2, 3 (2020): 032004. https://doi.org/10.1088/2515-7655/ab9b1d
22. Benchakar, Mohamed, Lola Loupias, Cyril Garnero, Thomas Bilyk, Claudia Morais, Christine Canaff, and Stéphane Célérier. "One MAX Phase, Different MXenes: A Guideline to Understand the Crucial Role of Etching Conditions on $Ti_3C_2T_x$ Surface Chemistry." *Applied Surface Science* 530 (2020): 147209. https://doi.org/10.1016/j.apsusc.2020.147209
23. Malaki, Massoud, Aziz Maleki, and Rajender Singh Varma. "MXenes and Ultrasonication." *Journal of Materials Chemistry A* 7, 18 (2019): 10843–10857. https://doi.org/10.1039/C9TA01850F
24. Hu, Minmin, Hui Zhang, Tao Hu, Bingbing Fan, Xiaohui Wang, and Zhenjiang Li. "Emerging 2D MXenes for Supercapacitors: Status, Challenges and Prospects." *Chemical Society Reviews* 49, 18 (2020): 6666–6693. https://doi.org/10.1039/D0CS00175A
25. Ali, Salamat, Muhammad Ahsan Farooq Qaisar, Muhammad Sufyan Javed, Khalid Umer, Saleh S. Alarfaji, Muhammad Mateen, Muhammad Chhattal, Shafqat Ali, Anand Parkash, Tensangmu Lama Tamang, and Jing Qi. "Two-Dimensional MXene Based Innovative Electrode Materials for Supercapacitors: Recent Advances and Prospects." *Fuel* 377 (2024): 132783. https://doi.org/10.1016/j.fuel.2024.132783
26. An, Yonglingn, Yuan Tian, Hengtao Shen, Quanyan Man, Shenglin Xiong, and Jinkui Feng. "Two-Dimensional MXenes for Flexible Energy Storage Devices." *Energy & Environmental Science* 16, 10 (2023): 4191–4250. https://doi.org/10.1039/D3EE01841E
27. Liang, Chenming, Yuntong Meng, Yu Zhang, Haifeng Zhang, Weixue Wang, Min Lu, and Guangsheng Wang. "Insights into the Impact of Interlayer Spacing on MXene-Based Electrodes for Supercapacitors: A Review." *Journal of Energy Storage* 65 (2023): 107341. https://doi.org/10.1016/j.est.2023.107341
28. Bilibana, Mawethu Pascoe. "Electrochemical Properties of MXenes and Applications." *Advanced Sensor and Energy Materials* (2023): 100080. https://doi.org/10.1016/j.asems.2023.100080
29. Shi, Haodong, Ming Yue, Chuanfang John Zhang, Yanfeng Dong, Pengfei Lu, Shuanghao, Zheng, Huijuan Huang, Jie Chen, Pengchao Wen, Zhaochao Xu, Qiong Zheng, Xianfeng Li, Yan Yu, Zhong-Shuai Wu. "3D Flexible, Conductive, and Recyclable $Ti_3C_2T_x$ MXene-Melamine Foam for High-Areal-Capacity and Long-Lifetime Alkali-Metal Anode." *ACS Nano* 14, 7 (2020): 8678–8688. https://doi.org/10.1021/acsnano.0c03042
30. Siddu N. K. Pavithra, Jeong, Sang Mun, and Chandra Sekhar Rout. "MXene–Carbon Based Hybrid Materials for Supercapacitor Applications." *Energy Advances* 3, 2 (2024): 341–365. https://doi.org/10.1039/D3YA00502J

31. Jangra, Sahil, Bhushan Kumar, Jaishree Sharma, Shilpi Sengupta, Subhankar Das, R. K. Brajpuriya, Anil Ohlan, Yogendra Kumar Mishra, and Manjeet Singh Goyat. “A Review on Overcoming Challenges and Pioneering Advances: MXene-Based Materials for Energy Storage Applications.” *Journal of Energy Storage* 101 (2024): 113810. https://doi.org/10.1016/j.est.2024.113810
32. Akhter, Raheela, and Shrikant S. Maktedar. “MXenes: A Comprehensive Review of Synthesis, Properties, and Progress in Supercapacitor Applications.” *Journal of Materiomics* 9, 6 (2023): 1196–1241. https://doi.org/10.1016/j.jmat.2023.08.011
33. Vellacheri, Ranjith, Ahmed Al-Haddad, Huaping Zhao, Wenxin Wang, Chengliang Wang, and Yong Lei. “High Performance Supercapacitor for Efficient Energy Storage under Extreme Environmental Temperatures.” *Nano Energy* 8 (2014): 231–237. https://doi.org/10.1016/j.nanoen.2014.06.015
34. Li, Jian, Xiaotao Yuan, Cong Lin, Yanquan Yang, Le Xu, Xin Du, Jinglin Xie, Jianhua Lin, and Junliang Sun. “Achieving High Pseudocapacitance of 2D Titanium Carbide (MXene) by Cation Intercalation and Surface Modification.” *Advanced Energy Materials* 7, 15 (2017): 1602725. https://doi.org/10.1002/aenm.201602725
35. Jayakumar, Suresh, P. Chinnappan Santhosh, Sreeram Anantha Ramakrishna, and Adiyodi Veettil Radhamani. “2D ($Ti_3C_2T_x$) MXene: A Comprehensive Review of Advancements in Synthesis Protocols, Applications in Supercapacitors, Sustainability Targets and Future Prospects.” *Journal of Energy Storage* 97 (2024): 112741. https://doi.org/10.1016/j.est.2024.112741
36. Chen, Yucheng, Huachao Yang, Zhaojun Han, Zheng Bo, Jianhua Yan, Kefa Cen, and Kostya Ken Ostrikov. “MXene-Based Electrodes for Supercapacitor Energy Storage.” *Energy & Fuels* 36, 5 (2022): 2390–2406. https://doi.org/10.1021/acs.energyfuels.1c04104
37. Thomas, Susmi Anna, Abhinandan Patra, Badria M. Al-Shehri, Manickam Selvaraj, Arun Aravind, and Chandra Sekhar Rout. “MXene-Based Hybrid Materials for Supercapacitors: Recent Developments and Future Perspectives.” *Journal of Energy Storage* 55 (2022): 105765. https://doi.org/10.1016/j.est.2022.105765
38. Zhan, Cheng, Michael Naguib, Maria Lukatskaya, Paul RC Kent, Yury Gogotsi, and De-en Jiang. “Understanding the MXene Pseudocapacitance.” *The Journal of Physical Chemistry Letters* 9, 6 (2018): 1223–1228. https://doi.org/10.1021/acs.jpclett.8b00200
39. Etman, Ahmed S, Joseph Halim, and Johanna Rosen. “MXene-Based Zn-Ion Hybrid Supercapacitors: Effects of Anion Carriers and MXene Surface Coatings on the Capacities and Life Span.” *Journal of Energy Storage* 52 (2022): 104823. https://doi.org/10.1016/j.est.2022.104823
40. Panda, Subhasree, Kalim Deshmukh, SK Khadheer Pasha, Jayaraman Theerthagiri, Sivakumar Manickam, and Myong Yong Choi. “MXene Based Emerging Materials for Supercapacitor Applications: Recent Advances, Challenges, and Future Perspectives.” *Coordination Chemistry Reviews* 462 (2022): 214518. https://doi.org/10.1016/j.ccr.2022.214518
41. Shimada, Tatau, Norio Takenaka, Yasunobu Ando, Minoru Otani, Masashi Okubo, and Atsuo Yamada. “Relationship between Electric Double-Layer Structure of MXene Electrode and Its Surface Functional Groups.” *Chemistry of Materials* 34, 5 (2022): 2069–2075. https://doi.org/10.1021/acs.chemmater.1c03328
42. Xu, Peng, Hongyu Xiao, Xiao Liang, Tengfei Zhang, Fanchao Zhang, Chenhao Liu, Beibei Lang, and Qiuming Gao. “A MXene-Based EDA-$Ti_3C_2T_x$ Intercalation Compound with Expanded Interlayer Spacing as High Performance Supercapacitor Electrode Material.” *Carbon* 173 (2021): 135–144. https://doi.org/10.1016/j.carbon.2020.12.043

43. Yedluri, Anil Kumar, Chaitany Jayprakash Raorane, Hosameldin Helmy Hegazy, Tholkappiyan Ramachandran, Seong Cheol Kim, and Md Moniruzzaman. "2D MXene-Based Supercapacitors: A Promising Path Towards High-Performance Energy Storage." *Journal of Energy Storage* 72 (2023): 108433. https://doi.org/10.1016/j.est.2023.108433
44. Yu, Jinyou Qingwenan Fan, Zhiyu Li, and Peng Fu. "MXene-Based Electrode Materials for Supercapacitors: Synthesis, Properties, and Optimization Strategies." *Materials Today Sustainability* 24 (2023): 100551. https://doi.org/10.1016/j.mtsust.2023.100551
45. Aziz, Shujahadeen B, Mohamad A. Brza, Rebar T. Abdulwahid, Jamal Hassan, Hawzhin B. Tahir, Sameerah I. Al-Saeedi, Ranjdar M. Abdullah, and Jihad M. Hadi. "Electrochemical Properties of a Novel EDLC Derived from Plasticized Biopolymer-Based Electrolytes with Valuable Energy Density Close to NiMH Batteries." *Scientific Reports* 13, 1 (2023): 21139. https://doi.org/10.1038/s41598-023-48417-6
46. Armelin, Elaine, Maria M. Pérez-Madrigal, Carol. Alemán, and David Diaz Díaz. "Current Status and Challenges of Biohydrogels for Applications as Supercapacitors and Secondary Batteries." *Journal of Materials Chemistry A* 4, 23 (2016): 8952–8968. https://doi.org/10.1039/c6ta01846g
47. Kim, Eunji, Seongbeen Kim, Hyeong Min Jin, Gyungtae Kim, Hwi-Heon Ha, Yunhui Choi, Kyoungha Min, Su-Ho Cho, Hee Han, Chi Won Ahn, Jaewoo Roh, Il-Kwon Oh, Jinwoo Lee and Yonghee Lee. "Unlocking Novel Functionality: Pseudocapacitive Sensing in MXene-Based Flexible Supercapacitors." *Nano-Micro Letters* 17, 1 (2025): 86. https://doi.org/10.1007/s40820-024-01567-2
48. Faruk, Omer, and Bapan Adak. "Recent Advances in PEDOT:PSS Integrated Graphene and MXene-Based Composites for Electrochemical Supercapacitor Applications." *Synthetic Metals* 297 (2023): 117384. https://doi.org/10.1016/j.synthmet.2023.117384
49. Bhojane, Prateek. "Recent Advances and Fundamentals of Pseudocapacitors: Materials, Mechanism, and Its Understanding." *Journal of Energy Storage* 45 (2022): 103654. https://doi.org/10.1016/j.est.2021.103654
50. Fleischmann, Simon, James B. Mitchell, Ruocun Wang, Cheng Zhan, De-En Jiang, Volker Presser, and Veronica Augustyn. "Pseudocapacitance: From Fundamental Understanding to High Power Energy Storage Materials." *Chemical Reviews* 120, 14 (2020): 6738–6782. https://doi.org/10.1021/acs.chemrev.0c00170
51. Kayali, Emre, Armin VahidMohammadi, Jafar Orangi, and Majid Beidaghi. "Controlling the Dimensions of 2D MXenes for Ultrahigh-Rate Pseudocapacitive Energy Storage." *ACS Applied Materials & Interfaces* 10, 31 (2018): 25949–25954. https://doi.org/10.1021/acsami.8b07397
52. Wu, Jiabin, Qun Li, Christopher E. Shuck, Kathleen Maleski, Husam N. Alshareef, Jun Zhou, Yury Gogotsi, and Liang Huang. "An Aqueous 2.1 V Pseudocapacitor with MXene and V-MnO_2 Electrodes." *Nano Research* 15, 1 (2022): 535–541. https://doi.org/10.1007/s12274-021-3704-5
53. Wang, Xianfen, Satoshi Kajiyama, Hiroki Iinuma, Eiji Hosono, Shinji Oro, Isamu Moriguchi, Masashi Okubo, and Atsuo Yamada. "Pseudocapacitance of MXene Nanosheets for High-Power Sodium-Ion Hybrid Capacitors." *Nature Communications* 6, 1 (2015): 6544. https://doi.org/10.1038/ncomms7544
54. Wu, You, Xiaoming Li, Hui Zhao, Fubing Yao, Jiao Cao, Zhuo Chen, Xiaoding Huang, Dongbo Wang, and Qi Yang. "Recent Advances in Transition Metal Carbides and Nitrides (MXenes): Characteristics, Environmental Remediation and Challenges." *Chemical Engineering Journal* 418 (2021): 129296. https://doi.org/10.1016/j.cej.2021.129296
55. Tamirat, Andebet Gedamu, Xuze Guan, Jingyuan Liu, Jiayan Luo, and Yongyao Xia. "Redox Mediators as Charge Agents for Changing Electrochemical Reactions." *Chemical Society Reviews* 49, 20 (2020): 7454–7478. https://doi.org/10.1039/d0cs00489h

56. Okubo, Masashi, Akira Sugahara, Satoshi Kajiyama, and Atsuo Yamada. "MXene as a Charge Storage Host." *Accounts of Chemical Research* 51, 3 (2018): 591–599. https://doi.org/10.1021/acs.accounts.7b00481
57. Guo, Yanpeng, Yaqing Wei, Huiqiao Li, and Tianyou Zhai. "Layer Structured Materials for Advanced Energy Storage and Conversion." *Small* 13, 45 (2017): 1701649. https://doi.org/10.1002/smll.201701649
58. Li, Kangle, Jiapeng Li, Qizhen Zhu, and Bin Xu. "Three-Dimensional MXenes for Supercapacitors: A Review." *Small Methods* 6, 4 (2022): 2101537. https://doi.org/10.1002/smtd.202101537
59. Bandpey, Mohammad, and Dominik P. J. Barz. "Effects of Interlayer Space Engineering and Surface Modification on the Charge Storage Mechanisms of MXene Nanomaterials: A Review on Recent Developments." *Nanoscale*. (2024). https://doi.org/10.1039/D4NR01317D
60. Ansari, Shagufi Naz, Mohit Saraf, Zahir Abbas, and Shaikh M. Mobin. "Heterostructures of MXenes and Transition Metal Oxides for Supercapacitors: An Overview." *Nanoscale* (2023). https://doi.org/10.1039/D3NR01755A
61. Tomar, Anuj Kumar, Tolendra Kshetri, Nam Hoon Kim, and Joong Hee Lee. "Cation and Anion (De) Intercalation into MXene/Perovskite Oxides for High-Rate Intercalation Pseudocapacitance." *Energy Storage Materials* 50 (2022): 86–95. https://doi.org/10.1016/j.ensm.2022.05.026
62. Das, K Sushanta, Lingaraj Pradhan, Jongwoo Hong, Seong Chan Jun, and Bikash Kumar Jena. "Intercalation Pseudocapacitance of Monolithic Microporous Carbon Thin Film Scrolls and Their Battery Type Behavior." *Available at SSRN 5054645*. https://doi.org/10.1016/j.jpowsour.2025.237174
63. Javed, Muhammad Sufyan, Abdul Mateen, Iftikhar Hussain, Awais Ahmad, Muhammad Mubashir, Shaukat Khan, Mohammed A. Assiri, Sayed M. Eldin, Syed Shoaib Ahmad Shah, and Weihua Han. "Recent Progress in the Design of Advanced MXene/Metal Oxides-Hybrid Materials for Energy Storage Devices." *Energy Storage Materials* 53 (2022): 827–872. https://doi.org/10.1016/j.ensm.2022.09.015
64. Wang, Yitong, and Yuhua Wang. "Recent Progress in MXene Layers Materials for Supercapacitors: High-Performance Electrodes." *SmartMat* 4, 1 (2023): e1130. https://doi.org/10.1002/smm2.1130
65. Wang, Zhiyu, Si Qin, Shayan Seyedin, Jizhen Zhang, Jiangting Wang, Ariana Levitt, Na Li, Carter Haines, Raquel Ovalle-Robles, Weiwei Lei, Yury Gogotsi, Ray H. Baughman, and Joselito M. Raza. "High-Performance Biscrolled MXene/Carbon Nanotube Yarn Supercapacitors." *Small* 14, 37 (2018): 1802225. https://doi.org/10.1002/smll.201802225
66. Lee, Jeong Bok, Gwan Hyun Choi, and Pil J. Yoo. "Oxidized-Co-Crumpled Multiscale Porous Architectures of MXene for High Performance Supercapacitors." *Journal of Alloys and Compounds* 887 (2021): 161304. https://doi.org/10.1016/j.jallcom.2021.161304
67. Pınar, Pınar Talay, Mehmet Gülcan, and Yavuz Yardım. "Synthesis and Characterization of a Nanocomposite Consisting of $Ti_3C_2T_x$ (MXene) and WS_2 Nanosheets for Potential Use in Supercapacitors." *Journal of Alloys and Compounds* 1010 (2025): 177656. https://doi.org/10.1016/j.jallcom.2024.177656
68. Abdah, Muhammad Amirul Aizat Mohd, Hafiz Taimoor Ahmed Awan, Maida Mehar, Muhammad Norhaffis Mustafa, Rashmi Walvekar, Mir Waqas Alam, Mohammad Khalid, Reddicherla Umapathi, and Vishal Chaudhary. "Advancements in MXene-Polymer Composites for High-Performance Supercapacitor Applications." *Journal of Energy Storage* 63 (2023): 106942. https://doi.org/10.1016/j.est.2023.106942
69. Patil, Amar M, Nutthaphak Kitiphatpiboon, Xiaowei An, Xiaoqiong Hao, Shasha Li, Xiaogang Hao, Abuliti Abudula, and Guoqing Guan. "Fabrication of a High-Energy Flexible All-Solid-State Supercapacitor Using Pseudocapacitive 2D-$Ti_3C_2T_x$-MXene

and Battery-Type Reduced Graphene Oxide/Nickel–Cobalt Bimetal Oxide Electrode Materials." *ACS Applied Materials & Interfaces* 12, 47 (2020): 52749–52762. https://doi.org/10.1021/acsami.0c16221

70. Luo, Yangyang, Chenhui Yang, Yapeng Tian, Yi Tang, Xingtian Yin, and Wenxiu Que. "A Long Cycle Life Asymmetric Supercapacitor Based on Advanced Nickel-Sulfide/Titanium Carbide (MXene) Nanohybrid and MXene Electrodes." *Journal of Power Sources* 450 (2020): 227694. https://doi.org/10.1016/j.jpowsour.2019.227694
71. Niu, Shanshan, Zhiyu Wang, Mingliang Yu, Mengzhou Yu, Luyang Xiu, Song Wang, Xianhong Wu, and Jieshan Qiu. "MXene-Based Electrode with Enhanced Pseudocapacitance and Volumetric Capacity for Power-Type and Ultra-Long Life Lithium Storage." *ACS Nano* 12, 4 (2018): 3928–3937. https://doi.org/10.1021/acsnano.8b01459
72. Manoharan, Sindhuja, Karthikeyan Krishnamoorthy, Arunprasath Sathyaseelan, and Sang-Jae Kim. "High-Power Graphene Supercapacitors for the Effective Storage of Regenerative Energy during the Braking and Deceleration Process in Electric Vehicles." *Materials Chemistry Frontiers* 5, 16 (2021): 6200–6211. https://doi.org/10.1039/D1QM00465D
73. Liu, Xinlong, Feng Xu, Zhiyong Li, Zhiting Liu, Wei Yang, Yufei Zhang, Haosen Fan, and Hui Ying Yang. "Design Strategy for MXene and Metal Chalcogenides/Oxides Hybrids for Supercapacitors, Secondary Batteries and Electro/Photocatalysis." *Coordination Chemistry Reviews* 464 (2022): 214544. https://doi.org/10.1016/j.ccr.2022.214544
74. Feng, Mengyue, Yang Zhang, Xiaolin Zhu, Wenxing Chen, Wangyang Lu, and Guan Wu. "Interface-Anchored Covalent Organic Frameworks@Amino-Modified $Ti_3C_2T_x$ MXene on Nylon 6 Film for High-Performance Deformable Supercapacitors." *Angewandte Chemie* 135, 33 (2023): e202307195. https://doi.org/10.1002/ange.202307195
75. Shi, Chenjing, Zijie Liu, Zhen Tian, Dan Li, Yanjun Chen, Li Guo, and Yanzhong Wang. "Fabrication of 3D MXene@Graphene Hydrogel with High Ion Accessibility via Al-Induced Self-Assembly and Reduction for High-Performance Supercapacitors." *Electrochimica Acta* 464 (2023): 142892. https://doi.org/10.1016/j.electacta.2023.142892
76. Sharma, Rahul, Harish Kumar, Gaman Kumar, Saloni Sharma, Ranjan Aneja, Ashok K. Sharma, Ramesh Kumar, and Parvin Kumar. "Progress and Challenges in Electrochemical Energy Storage Devices: Fabrication, Electrode Material, and Economic Aspects." *Chemical Engineering Journal* 468 (2023): 143706. https://doi.org/10.1016/j.cej.2023.143706
77. Stephen, Ikpatim, Lawal Usman, Amadi I. Juliet, and James T. Ugye. "Advancing Chemical Sciences for Promoting Food Security and Human Empowerment in North–Eastern Nigeria." *ChemTrends* (2024): 193.
78. Vaishag, Pushpalatha Vijayakumar, Sabarija A. Mohandas, Mufeeda Mufeeda, Parveena Gangadharan, and Pathath Abdul Rasheed. "Conversion of Electronic Waste to an Electrochemical Sensor for Dopamine: Using MXene-Modified Liquid Crystal Display Panels." *ACS Sustainable Chemistry & Engineering* 11, 34 (2023): 12771–12779. https://doi.org/10.1021/acssuschemeng.3c03173
79. Hussain, Zahid, Muhammad Imran, Amir Muhammad Afzal, Muhammad Waqas Iqbal, Sohail Mumtaz, Shaik Abdul Munnaf, Saikh Mohammad Wabaidur, Wajeeha Fatima, Samia Safdar, Muhammad Azhar Mumtaz, Muhammad Hamza Waris and Zubair Ahmad. "High Performance MnNbS@MXene Hybrid Electrode for Battery-Supercapacitor Hybrid Device and Biomedical Applications." *Materials Science in Semiconductor Processing* 173 (2024): 108091. https://doi.org/10.1016/j.mssp.2023.108091

80. Liu, Yang, Hui Zhou, Weixiao Zhou, Si Meng, Cheng Qi, Zhou Liu, and Tiantian Kong. "Biocompatible, High-Performance, Wet-Adhesive, Stretchable All-Hydrogel Supercapacitor Implant Based on PANI@rGO/MXenes Electrode and Hydrogel Electrolyte." *Advanced Energy Materials* 11, 30 (2021): 2101329. https://doi.org/10.1002/aenm.202101329
81. Zhang, Jian, Xiuchen Wang, Gege Hang, Wenjing Zhang, Zhifang Zheng, Jin Duan, and Zhe Liu. "Polypyrrole In-Situ Polymerized MXene/TPU Fiber Electrode for Flexible Supercapacitors." *Composites Communications* 45 (2024): 101817. https://doi.org/10.1016/j.coco.2024.101817
82. Zhu, Xiaolin, Yang Zhang, Zengming Man, Wangyang Lu, Wei Chen, Jianhong Xu, Ningzhong Bao, Wenxing Chen, and Guan Wu. "Microfluidic-Assembled Covalent Organic Frameworks@$Ti_3C_2T_x$ MXene Vertical Fibers for High-Performance Electrochemical Supercapacitors." *Advanced Materials* 35, 46 (2023): 2307186. https://doi.org/10.1002/adma.202307186
83. Xin, Wei, Guo-Qiang Xi, Wen-Tao Cao, Chang Ma, Tong Liu, Ming-Guo Ma, and Jing Bian. "Lightweight and Flexible MXene/CNF/Silver Composite Membranes with a Brick-Like Structure and High-Performance Electromagnetic-Interference Shielding." *RSC Advances* 9, 51 (2019): 29636–29644. https://doi.org/10.1039/C9RA06399D
84. Wang, Wei, Guohui Chen, Weiqi Kong, Junshu Chen, Linyu Pu, Jiaxu Gong, Huan Zhang, and Yatang Dai. "Sandwich-Like High-Performance $Ti_3C_2T_x$ MXene/$NiCo_2O_4$ Nanosphere Composites for Asymmetric Supercapacitor Application." *Journal of Energy Storage* 86 (2024): 111097. https://doi.org/10.1016/j.est.2024.111097
85. Xu, Xiaoxu, Zhaoqing Lu, Li Hua, Yajie Kou, Jiayue Dong, Jinlong Wu, and Yuanming Wang. "All-Pseudocapacitive Two-Dimensional Materials with Covalent Active Groups and Reinforcement of Aramid Nanofibers for Lightweight, High-Strength, and Stable Asymmetric Supercapacitors." *Journal of Energy Storage* 108 (2025): 115179. https://doi.org/10.1016/j.est.2024.115179
86. Zhou, Zehang, Jize Liu, Xinxing Zhang, Dong Tian, Zeying Zhan, and Canhui Lu. "Ultrathin MXene/Calcium Alginate Aerogel Film for High-Performance Electromagnetic Interference Shielding." *Advanced Materials Interfaces* 6, 6 (2019): 1802040. https://doi.org/10.1002/admi.201802040
87. Vijayakumar, Manavalan, George Elsa, Aamani Nirogi, Rajendran Navaneethan, Ammaiyappan Bharathi Sankar, and Mani Karthik. "MXenes and Their Composites for Hybrid Capacitors and Supercapacitors: a Critical Review." *Emergent Materials* 4, 3 (2021): 655–672. https://doi.org/10.1007/s42247-021-00204-7
88. Patra, Abhinandan, Pratap Mane, Shyam. R. Polaki, Brahmananda Chakraborty, and Chandra Sekhar Rout. "Enhanced Charge Storage Performance of MXene-Based All-Solid-State Supercapacitor with Vertical Graphene Arrays as the Current Collector." *Journal of Energy Storage* 54 (2022): 105355. https://doi.org/10.1016/j.est.2022.105355
89. Adekoya, Gbolahan Joseph, Oluwaseun Chijioke Adekoya, Rotimi Emmanuel Sadiku, Yskandar Hamam, and Suprakas Sinha Ray. "Applications of MXene-Containing Polypyrrole Nanocomposites in Electrochemical Energy Storage and Conversion." *ACS Omega* 7, 44 (2022): 39498–39519. https://doi.org/10.1021/acsomega.2c02706
90. Wang, Yongbin, Ningjun Chen, Bin Zhou, Xuefeng Zhou, Ben Pu, Jia Bai, Qi Tang, Yan Liu, and Weiqing Yang. "NH_3-Induced In Situ Etching Strategy Derived 3D-Interconnected Porous MXene/Carbon Dots Films for High-Performance Flexible Supercapacitors." *Nano-Micro Letters* 15, 1 (2023): 231. https://doi.org/10.1007/s40820-023-01204-4
91. Ghaemi, S. P., S. Morteza Masoudpanah, and Parham Heidari. "Facile Synthesis of $Ni_{0.5}Zn_{0.5}Fe_2O_4$/MXene Composite Powders for High-Performance Asymmetric Supercapacitors." *Journal of Energy Storage* 72 (2023): 108439. https://doi.org/10.1016/j.est.2023.108439

92. Akbar, Sawera, Mirza Mahmood Baig, Muhammad Naziruddin Khan, Zeid A. AlOthman, Imran Shakir, Seung-Goo Lee, and Muhammad Farooq Warsi. "Ag-Doped $MgCo_2O_4$/MXene Composite: Tailoring the Electrochemical Properties via 2D Layered Material for Next Generation Supercapacitor Study." *Journal of Alloys and Compounds* 1009 (2024): 176813. https://doi.org/10.1016/j.jallcom.2024.176813
93. Singh, Uday Karanbir, Harpreet Kaur, Pushpinder Kaur, and Richa Rastogi. "Unlocking Potential: Recent Advances in MXene Supercapacitors for Flexible Energy Storage Devices." *Nano-Structures & Nano-Objects* 39 (2024): 101290. https://doi.org/10.1016/j.nanoso.2024.101290.
94. Pang, Jinbo, Rafael G. Mendes, Alicja Bachmatiuk, Liang Zhao, Huy Q. Ta, Thomas Gemming, Hong Liu, Zhongfan Liu, and Mark H. Rummeli. "Applications of 2D MXenes in Energy Conversion and Storage Systems." *Chemical Society Reviews* 48, 1 (2019): 72–133. https://doi.org/10.1039/C8CS00324F
95. Boota, Muhammad, Murugesan Rajesh, and Matthieu Bécuwe. "Multi-Electron Redox Asymmetric Supercapacitors Based on Quinone-Coupled Viologen Derivatives and $Ti_3C_2T_x$ MXene." *Materials Today Energy* 18 (2020): 100532. https://doi.org/10.1016/j.mtener.2020.100532
96. Tyagi, Nahid, Vipul Bhardwaj, Sudheshna Moka, Manoj Kumar Singh, Manika Khanuja, and Gaurav Sharma. "An Overview on Synthesis of MXene and MXene Based Nanocomposites for Supercapacitors." *Materials Today Communications* (2024): 110223. https://doi.org/10.1016/j.mtcomm.2024.110223
97. Alikhani, Mohammadreza, Soheil Hajibaba, Mohsen Moayedi, and Yaser Abdi. "Borophene–MXene Heterostructures as Supercapacitor." *Journal of Energy Storage* 99 (2024): 113301. https://doi.org/10.1016/j.est.2024.113301
98. Benzigar, Mercy R, Venkata DBC Dasireddy, Xinwei Guan, Tom Wu, and Guozhen Liu. "Advances on Emerging Materials for Flexible Supercapacitors: Current Trends and Beyond." *Advanced Functional Materials* 30, 40 (2020): 2002993. https://doi.org/10.1002/adfm.202002993
99. Tareen, Ayesha Khan, Karim Khan, Muhammad Iqbal, Ye Zhang, Jianyu Long, Asif Mahmood, Nasir Mahmood, Zhongjian Xie, Chuan Li, and Han Zhang. "Recent Advance in Two-Dimensional MXenes: New Horizons in Flexible Batteries and Supercapacitors Technologies." *Energy Storage Materials* 53 (2022): 783–826. https://doi.org/10.1016/j.ensm.2022.09.030
100. Watanabe, Masayoshi, Morgan L. Thomas, Shiguo Zhang, Kazuhide Ueno, Tomohiro Yasuda, and Kaoru Dokko. "Application of Ionic Liquids to Energy Storage and Conversion Materials and Devices." *Chemical Reviews* 117, 10 (2017): 7190–7239. https://doi.org/10.1021/acs.chemrev.6b00504
101. Ma, Rui, Lingyun Cao, Jingting Zhuo, Jintao Lu, Jiaxiang Chen, Jue Huang, Guowei Yang, and Fang Yi. "Designed Redox-Electrolyte Strategy Boosted with Electrode Engineering for High-Performance $Ti_3C_2T_x$ MXene-Based Supercapacitors." *Advanced Energy Materials* 13, 34 (2023): 2301219. https://doi.org/10.1002/aenm.202301219
102. Ma, Rui, Xujing Zhang, Jingting Zhuo, Lingyun Cao, Yutong Song, Yajiang Yin, Xiaofeng Wang, Guowei Yang, and Fang Yi. "Self-Supporting, Binder-Free, and Flexible $Ti_3C_2T_x$ MXene-Based Supercapacitor Electrode with Improved Electrochemical Performance." *ACS Nano* 16, 6 (2022): 9713–9727. https://doi.org/10.1021/acsnano.2c03351
103. Prabhakar Vattikuti, Surya V, Jaesool Shim, Pitcheri Rosaiah, Alain Mauger, and Christian M. Julien. "Recent Advances and Strategies in MXene-Based Electrodes for Supercapacitors: Applications, Challenges and Future Prospects." *Nanomaterials* 14, 1 (2023): 62. https://doi.org/10.3390/nano14010062

104. Zhou, Jingyuan, Zhaoyang Lin, Huaying Ren, Xidong Duan, Imran Shakir, Yu Huang, and Xiangfeng Duan. "Layered Intercalation Materials." *Advanced Materials* 33, 25 (2021): 2004557. https://doi.org/10.1002/adma.202004557
105. Guo, Tiezhu, Di Zhou, Min Gao, Shungui Deng, Mohammad Jafarpour, Jonathan Avaro, Antonia Neels Erwin Hack, Jing Wang, Jakob Heier, and Chuanfang Zhang. "Large-Area Smooth Conductive Films Enabled by Scalable Slot-Die Coating of $Ti_3C_2T_x$ MXene Aqueous Inks." *Advanced Functional Materials* 33, 15 (2023): 2213183. https://doi.org/10.1002/adfm.202213183
106. Zheng, Wei, Li Yang, Peigen Zhang, Varun Natu, Zhengming Sun, Johanna Rosen, and Michel W. Barsoum. "Mass Loading and Self-Discharge Challenges for MXene-Based Aqueous Supercapacitors." *Energy Storage Materials* (2023): 103037. https://doi.org/10.1016/j.ensm.2023.103037
107. Li, Pei-Xuan, GuoZhen Guan, Xin Shi, Lei Lu, Yu Chao Fan, Jie Xu, Yuan-Yuan Shang, Ying-Jiu Zhang, Jin-Quan Wei, and Feng Mei Guo. "Bidirectionally Aligned MXene Hybrid Aerogels Assembled with MXene Nanosheets and Microgels for Supercapacitors." *Rare Metals* 42, 4 (2023): 1249–1260. https://doi.org/10.1007/s12598-022-02189-6
108. Liu, Wenen, Dan Luo, Maiwen Zhang, Jiangtao Chen, Mathew Li, Anna Chen, Shibo Xi and Aiping Yu. "Engineered MXene Quantum Dots for Micro-Supercapacitors with Excellent Capacitive Behaviors." *Nano Energy* 122 (2024): 109332. https://doi.org/10.1016/j.nanoen.2024.109332
109. Feng, Shuxuan, Xin Wang, Menglu Wang, Chong Bai, Shitai Cao, and Desheng Kong. "Crumpled MXene Electrodes for Ultrastretchable and High-Area-Capacitance Supercapacitors." *Nano Letters* 21, 18 (2021): 7561–7568. https://doi.org/10.1021/acs.nanolett.1c02071
110. Le, Thi Anh, Ngoc Quang Tran, Yeseul Hong, and Hyoyoung Lee. "Intertwined Titanium Carbide MXene within a 3D Tangled Polypyrrole Nanowires Matrix for Enhanced Supercapacitor Performances." *Chemistry–A European Journal* 25, 4 (2019): 1037–1043. https://doi.org/10.1002/chem.201803748
111. Cui, Haonan, Hongyu Mi, Chenchen Ji, Fengjiao Guo, Yanna Chen, Dandan Wu, Jieshan Qiu, and Haijiao Xie. "A Durable MXene-Based Zinc Ion Hybrid Supercapacitor with Sulfated Polysaccharide Reinforced Hydrogel/Electrolyte." *Journal of Materials Chemistry A* 9, 42 (2021): 23941–23954. https://doi.org/10.1039/D1TA06974H
112. Yang, Mengdan, Huiqing Lou, Xiaobing Kong, Rui Pang, Ding Zhang, Weixue Meng, Meng Li, Xinguang Huang, Shipeng Zhang, Yuanyuan Shang, and Anyuan Cao. "Recent Advances in MXene-Based Fibers: Fabrication, Performance, and Application." *Small Methods* 7, 10 (2023): 2300518. https://doi.org/10.1002/smtd.202300518
113. Zhuang, Xiaoli, Songtao Zhang, Yijian Tang, Feng Yu, Zhaomin Li, and Huan Pang. "Recent Progress of MOF/MXene-Based Composites: Synthesis, Functionality and Application." *Coordination Chemistry Reviews* 490 (2023): 215208. https://doi.org/10.1016/j.ccr.2023.215208
114. Qiu, Hui, Xiaowei Qu, Yujiao Zhang, Su Chen, and Yizhong Shen. "Robust PANI@MXene/GQDs-Based Fiber Fabric Electrodes via Microfluidic Wet-Fusing Spinning Chemistry." *Advanced Materials* 35, 38 (2023): 2302326. https://doi.org/10.1002/adma.202302326
115. Wang, Yesheng, Yongpeng Cui, Dongqing Kong, Xiaoning Wang, Bin Li, Tonghui Cai, Xuejin Li, Jing Xu, Yanpeng Li, Youguo Yan, Han Hu, Mingbo Wu, Qingzhong Xue, Zifeng Yan, Lianming Zhao, and Wei Xing. "Stimulation of Surface Terminating Group by Carbon Quantum Dots for Improving Pseudocapacitance of $Ti_3C_2T_x$ MXene Based Electrode." *Carbon* 180 (2021): 118–126. https://doi.org/10.1016/j.carbon.2021.04.050

116. Zhang, Haonan, Yunchuan Luo, Jie Zhou, Longsen Wang, Leilei Shu, Wen He, Qi Zhang, and Peihong Wang. "One-Step Wet-Spinning of High-Energy Density Coaxial Fibrous Supercapacitors Based on In Situ Carbon-Modified Nitrogen-Doped MXene Nanosheets." *Nano Letters* 24, 33 (2024): 10131–10138. https://doi.org/10.1021/acs.nanolett.4c02266.
117. Garg, Ruby, Alpana Agarwal, and Mohit Agarwal. "A Review on MXene for Energy Storage Application: Effect of Interlayer Distance." *Materials Research Express* 7, 2 (2020): 022001. https://doi.org/10.1088/2053-1591/ab750d
118. Sagadevan, Suresh, Is Fatimah, Anita Lett, Babak Kakavandi, Tetsuo Soga, Won-Chun Oh, and Hyacinthe Randriamahazaka. "Exploring the Potential of MXene-Based Aerogels and Hybrid Nanocomposites for Supercapacitor Applications." *Journal of Energy Storage* 99 (2024): 113269. https://doi.org/10.1016/j.est.2024.113269.
119. Jaberi, S. Y. Shajaripour, Parvin Asen, Ali Esfandiar, and Valeri P. Tolstoy. "MXene/Carbon Hybrid Nanostructures and Heteroatom-Doped Derivatives for Enhanced Electrochemical Energy Storage." *Journal of Energy Storage* 90 (2024): 111751. https://doi.org/10.1016/j.est.2024.111751
120. Solangi, Nadeem Hussain, Aumber Abbas, Nabisab Mujawar Mubarak, Rama Rao Karri, Shrouq H. Aleithan, Jamal Kazmi, Waqas Ahmad, and Karim Khan. "Insight Mechanism of MXene for Future Generation Highly Efficient Energy Storage Device." *Materials Today Sustainability* (2024): 100896. https://doi.org/10.1016/j.mtsust.2024.100896
121. Hepel, Maria. "Advances in Micro-Supercapacitors (MSCs) with High Energy Density and Fast Charge-Discharge Capabilities for Flexible Bioelectronic Devices—A Review." *Electrochemical Science Advances* 3, 3 (2023): e2100222. https://doi.org/10.1002/elsa.202100222
122. Awan, Hafiz Taimoor Ahmed, Muhammad Amirul Aizat Mohd Abdah, Maida Mehar, Rashmi Walvekar, Vishal Chaudhary, Mohammad Khalid, and Ajit Khosla. "MXene-Polymer Hybrid Composites for Advanced Energy Storage: Insights into Supercapacitors and Batteries." *Journal of Energy Storage* 95 (2024): 112449. https://doi.org/10.1016/j.est.2024.112449
123. Luo, Wenlong, Yong. Ma, Tingxi Li, Hamdy Khamees Thabet, Chunping Hou, Mohamed M. Ibrahim, Salah M. El-Bahy, Ben Bin Xu, and Zhanhu Guo. "Overview of MXene/Conducting Polymer Composites for Supercapacitors." *Journal of Energy Storage* 52 (2022): 105008. https://doi.org/10.1016/j.est.2022.105008
124. Ezika, Anthony Chidi, Christopher Chiedozie Obi, Henry Chukwuka Oyeoka, Uzoma Ebenezer Enwerem, Ugwu Shadrack Chukwuemeka, and Ibenta Martin Emeka. "Insight on Electronic and Thermal Behaviors of Conductive MXene-$Ti_3C_2T_x$-Based Polymeric Hybrid Material and Their Capacitive Energy Storage Applications: A Review." *Advanced Energy Conversion Materials* (2024): 156–194. https://doi.org/10.37256/aecm.5220244550
125. Wu, Zhitan, Tongxin Shang, Yaqian Deng, Ying Tao, and Quan-Hong Yang. "The Assembly of MXenes from 2D to 3D." *Advanced Science* 7, 7 (2020): 1903077. https://doi.org/10.1002/advs.201903077
126. Li, Lingfang, Bin Zeng, Chuang Xiang, and Wen Liu. "Novel Design Strategies of Three-Dimensional MXene Structures and Their Applications in Metal-Ion Hybrid Capacitors." *Australian Journal of Chemistry* (2023). https://doi.org/10.1071/CH23090.
127. Chetana, S., Muhammad Amirul Aizat Mohd Abdah, Vikas N. Thakur, M. S. Govinda Gowda, Priyvart Choudhary, Jagadeesh Babu Sriramoju, Dinesh Rangappa, Sumira Malik, Sarvesh Rustagi, and Mohammad Khalid. "Progress and Prospects of MXene-Based Hybrid Composites for Next-Generation Energy Technology." *Journal of The Electrochemical Society* 170, 12 (2023): 120530. https://doi.org/10.1149/1945-7111/ad0c64

128. Srikrajang, Siwaluk, Latiful Kabir, Suresh Sagadevan, Karna. Wijaya, and Won-Chun. Oh. "Representative Modeling of Biocompatible MXene Nanocomposites for Next-Generation Biomedical Technologies and Health Care." *Journal of Materials Chemistry B* (2025). https://doi.org/10.1039/D4TB02478H
129. Tian, Yapeng, Chenhui Yang, Yangyang Luo, Hongyang Zhao, Yaping Du, Ling Bing Kong, and Wenxiu Que. "Understanding MXene-Based 'Symmetric' Supercapacitors and Redox Electrolyte Energy Storage." *ACS Applied Energy Materials* 3, 5 (2020): 5006–5014. https://doi.org/10.1021/acsaem.0c00527
130. Buhari, Salahuddeen, Manuj Ahuja, Binson Babu, and Ashish Chalana. "Advances in MXenes: Synthesis and Electrochemical Applications." *Advanced Sustainable Systems* (2025): e00419. https://doi.org/10.1002/adsu.202500419
131. Zhang, Peng, Yanmeng Peng, Qizhen Zhu, Razium Ali Soomro, Ning Sun, and Bin Xu. "3D Foam-Based MXene Architectures: Structural and Electrolytic Engineering for Advanced Potassium-Ion Storage." *Energy & Environmental Materials* 7, 4 (2024): e12379. https://doi.org/10.1002/eem2.12379
132. Mohanty, Ritik, Amtul Nashim, Koushik Parida, and Kulamani Parida. "Can Doped-MXene-Based Supercapacitors Be the Game-Changer for Future Energy Landscape? A Critical Perspective." *Langmuir* 40, 28 (2024): 14197–14213. https://doi.org/10.1021/acs.langmuir.4c01242
133. Sun, Li, Xiao Su, Yujuan Chen, Kelei Zhuo, Huanhuan Li, Dong Sun, and Jianji Wang. "Ferric Ion-Assisted Assembly of MXene/TiO_2-Graphene Aerogel for Ionic Liquid-Based Supercapacitors." *Chemical Engineering Journal* 476 (2023): 146731. https://doi.org/10.1016/j.cej.2023.146731
134. Senthilkumar, S. Thangavelu, Ramakrishnan Kalai Selvan, and J. S. Melo. "Redox Additive/Active Electrolytes: A Novel Approach to Enhance the Performance of Supercapacitors." *Journal of Materials Chemistry A* 1, 40 (2013): 12386–12394. https://doi.org/10.1039/C3TA11959A
135. Zhu, Yuanyuan, Shuanghao Zheng, Pengfei Lu, Jiaxin Ma, Pratteek Das, Feng Su, Hui-Ming Cheng, and Zhong-Shuai Wu. "Kinetic Regulation of MXene with Water-in-LiCl Electrolyte for High-Voltage Micro-Supercapacitors." *National Science Review* 9, 7 (2022): nwac024. https://doi.org/10.1093/nsr/nwac024
136. Nahirniak, Svitlana, Apurba Ray, and Bilge Saruhan. "Challenges and Future Prospects of the MXene-Based Materials for Energy Storage Applications." *Batteries* 9, 2 (2023): 126. https://doi.org/10.3390/batteries9020126
137. Shoeb, Mohd, Fouzia Mashkoor, Javed Alam Khan, Mohammad Naved Khan, Mohammed Ashraf Gondal, and Changyoon Jeong. "Effect of Mass Loading and Fabrication of VARTM-Based High-Performance Solid-State Supercapacitor Device with MXene-$NiCo_3S_4$ Nanocomposite." *Journal of Energy Storage* 97 (2024): 112654. https://doi.org/10.1016/j.est.2024.112654
138. Zhou, Yuxiao, Yali Zhang, Kunpeng Ruan, Hua Guo, Mukun He, Hua Qiu, and Junwei Gu. "MXene-Based Fibers: Preparation, Applications, and Prospects." *Science Bulletin* (2024). https://doi.org/10.1016/j.scib.2024.03.017
139. Yang, Qiuyan, Zhen Xu, Bo Fang, Tieqi Huang, Shengying Cai, Hao Chen, Yingjun Liu, Karthikeyan Gopalsamy, Weiwei Gao, and Chao Gao. MXene/Graphene Hybrid Fibers for High Performance Flexible Supercapacitors." *Journal of Materials Chemistry A* 5, 42 (2017): 22113–22119. https://doi.org/10.1039/C7TA07999K
140. Zhang, Chuanfang, Lorcan McKeon, Matthias P. Kremer, Sang-Hoon Park, Oskar Ronan, Andrés Seral-Ascaso, Sebastian Barwich, Cormac Ó Coileáin, Niall McEvoy, Hannah C. Nerl, Babak Anasori, Jonathan N. Coleman, Yury Gogotsi, and Valeria Nicolosi. "Additive-Free MXene Inks and Direct Printing of Micro-Supercapacitors." In *MXenes*, 463–485. Jenny Stanford Publishing, 2023. https://doi.org/10.1038/s41467-019-09398-1

141. Ma, Chang, Ming-Guo Ma, Chuanling Si, Xing-Xiang Ji, and Pengbo Wan. "Flexible MXene-Based Composites for Wearable Devices." *Advanced Functional Materials* 31, 22 (2021): 2009524. https://doi.org/10.1002/adfm.202009524
142. Qiongqiong Lu, Congcong Liu, Yirong Zhao, Wengao Pan, Kun Xie, Pengfei Yue, Guoshang Zhang, Ahmad Omar, Lixiang Liu, Minghao Yu, and Daria Mikhailova. "Freestanding MXene-Based Macroforms for Electrochemical Energy Storage Applications." *SusMat* 3, 4 (2023): 471–497. https://doi.org/10.1002/sus2.151
143. Xu, Shuaikai, Yubing Li, Kunhua Quan, Guodong Wei, Tangming Mo, Dan Huang, Haifu Huang, Xianqing Liang, and Ya Yang. "Facilitating Ion Transport in Porous Chemically Bonded Black Phosphorene/MXene Heterostructured Films for Flexible High-Rate Supercapacitors." *Energy Storage Materials* 72 (2024): 103721. https://doi.org/10.1016/j.ensm.2024.103721
144. Aghayar, Zahra, Massoud Malaki, and Yizhou Zhang. "MXene-Based Ink Design for Printed Applications." *Nanomaterials* 12, 23 (2022): 4346. https://doi.org/10.3390/nano12234346
145. Sohan, Arya, Pravallika Banoth, Mariya Aleksandrova, Andrews Nirmala Grace, and Pratap Kollu. "Review on MXene Synthesis, Properties, and Recent Research Exploring Electrode Architecture for Supercapacitor Applications." *International Journal of Energy Research* 45, 14 (2021): 19746–19771. https://doi.org/10.1002/er.7075
146. Wang, Zixing, Zhong Xu, Haichao Huang, Xiang Chu, Yanting Xie, Da Xiong, Cheng Yan, Haibo Zhao, Haitao Zhang, and Weiqing Yang. "Unraveling and Regulating Self-Discharge Behavior of $Ti_3C_2T_x$ MXene-Based Supercapacitors." *ACS Nano* 14, 4 (2020): 4916–4924. https://doi.org/10.1021/acsnano.0c01056
147. Otgonbayar, Zambaga, Sunhye Yang, Ick-Jun Kim, and Won-Chun Oh. "Recent Advances in 2D MXene and Solid State Electrolyte for Energy Storage Applications: Comprehensive Review." *Chemical Engineering Journal* (2023): 144801. https://doi.org/10.1016/j.cej.2023.144801
148. Tripathy, Dipti, and Sanjeev Moharana. "MXene-Based Materials for Fuel Cell Applications." In *MXenes*, 280–302. CRC Press, 2025. https://doi.org/10.1201/9781003366225-18
149. Kanti, Praveen Kumar, K. Deepthi Jayan, V. Vicki Wanatasanappan, Jhilmil Swapnalin, Prabhakar Sharma, and Prasun Banerjee. "A State-of-the-Art Review on MXene-Based Hybrid Nanomaterial for Energy Storage Applications." *Journal of Alloys and Compounds* (2024): 174786. https://doi.org/10.1016/j.jallcom.2024.174786
150. Khatami, Mehrdad, and Siavash Iravani. "MXenes and MXene-Based Materials for the Removal of Water Pollutants: Challenges and Opportunities." *Comments on Inorganic Chemistry* 41, 4 (2021): 213–248. https://doi.org/10.1080/02603594.2021.1922396
151. Gopi, Rajakannu Ravi, Avadhani Kadathamana, Jinesh Varghese, Sandra L. Loredo, Ana Cerdán-Pasarán, Santhosh Pushpan, and K. C. Sanal. "Electrochemical Performance and Design Strategies of MXene-Based Supercapacitors." *Journal of Energy Storage* (2024). https://doi.org/10.5772/intechopen.1008245
152. Xu, Shuaikai, Yubing Li, Tangming Mo, Guodong Wei, and Ya Yang. "Highly Matched Porous MXene Anodes and Graphene Cathodes for High-Performance Aqueous Asymmetric Supercapacitors." *Energy Storage Materials* 69 (2024): 103379. https://doi.org/10.1016/j.ensm.2024.103379
153. Gandla, Dayakar, Zilong Zhuang, Vijaykumar V. Jadhav, and Daniel Q. Tan. "Lewis Acid Molten Salt Method for 2D MXene Synthesis and Energy Storage Applications: A Review." *Energy Storage Materials* (2023): 102977. https://doi.org/10.1016/j.ensm.2023.102977
154. Ye Zar Ni Htwe, Suriani Abu Bakar, Azmi Mohamed, Muqoyyanah, Mohd Hafiz Dzarfan Othman, Mohamad Hafiz Mamat, Mohd Khairul Ahmad, Muhammad Noorazlan Abd Azis, Ratno Nuryadi, Seeram Ramakrishna, Numan Salah, and Ahmed

Alshahrie "Progress in Etching-Driven MXene Synthesis and Utilization in Conductive Inks for Printed Electronics Applications: A Comprehensive Review." *Synthetic Metals* 306 (2024): 117631. https://doi.org/10.1016/j.synthmet.2024.117631

155. Li, Hongpeng, Xiran Li, Jiajie Liang, and Yongsheng Chen. "Hydrous RuO_2-Decorated MXene Coordinating with Silver Nanowire Inks Enabling Fully Printed Micro-Supercapacitors with Extraordinary Volumetric Performance." *Advanced Energy Materials* 9, 15 (2019): 1803987. https://doi.org/10.1002/aenm.201803987
156. Ibrahim, Yassmin, Ahmed Mohamed, Ahmed M. Abdelgawad, Kamel Eid, Aboubakr M. Abdullah, and Ahmed Elzatahry. "The Recent Advances in the Mechanical Properties of Self-Standing Two-Dimensional MXene-Based Nanostructures: Deep Insights into the Supercapacitor." *Nanomaterials* 10, 10 (2020): 1916. https://doi.org/10.3390/nano10101916
157. Attia, Sayed Y, Saad G. Mohamed, Yosry F. Barakat, Hamdy H. Hassan, and Wail Al Zoubi. "Supercapacitor Electrode Materials: Addressing Challenges in Mechanism and Charge Storage." *Reviews in Inorganic Chemistry* 42, 1 (2022): 53–88. https://doi.org/10.1515/revic-2020-0022
158. Cai, Chenyang, Zechang Wei, Leixin Deng, and Yu Fu. "Temperature-Invariant Superelastic Multifunctional MXene Aerogels for High-Performance Photoresponsive Supercapacitors and Wearable Strain Sensors." *ACS Applied Materials & Interfaces* 13, 45 (2021): 54170–54184. https://doi.org/10.1021/acsami.1c16318
159. Behl, Shilpa, Vikas Lahariya, Pratima Parashar Pandey, and Rajesh Kumar. "Recent Developments in V_2C MXene as Energy Storage Materials: Promises, Challenges, and Future Prospects." *Journal of Energy Storage* 92 (2024): 112176. https://doi.org/10.1016/j.est.2024.112176
160. Qin, Leiqiang, Jianxia Jiang, Quanzheng Tao, Chuanfei Wang, Ingemar Persson, Mats Fahlman, Per OÅ Persson, Lintao Hou, Johanna Rosen, and Fengling Zhang. "A Flexible Semitransparent Photovoltaic Supercapacitor Based on Water-Processed MXene Electrodes." *Journal of Materials Chemistry A* 8, 11 (2020): 5467–5475. https://doi.org/10.1039/D0TA00687D
161. Xie, Yanting, Haitao Zhang, Haitao Hu, and Zhengyou He. "Large-Scale Production and Integrated Application of Micro-Supercapacitors." *Chemistry – A European Journal* 30, 21 (2024): e202304160. https://doi.org/10.1002/chem.202304160
162. Yu, LePing, Lu Lu, XiaoHong Zhou, Lyu Xu, Zahrah Alhalili, and FengJun Wang. "Strategies for Fabricating High-Performance Electrochemical Energy-Storage Devices by MXenes." *ChemElectroChem* 8, 11 (2021): 1948–1987. https://doi.org/10.1002/celc.202100385
163. Barik, Rasmita, Vaishali Tanwar, and Pravin P. Ingole. "Supercapacitors: Future Direction and Challenges." In *Nanostructured Materials for Supercapacitors*, 619–644. Cham: Springer International Publishing, 2022. https://doi.org/10.1007/978-3-030-99302-3_27

13 Ionic Liquid Electrolytes for Supercapacitors

Sreeja P C and Sapna Nehra

13.1 INTRODUCTION

Supercapacitors (SCs) are advanced energy storage devices that bridge batteries and conventional capacitors. Unlike batteries, which use chemical reactions, they store charge in the electric double layer at the electrode–electrolyte interface. This capability allows for millions of charge–discharge cycles with minimal aging and high power density for fast discharge [1, 2]. While their energy densities are lower than those of batteries, advancements in new materials and ILEs are improving this aspect. SCs are utilized in energy conversion, transportation, and information technology, where high energy demand, rapid response times, and longevity are crucial. They are found in renewable energy systems, electric vehicles, consumer electronics, and backup power supplies [3–5].

Energy storage devices like SCs require electrolytes to move ions between electrodes during charge and discharge cycles. Three critical indicators for electrolyte choice are energy density, power density, and operational stability. Ionic liquids and organic electrolytes are new-generation options that enhance SCs' voltage range and thermal stability, improving efficiency for various markets [6, 7]. Scientists are refining electrolyte formulations to improve energy density and complement next-gen electrode materials.

Electricity from ionic liquids is widely used in energy storage due to their unique characteristics. While traditional electrolytes have drawbacks, ionic liquid electrolytes (ILEs) perform well, making them promising for future SCs and batteries thanks to their thermal stability, broad electrochemical stability windows, and high ionic conductivity [8, 9]. Additionally, ionic liquids are safer than conventional electrolytes in harsh environments as they are non-flammable and non-volatile. The adaptability of ionic liquids through varying cation–anion combinations is valuable for targeted applications. Adopting ILEs improves energy density, operating voltage, and device lifespan, aiding the development of portable devices, electric vehicles, and renewable energy systems.

13.2 COMPOSITION OF ILEs

Based solely on cations and anions, the ILE consists of salts in a liquid state at or near room temperature. Although they share the same title as electrolytes, traditional ones differ from ionic liquids because they are composed of salts dissolved in a solvent [10]. Typically, this solvent is an inorganic or organic anion, such as tetrafluoroborate, hexafluorophosphate, or bis(trifluoromethylsulfonyl)imide (TFSI), paired with an organic

DOI: 10.1201/9781003535836-13

cation, such as imidazolium, pyridinium, or ammonium. The unique structure of ionic liquids fundamentally imparts them extraordinary properties, including a wide electrochemical stability range, low vapor pressure, and good thermal stability, making them highly suitable for modern energy storage applications [11–13].

13.3 KEY CHARACTERISTICS OF ILEs

ILEs have several key characteristics that make them highly beneficial for energy storage and other advanced applications.

13.3.1 Wide Electrochemical Stability Window

The electrolyte's broad electrochemical stability window, which indicates the voltage range without rupturing, is a key factor distinguishing ILEs. This property is crucial for enhancing energy storage capacity in devices like SCs and batteries, as these electrolytes operate at higher voltages than traditional ones [14]. Aqueous electrolytes (~1.23 V) and organic solvents (~3 V) are much less stable. ILEs with a stability range of 4–6 V show significant performance differences. Key factors for stability include the strong covalent bonds between cations and anions and the absence of volatile solvents. ILEs can efficiently store energy while minimizing adverse reactions that decrease the device's performance and lifespan at higher voltages. Thus, they are ideal for applications such as grid storage, electric vehicles, and aeronautical technology [15].

13.3.1.1 High Ionic Conductivity

ILEs possess high ionic conductivity, which makes them ideal for energy storage and electrochemical applications. Unlike regular electrolytes, ionic liquids consist only of free cations and anions, facilitating ion transfer between electrodes. This conductivity enables rapid charging and discharging, suitable for high power density applications like SCs and advanced batteries. Additionally, their ionic conductivity remains stable over a wide temperature range, improving performance in extreme environments. By reducing internal resistance and boosting energy efficiency, ILEs aid in developing faster and more reliable energy storage systems [16–19].

13.3.2 Non-Volatility and Non-Flammability

The advantages of safety and dependability in energy storage systems stem from ILEs' non-volatile and non-flammable properties. Unlike traditional organic electrolytes, ionic liquids do not evaporate at high temperatures due to their low vapor pressure. This makes them suitable for high-temperature and vacuum systems, eliminating health and safety risks from fluid leakage and pressure swelling. Their ionic nature and thermal stability provide unignitability, meaning they resist ignition. Thus, these features make ILEs highly appealing for electric vehicles, airplanes, and grid energy storage, where durability and safety are crucial. Their stability also leads to less frequent maintenance and promotes more resilient and sustainable energy solutions [20–22].

13.3.3 Thermal and Chemical Stability

The electric transfer process of ionic liquid systems is thermally and chemically stable, making them desirable for challenging energy tasks. They can operate in extremely high temperatures, far exceeding what organic cells can endure, and are reported to be more reliable than current systems. Consequently, ionic liquids sustain demanding electrochemical conditions and high voltages, ensuring that electrode materials do not oxidize or reduce. This implies longer lifespan and minimal wear of power storage units. Furthermore, they can be utilized in energy-efficient applications such as e-transport, e-home, and aerospace. These qualities also reduce the risk of side reactions and thus increasing the equipment's overall efficiency and safety of service to the human being [23, 24].

13.3.4 Customizability

The ability to modify ILE is a remarkable feature that enables adjustment of its properties for specific applications. It is possible to control key power parameters such as viscosity, ionic conductivity, thermal stability, and electrochemical stability by selecting combinations of cations and anions. Thus, changes in the size, structure, or functional groups of the cation can effectively enhance ion transport, while adjustments to the anion can improve voltage stability and reduce reactivity. Furthermore, due to their molecular versatility, ionic liquids can be designed for various purposes, such as high-temperature stability for aerospace applications, non-flammability for electric vehicles, and the environmentally friendly character of these new liquids (biodegradability) for sustainable energy solutions. The unique features of ILEs position them as a versatile and promising component of modern energy storage technologies and electrochemical systems [25, 26].

13.4 TYPES OF ILEs USED IN SCs

Ionic liquids are categorized by chemical structure into three main types: aprotic, protic, and zwitterionic ILs. Aprotic and protic ILs share similarities in solubility, thermal stability, and electric conductivity, but protic ILs have a cation with a free proton. Aprotic ILs offer a large, stable electrochemical potential window, making them suitable for energy storage in SCs and lithium ion batteries (LIB) electrolytes. Current applications of protic ILs in fuel cells reflect ongoing energy conversion and storage research, while issues like zeta potential are encountered in ionic liquid-based membranes [27, 28]. Various ionic liquids for energy storage are based on cationic salts. Figure 13.1 shows the types of cations in ionic liquids [29].

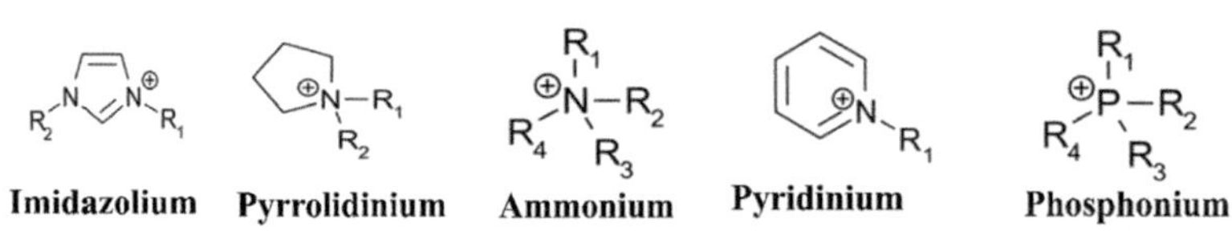

FIGURE 13.1 Different types of cations in ILEs.

13.4.1 Imidazolium-Based Ionic Liquids

Imidazolium-based ionic liquids are a class of ionic liquids that have cations derived from the imidazole ring, typically substituted with alkyl groups, and combined with a variety of anions, including bis(trifluoromethylsulfonyl)imide (Tf_2N), tetrafluoroborate (BF_4), or hexafluorophosphate (PF_6). An epitome of a product in this category is 1-ethyl-3-methylimidazolium bis(trifluoromethylsulfonyl)imide ([EMIm][Tf_2N]). These ionic liquids are extensively employed in energy storage devices like SCs due to their exceptional features such as high ionic conductivity, unmatched electrochemical stability, and wide liquid temperature range [30]. Their outstanding stability makes them suitable for applications that require rapid charging and discharging, while their minimal volatility contributes to enhanced safety and durability. In the case of ethyl-methyl imidazolium tetrafluoroborate ($EMIMBF_4$) electrolyte, a maximum potential window of 3.5 V is available with excellent electrochemical stability. When an AC-based SC is used, the material has also been found to perform exceptionally well with a maximum specific capacitance of 198.15 F/g, an energy density of 82.93 Wh/kg, and a power density of 3487 W/kg [31]. For ultrafast, compact SCs, $BMIMBF_4$ electrolytes in CNT-bridged graphene 3D structures not only have a wider potential window of 4.0 V but also exhibit a power density of 400,000 W/kg, an energy density of 110.6 Wh/kg, and a specific capacitance of 49 F/g at 0.5 A/g [32]. Imidazolium-based ionic liquids represent a group of materials that offer the added advantage of modifiability in terms of viscosity and conductivity through chain length modification or the choice of various anions, which is key to SC performance [33, 34].

13.4.2 Pyrrolidinium-Based Ionic Liquids

Ionic liquids containing pyrrolidinium contain a cation featuring a nitrogen atom in a five-membered ring [35]. Such ionic liquids often include tetrafluoroborate (BF_4), hexafluorophosphate (PF_6), or bis(trifluoromethylsulfonyl)imide (Tf_2N) as their anion. Ionic liquids based on pyrrolidinium exhibit the fastest charge and discharge capacities, thanks to high conductivity (due to low viscosity), facilitating rapid ion transport. Furthermore, these ionic liquids are less toxic to the environment and can withstand high temperatures; therefore, they are suitable for producing green energy [36]. Such ionic liquids can be modified to meet energy storage and conversion device performance requirements by altering the pyrrolidinium cation structure or selecting different anions. One of the electrolytes is 1-ethyl-1-methylpyrrolidinium bis(trifluoromethylsulfonyl)imide ([EMpyr][Tf_2N]). These highly conductive pyrrolidinium-based polymeric ILEs are well-suited for use with composite SC electrodes. The electrolyte functioning at 10–35 wt.% concentration in Dimethyl formamide (DMF) demonstrates stable capacitance and a long cycling life. The 87% capacitance is retained over 7000 cycles without performance degradation [37].

13.4.3 Phosphonium-Based Ionic Liquids

Phosphonium-based ionic liquids are a type of liquid wherein the cation results from a phosphonium ion, featuring a phosphorus atom bonded to four organic groups (usually alkyl or aryl groups) and an anion such as bis(trifluoromethylsulfonyl)imide (Tf_2N), tetrafluoroborate (BF_4), or hexafluorophosphate (PF_6) [38]. Ionic liquids derived from phosphorus are less viscous than other types of ionic liquids, when referring to viscosity as a fluid's resistance to deformation. This allows ions to move more freely, enabling SCs to achieve faster charge/discharge rates. Moreover, phosphorus-based ionic liquids facilitate the smooth movement of ions within their environmental context. Given that the components of this substance are greener and biodegradable, they could potentially have a lower ecological impact than other ionic liquids in the future.

Additionally, these ionic liquids exhibit remarkable resistance to chemical decay and are effective across a broad range of applications. Thus, phosphonium-based ionic liquid [P666,14] [Tf_2N] contains the cation tri-n-hexyl(tetradecyl)phosphonium and the anion bis(trifluoromethylsulfonyl)imide (Tf_2N). Despite their advantages, phosphonium-based ionic liquids have not yet achieved widespread use, primarily due to their low ionic conductivities and bulky cations, which complicate transportation. However, they may find applications with the advent of multicomponent electrolyte types, as their research potential has been recognized for future electrolytes. They can also serve as promising substitute electrolytes when combined with other ionic liquids, distinguished by their excellent electron transfer rates and outstanding temperature stability (up to 400°C in some cases) compared to imidazolium and ammonium-based ionic liquids [39, 40].

13.4.4 Ammonium-Based Ionic Liquids

Ionic liquids containing cations derived from ammonium ions (NH_4^+), often associated with organic groups like alkyl or aryl chains, are known as ammonium-based ionic liquids. These ionic liquids typically combine with various anions such as chloride, bis(trifluoromethylsulfonyl)imide, or tetrafluoroborate. Modifying the structure of ammonium cations allows them to accept diverse alkyl and functional groups, thereby enabling the adjustment of their physical and chemical properties to meet energy device requirements. Among potential electrolyte types, quaternary ammonium-based ionic liquids featuring short alkyl chains show great promise, with examples including N,N-diethyl-N-methyl-N-(2-methoxyethyl) ammonium TFSA (DEME-TFSA) and N,N-diethyl-N-methyl-N-(2-methoxyethyl) ammonium tetrafluoroborate (DEME-BF_4), which are noted as exemplary representatives of this category [41, 42]. Notably, the DEME-BF_4 electrolyte can extend the potential window to 6.0 V and enhance electrochemical performance, boasting a notable ionic conductivity of 4.8 mS/cm at 25°C that surpasses that of $EMIMBF_4$ and other ionic liquids [43]. However, their application encounters limitations due to issues such as larger cation sizes, higher viscosity, and ionic conductivity lower than that of other ionic liquids, which are related to the weak van der Waals forces arising from cation–anion interactions [44].

13.4.5 Pyridinium-Based ILs

Pyridinium-based ILEs consist of ionic liquids made up of pyridinium cations, which are well-known for their unique chemical and electrochemical properties. In addition to their good thermal stability, pyridinium-based ionic liquids can also be utilized in high-temperature applications. However, their higher viscosity and moderate ionic conductivity compared to other ionic liquids, such as imidazolium-based ones, present major challenges in achieving optimal ion transport and power densities [45]. These are mostly either tailor-made or custom-made through molecular modifications to exploit the properties of the electrolytes. For instance, introducing functional groups on the pyridinium ring to enhance ionic mobility or to generate compatible electrode materials is a typical example of the chemical modifications performed. These ILEs represent a developing group of electrolytes and could serve as a future power source in the renewable energy sector. Although the concept of pyridinium-based ILE is not as widely recognized as imidazolium or phosphonium-based ionic liquids, they hold potential for specific SC applications, particularly when dual stability and performance are key criteria for a specific application. Some common pyridinium-based ILEs include 1-butyl-3-methylpyridinium hydrogen sulphate ((BMPy)(HSO_4)). Another common pyridinium-based ILE is 1-butyl-3-methylpyridinium ethylsulfate ((BMPy)(ESO4)), both of which have demonstrated outstanding performance in energy research [46–48].

13.5 MECHANISMS OF ILEs IN SCs

The mechanisms by which ILEs operate in SCs are primarily related to their high-performance energy storage functionalities, involving both electrochemical double-layer formation and pseudocapacitive contributions. The efficiency of these processes heavily depends on effective ion transport and adaptable charge storage, which is facilitated by the unique characteristics of ionic liquids, including their high ionic conductivity and broad electrochemical stability [49].

13.5.1 Electrochemical Double-Layer Formation

When voltage is applied to the electrodes, the cations and anions in the ionic liquid migrate to the electrodes with the opposite charge. This situation prompts the emergence of two different layers: the first comprises the ions directly attached to the electrode and is referred to as the inner Helmholtz layer, while the second consists of the ions from the bulk electrolyte (diffuse layer). The charge separation at the electrode–electrolyte interface facilitates energy storage through electrostatic forces, meaning no chemical reactions occur. This process properly accounts for its high reversibility, which minimizes degradation and allows it to withstand millions of charge–discharge cycles [50, 51]. The size and shape of the ions, along with the dimensions and morphology of the capacitor electrode, not only depend on these factors but also directly impact the electric double layer's thickness and capacitance. ILEs containing ionic liquids exhibit higher ionic conductivity and better electrochemical stability, leading to remarkable enhancements in the efficiency and energy density of this process; thus, this type of electrolyte is suitable for high-performance SCs [52].

Figure 13.2(a) illustrates the integral capacitance of the positive, negative, and total electrodes when the operating potential window (OPW) is set at 3.0 V. These are

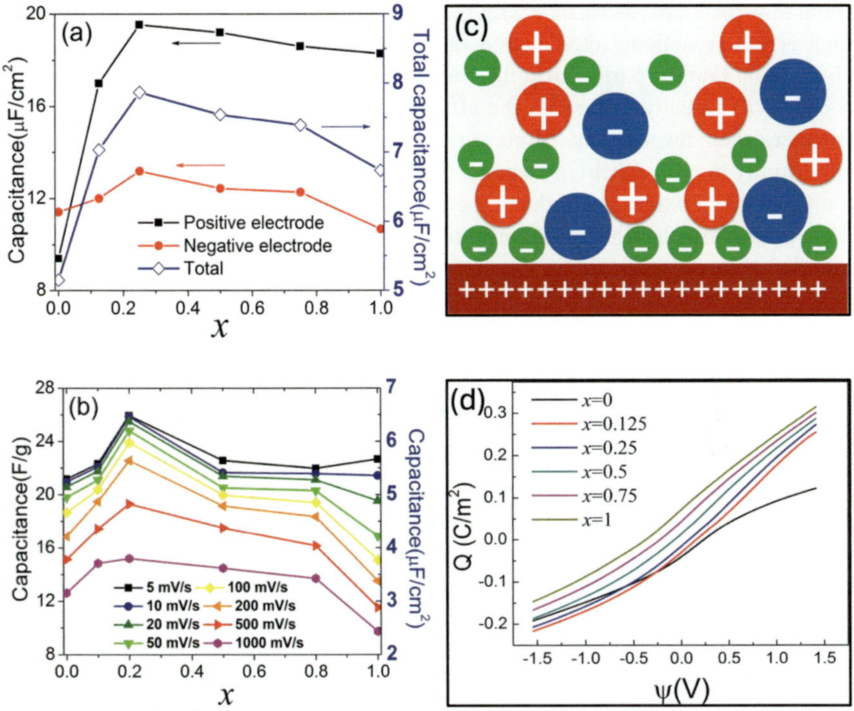

FIGURE 13.2 (a) Calculated integral capacitance for the positive, negative, and total electrodes at 3.0 V operating potential window (OPW). (b) A symmetric supercapacitor operating at 3.0 V as a function of the concentration of the RTIL mixture for different scan rates. (c) An RTIL mixture near the electrode surface, and (d) charge density as a function of the electrode potential at several compositions of the EMI-TFSI-BF4 mixture. (Adapted with permission [53]. © ACS 2016.)

the experimental results for a symmetric SC functioning at 3.0 V. The capacitance of the electrodes is presented as a function of the concentration of the room temperature ionic liquid (RTIL) mixture at varying scan rates, as illustrated in Figure 13.2(b). Figure 13.2(c) illustrates an RTIL mixture near the electrode surface. In the coarse-grained model, cations and anions are modeled as charged hard spheres of different sizes, and a planar geometry characterizes the electrode surface. Furthermore, it shows the charge density as a function of the electrode potential across several compositions of the EMI-TFSI-BF4 mixture. Here, x symbolizes the mole fraction of EMI-BF4 in the RTIL mixture; x = 0 corresponds to pure EMI-TFSI, while x = 1 indicates pure EMI-BF4, as shown in Figure 13.2(d) [53].

13.5.2 Pseudocapacitance Contributions

Formally, they are used in electric double-layer capacitors with ILEs that, in turn, promote pseudocapacitance [54, 55] through a fast and reversible redox reaction process at the electrode surface. Pseudocapacitance involves ions from the ionic liquid being absorbed and interacting with the electrode materials via faradaic processes, such as redox reactions, intercalation, or surface adsorption. The unique composition of the ionic liquids, particularly their wide electrochemical window and high ionic conductivity, greatly influences the efficiency of the charge transfer reactions and enhances their reversibility. Moreover, the tunable properties of ILEs provide the opportunity to utilize them in conjunction with the latest electrode materials, such as transition metal oxides or conducting polymers, which are designed to achieve pseudocapacitive effects. This mechanism reduces the burden on energy storage without sacrificing the properties of SCs, such as rapid response time and extended cycle life, thus making capacitors, in fact, the link between the two [56, 57].

13.5.3 Ion Transport and Charge Storage

Efficient ion transport and charge storage processes in SCs using ILEs are essential prerequisites for high operational performance [49]. When a voltage is applied, the ionic liquid, which contains both cations and anions, directs the ions to the electrodes that possess charges opposite to those of the ions. The high ionic conduction of ionic liquids facilitates the movement of ions, ensuring that many charges travel rapidly throughout the system. The surface of the electrode can form the electrochemical double layer (EDL), where energy is transferred through electrostatic charge separation. Some ILEs provide charge storage by converting ions into electrode material via reversible faradaic reactions, which increases energy storage capacity [9]. The low viscosity and adjustable ionic structure of ionic liquids significantly enhance ion mobility and compatibility with various electrodes, revolutionizing the entire charge storage process. Effective ion transport and flexible charge storage systems position ILEs as leading candidates for high-performance SCs [58].

13.6 ADVANTAGES OF ILEsin SCs

The unique properties of ILEs, stemming from their physical and chemical characteristics, make them ideal candidates for SC applications. The enhanced performance, safety, and operational efficiency of energy storage systems are positive effects derived from these advantages.

13.7 HIGH VOLTAGE CAPABILITY

One of the main advantages of SCs powered by ILEs is that they can be used at high voltages, resulting in increased energy storage capability [59]. Ionic liquids are electrochemically stable, and thus their electrochemical stability window on the voltage scale is essentially in the range from 4 to 6 V, depending on the specific ionic liquid, which stands in striking contrast to conventional aqueous or organic electrolytes. Notably, the energy content scales directly with voltage; the higher the potential, the more stable the square, enabling SCs to store more energy by operating at higher voltages without the risk of electrolyte decomposition, thus achieving higher specific energy. Furthermore, ionic liquids' high resistance to various chemical and thermal agents allows the devices to reliably operate over much more extended periods and in diverse climatic conditions.

Additionally, ILE facilitates the easier construction of energy storage systems and simplifies transporting electricity from the grid to the vehicle and back when connected in series to produce high voltage. This property makes them the best choice for applications such as advanced electronics, grid energy storage, and electric vehicles that require high-energy, compact, and efficient solutions [60].

13.7.1 Safety Enhancements

One of the many practical applications of ILEs, then, is SCs, which benefit remarkably in terms of safety. They are non-flammable, eliminating any fire hazard of the commonly used organic electrolytes. Still, they also possess the lowest volatility and emit the least heat at high temperatures, even below the saturated vapor pressure. This results in the lowest possibility of issues related to leakage and the formation of hazardous gases during the device's operation. It has been indicated that SCs can be used extensively over a wide range of temperatures without any degradation or performance loss over time, due to their high thermal stability [61].

Additionally, ionic liquids are chemically inert, which decreases the number of undesired side reactions with the electrode materials, further extending the device's lifespan. Furthermore, having a wide electrochemical window minimizes the likelihood of thermal runaway or catastrophic failures, allowing the system to operate at higher voltages without electrolyte breakdown, thereby ensuring safety during operation. This combination of properties makes ILEs the preferred material in applications where high safety standards are essential, such as in electric vehicles, aerospace, and grid-scale energy storage systems.

13.7.2 Thermal Stability for Extreme Conditions

The thermal stability of ILEs makes them a suitable choice for SCs operating under extreme conditions. Also referred to as ultra-capacitors, these devices can function at high temperatures without performance losses, as the electrical solvents in the ionic liquid state possess high thermal decomposition temperatures, often exceeding 300°C, unlike traditional organic or aqueous electrolytes. Essentially, even at extremely high temperatures, ionic liquids do not readily evaporate into the atmosphere or lose their liquid phase, thus ensuring the stability and safety of the devices in extreme environments such as industrial, automotive, and aerospace systems. Ionic liquids behave like ordinary liquids even in very cold environments; they can maintain their liquid state over a wide temperature range. SCs with ILEs can be relied upon to deliver stable energy storage and power release due to their broad thermal stability window, even when subjected to rigorous thermal cycling. This innovation eliminates the issue of saying "necessary" because new thermal management systems are now superfluous, estimating that system complexity is reduced, costs are lower, and reliable performance in challenging conditions is guaranteed [62, 63].

13.7.3 Environmental Benefits

ILEs, because of their low flammability and non-volatility, are a superior choice for SCs in the pursuit of sustainable energy systems that reduce the occurrence of harmful volatile organic compounds (VOCs), leading to decreased atmospheric pollution and improved operational safety. Moreover, many ILEs are non-toxic and can be partially regenerated from natural resources. Compared to other chemicals, ionic liquids are more environmentally friendly, as a significant portion is derived from biodegradable, renewable sources, and a larger share is linked to their longer lifespan. Furthermore, they diminish the need for disposability due to reduced waste generation, resulting in fewer replacements. ILEs are even less hazardous and easier to dispose of, as their chemical structures can be modified to produce environmentally safe and non-toxic substances. While these attributes enhance the green image of SC systems, they also encourage the development of sustainable materials in energy storage technologies, thus aligning with the global sustainability agenda [64].

13.8 DESIGN STRATEGIES FOR OPTIMIZING IONIC LIQUID ELECTROLYTES

The chemical makeup and physical characteristics of ILEs must be carefully considered when designing them for optimal performance in SCs. The following tactics are commonly employed.

13.8.1 Molecular Engineering of Cations and Anions

The choice of cation structure affects several key properties, such as ionic conductivity, viscosity, and thermal stability [65]. Imidazolium and pyrrolidinium are the cations that provide expansive electrochemical stability windows and excellent conductivity [66]. Phosphorus and ammonium cations demonstrate excellent thermal stability and can be the best choice in harsh environmnts. When anion selection includes fluorinated anions, like bis(trifluoromethylsulfonyl)imide ($[Tf_2N]$), it influences the compatibility of the electrode materials, their stability, and solubility. The anions with high stability and low viscosity are typically fluorinated, such as bis(trifluoromethylsulfonyl)imide. Ideal anions, like phosphates or carboxylates, are preferred to create an environmentally friendly ionic liquid. Introducing functional groups to the cations or anions, such as hydroxyl, cyano, or ether groups, may be advantageous, improving compatibility with the electrode materials, lowering viscosity, and increasing ionic conductivity.

13.8.2 Incorporation of Additives

It is possible that adding small quantities of organic solvents, such as ethylene carbonate or acetonitrile, will enhance ionic conductivity and lower viscosity while not seriously affecting stability. Additionally, one can add thermal or oxidative stabilizers to give ionic liquids high-temperature or voltage resistance. For instance, by applying the method of additive engineering, i.e., the incorporation of cetyltrimethylammonium bromide-grafted Ti_3C_2MXene (Ti_3C_2-CTAB) into

1-ethyl-3-methylimidazolium tetrafluoroborate ($EMIMBF_4$), a widely used ILE, the rapid dynamics of ionic liquid-based SCs can be improved. When hydrogen bonding and electrostatic forces break apart ion pairs, thus promoting ion transfer efficiency, this change results in a drastic reduction in solid–liquid surface energy from 18.03 to 12.37 mN/m and raises ionic conductivity by 38% [67]. Furthermore, a series of substances that act on the electrode and prevent side reactions at the electrode–electrolyte interface can significantly enhance performance. By adding proton donors or proton acceptors, the conductivity and hydrogen bonding interactions in protic ionic liquids could be controlled [68, 69].

13.8.3 Hybrid Electrolyte Systems

Viscosity, cost, and electrochemical stability can be balanced using ionic liquids combined with organic solvents, without compromising conductivity or voltage range. Ionic liquids can be integrated into solid-state matrices or polymer electrolytes to create quasi-solid electrolytes, which are safer and more mechanically stable. In some applications, combining DES and ionic liquids in a hybrid system can provide similar benefits at lower costs. A new class of materials, known as "pseudo-solid-state electrolytes," including ionogels and ionic plastic crystals, has been developed through the IL modification of solid-state electrolytes. The impressive properties of these hybrid electrolytes are expected to significantly enhance battery performance and expand their application field [70]. The interface between solid polymer electrolytes and ILEs at the junction of the polymerized ionic liquid gel polymer electrolytes offers a new pathway. This approach allows these electrolytes to benefit from the polymer matrix's mechanical strength and leakage resistance while also enjoying the extensive electrochemical stability and high ionic conductivity of ionic liquids. Furthermore, due to their inherent elasticity, they are highly suitable for the development of wearable and flexible electronic gadgets in the future [71].

13.9 APPLICATIONS OF ILEs SCs

The excellent performance and tailor-made features of ILEs (i.e., high thermal stability, wide electrochemical stability window, non-flammable properties, and tunable properties) are vital components in improving the SC's performance in nearly any field of application. The most significant applications to mention are as follows.

13.9.1 EVs

EVs' regenerative braking systems utilize SCs with ILEs to harvest and deliver energy quickly. Their ability to operate over various temperatures and manage high voltages safeguards the vehicle's safety and enhances its efficiency [72, 73].

13.9.2 Renewable Energy Storage

Ionic SCs with liquid electrolytes offer a reliable method for rapidly storing and discharging energy in energy-producing systems, such as solar and wind power. This helps batteries maintain system stability and control power fluctuations [74, 75].

13.9.3 Consumer Electronics

Wearables, laptops, and smartphones now utilize ionic liquid-based SCs because of their high energy density, long lifespan, and compact size [76].

13.9.4 Industrial Equipment

These capacitors utilize ILEs and are primarily employed in industrial fields to meet safety and stability requirements, including large machinery, robots, and power backup systems [77].

13.9.5 Aerospace and Defense

SCs using ILEs are widely used and ideal candidates for aerospace and defense applications, including satellite systems, drones, and military weapon innovations, due to their broad working temperature range and long lifespan [78, 79].

13.9.6 Medical Devices

SCs utilizing ILEs for energy are the reason behind the latest medical devices, including implanted devices and portable diagnostic equipment, which require mobility, safety, and reliability [80].

13.10 CHALLENGES AND LIMITATIONS

13.10.1 High Viscosity and Ionic Mobility Issues

The high viscosity of many ILs is detrimental to ionic conductivity and ion mobility. This is particularly true for high-power applications, where the sluggish movement of ions restricts the battery's ability to charge or discharge quickly. Additionally, viscosity increases significantly at very low temperatures for instruments used in cold environments, creating serious challenges [81].

13.10.2 Cost and Scalability

Economic considerations are among the many factors that make the cost-effectiveness of alternatives crucial, serving as the most significant element in decision-making for a green policy. It is widely recognized that ionic liquids are typically much more expensive than traditional electrolytes, primarily due to the challenges in synthesizing them and the reliance on costly raw materials. Although these electrolytes are used commercially on a large scale, their supply remains relatively limited because of constraints in the production process and high economic costs. Conversely, ionic liquids produced from green processes or ethanol-based methods are the current efforts to lower manufacturing costs for ionic liquids [82, 83].

13.10.3 Compatibility with Electrode Materials

In most cases, ordinary electrode materials cannot be used with the range of ionic liquids, and they encounter other issues, such as the degradation of the electrode surface, which results in an unstable reaction. Meanwhile, the interface surface is affected by wetting and ion transport. Some distinctions between ionic liquids for specific electrodes arise from challenges in the production process and increased costs, which may be unattainable [84].

13.10.4 Environmental and Toxicological Concerns

Some ionic liquids are toxic, non-biodegradable, and environmentally persistent, raising concerns about their long-term environmental impact. Yet, numerous groups are investigating them as potential contributions to green technology. It is advisable to use ionic liquids cautiously to prevent environmental pollution. The catalysts used in biochemical processes must be decomposed for reuse, so more sustainable methods, such as green chemistry involving cascading reactions, must be developed [85].

13.10.5 Processing and Handling Challenges

A critical problem that often arises in ionic liquids is that impurities can completely shut down the conductivity and stability of the most commonly used liquids, necessitating stringent cleanliness control. In some instances, liquids that exhibit hygroscopicity have been reported to impair performance. These require special handling and storage [86].

The electrochemical window of various commercially available RTILs, featuring different cations and anions under dry conditions (at a relative humidity level below 1%), was determined using a cutoff current of 15 μA (~1.9 mA/cm^2 current density) (blue) and 1.5 μA (~0.19 mA/cm^2 current density) (red), as shown in Figure 13.3. Observations indicate that at higher cutoff currents, most cyclic voltammograms tend to be relatively flat, yielding expansive electrochemical windows ranging from 4.3 to 6.5 V. These values are comparable to those reported by other research groups studying dry RTILs. It is important to note that, due to the employment of a platinum pseudo-reference electrode, direct comparisons of anodic and cathodic potential limits among different ionic liquids are not feasible. Nonetheless, the unstable reference should not influence the absolute potential windows. The RTILs examined in this research were not subjected to vacuum drying; they were instead thoroughly purged with a nitrogen stream while maintaining a controlled humidity level of less than 1 RH%.

When comparing ionic liquids that share the same anion, specifically $[NTf_2]^-$, but have different cations as given in Figure 13.3a, b, d, e, and g, the following trend emerges regarding the electrochemical window and stability: $[C_2mim]^+ < [C_4mim]^+ < [S_{2,2,1}]^+ < [N_{4,1,1,1}]^+ \approx [C_4mpyrr]^+$. Cations based on imidazolium demonstrate greater ease of reduction than single-centered cations such as pyrolidinium or ammonium. Additionally, a longer alkyl chain in the RTIL correlates with improved electrochemical stability. For electrolytes containing the same cation, namely $[C_4mim]^+$, but differing in anions (refer to Figure 13.3c, e, f, and i), the stability trend is as follows: $[BF_4]^- < [PF_6]^- \approx [NTf_2]^- < [FAP]^-$. This contrasts with the stability trend derived

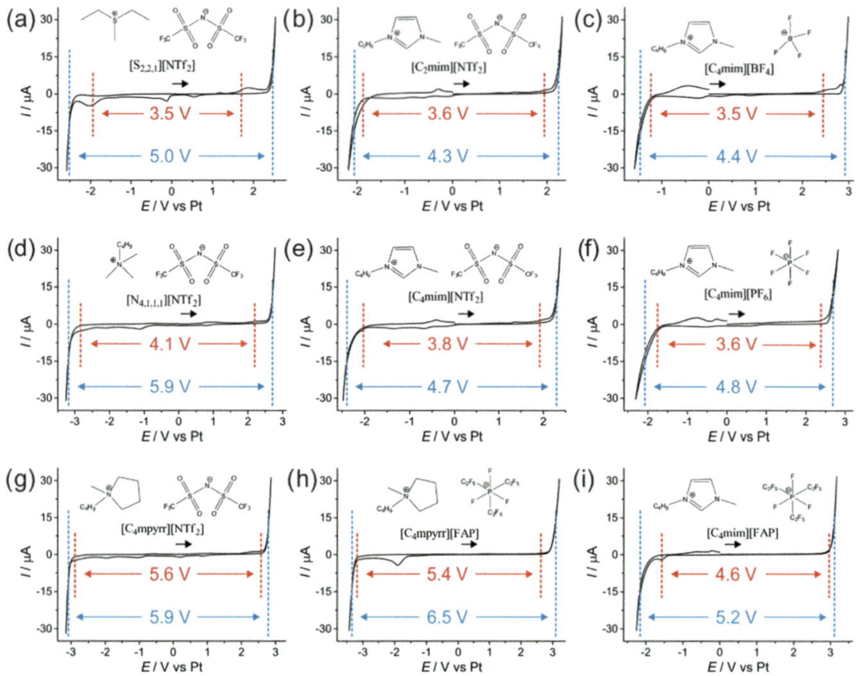

FIGURE 13.3 Electrochemical windows and chemical structures of (a) $[S_{2,2,1}][NTf_2]$, (b) $[C_2mim][NTf_2]$, (c) $[C_4mim][BF_4]$, (d) $[N_{4,1,1,1}][NTf_2]$, (e) $[C_4mim][NTf_2]$, (f) $[C_4mim][PF_6]$, (g) $[C_4mpyrr][NTf_2]$, (h) [C_4mpyrr][FAP], and (i) [C_4mim][FAP] on a platinum thin-film electrode at <1 RH%. The red-dashed lines represent a cutoff current of 1.5 μA, and the blue-dashed line shows a cutoff current of 15 μA. All scans were taken at 100 mV/s. The black arrows indicate the start potential and the direction of the scan. (Adapted with permission [87]. © ACS 2020.)

from computational calculations, which suggests an increasing order of $[NTf_2]^- < [BF_4]^- < [PF_6]^-$ [87].

13.10.6 Limited Data on Long-Term Performance

The long-term stability and performance of the ILEs in commercial SC applications remain unacknowledged in the existing literature. New experiments, such as those concerning ion quality, leakage prevention, and changes in the electrolyte over time, are highly anticipated [88].

13.11 FUTURE PERSPECTIVES

13.11.1 Advancements in Ionic Liquid Research

Recent developments in molecular engineering and computational simulation have enabled the targeted creation of ionic liquids with reduced viscosity, increased ion conductivity, and more expansive electrochemical stability windows [89, 90].

Ongoing exploration of dual-function ionic liquids allows them to act as both electrolytes and stabilizers by enhancing interfacial stability and reducing parasitic reactions. The combination of ionic liquids with other electrolytes, solvents, or additives creates hybrid systems known for their improved performance, cost-effectiveness, and scalability. Ideally, supporting the development of ionic liquids that can function at extremely high temperatures and pressures would benefit aeronautics and demanding industrial environments.

13.11.2 Integration with Emerging Electrode Materials

Energy density and power are optimized through ionic liquid modification for advanced electrode materials, including nanostructures like graphene, carbon nanotubes, and MXenes [91–93]. The emphasis is on ensuring consistent performance for pseudocapacitive materials such as conducting polymers or metal oxides, which require a strong and stable ionic interface. In the realm of flexible and elastic electrodes, researchers are employing ionic liquids to further energy storage systems, including wearable and portable applications [94]. The reduction in power consumption and the enhancement in cycling stability, stemming from the chemical interactions between ionic liquids and electrode surfaces, are the primary focus of researchers.

13.11.3 Development of Greener Ionic Liquids

The recent interest in using ionic liquids synthesized from renewable resources, such as biomass, represents a significant step toward environmental and energy sustainability [95]. Ecological and disposal issues have been addressed by creating ionic liquids that decompose into non-toxic elements after use. The goal of green chemistry in producing ionic liquids is to minimize environmental impact through solvent-free synthesis, repurposing waste for ionic liquids, and energy-efficient production processes. Safety, toxicity, and sustainability concerns have driven research in this field to ensure that the development of ionic liquids aligns with international standards [96]. In the future, it may be possible to develop new ionic liquids that are less harmful to the environment, which will at least comply with the 12 principles of green chemistry, supported by life cycle assessment studies and computational modeling research [97].

13.12 CONCLUSION

ILEs have undoubtedly elevated SC technology to a new level, solidifying their role in providing substantial advantages over conventional electrolytes. The unique properties of ionic liquids, such as their wide electrochemical stability windows, high thermal stability, non-flammability, and tunable features, enable SCs to operate at higher energy levels, function safely under harsh conditions, and ensure long-term durability. Their adoption is primarily hindered by high production costs, limited scalability, high viscosity, and negative environmental effects. The most critical aspect of their broader application will be tackling these challenges through innovative research and sustainable production methods.

Current developments in molecular engineering, hybrid systems, and green synthesis are paving the way for ILEs to become more cost-effective, efficient, and environmentally friendly. If the current challenges are overcome, ionic liquids could significantly transform related technologies and play a leading role in the energy revolution through sustainable energy systems.

13.13 AUTHOR AI DECLARATION

The authors hereby declare the use of generative AI and AI-assisted technologies in the writing process. While preparing this work, the authors used the ChatGPT tool/service to rephrase several statements in the manuscript. All AI-generated content was reviewed and edited by the authors as needed. The authors take full responsibility for the content of the publication.

ACKNOWLEDGMENT

The Editors acknowledge the use of Grammarly as a tool to assist in editing this book chapter. The suggestions made by Grammarly were carefully considered and incorporated where appropriate, but the author is solely responsible for the final content.

REFERENCES

1. Dutta, Arindam, Shirsendu Mitra, Mitali Basak, and Tamal Banerjee. "A comprehensive review on batteries and supercapacitors: Development and challenges since their inception." *Energy Storage* 5 (2023): e339. doi: 10.1002/est2.339.
2. Hasany, Syed Tanveer, Farhan Koray Bahadır, and Alam Dönmez. "Advancement in high energy density materials for energy storage systems." *Materials for Boosting Energy Storage. Advances in Sustainable Energy Technologies* 3 (2024): 101–122. doi: 10.1021/bk-2024-1488.ch005.
3. Kumar, Yogesh, Sangeeta Rawal, Bhawana Joshi, and S. A. Hashmi. "Background, fundamental understanding and progress in electrochemical capacitors." *Journal of Solid State Electrochemistry* 23 (2019): 667–692. doi: 10.1007/s10008-018-4160-3.
4. Volfkovich, Yu M. "Electrochemical supercapacitors (a review)." *Russian Journal of Electrochemistry* 57 (2021): 311–347. doi: 10.1134/S1023193521040108.
5. Wang, Yonggang, Yanfang Song, and Yongyao Xia. "Electrochemical capacitors: Mechanism, materials, systems, characterization and applications." *Chemical Society Reviews* 45 (2016): 5925–5950. doi: 10.1039/C5CS00580A.
6. Vishweswariah, Karthik, Anil Kumar Madikere Raghunatha Reddy, and Karim Zaghib. "Beyond organic electrolytes: An analysis of ionic liquids for advanced lithium rechargeable batteries." *Batteries* 10 (2024): 436. doi: 10.3390/batteries10120436.
7. Silva, Lucas de S., Henrique de A. Chagas, and G. Colherinhas. "Sustainable supercapacitors using advanced hydrated amino acid ionic liquids: A novel approach to biodegradable energy storage." *Journal of Molecular Liquids* 417 (2025): 126638. doi: 10.1016/j.molliq.2024.126638.

8. Yu, Linpo, and George Z. Chen. "Ionic liquid-based electrolytes for supercapacitor and supercapattery." *Frontiers in Chemistry* 7 (2019): 272. doi: 10.3389/fchem.2019.00272.
9. Feng, Jianze, Yan Wang, Yongtai Xu, Yinglun Sun, Yu Tang, and Xingbin Yan. "Ion regulation of ionic liquid electrolytes for supercapacitors." *Energy & Environmental Science* 14 (2021): 2859–2882. doi: 10.1039/D0EE04002A.
10. Galiński, Maciej, Andrzej Lewandowski, and Izabela Stępniak. "Ionic liquids as electrolytes." *Electrochimica Acta* 51 (2006): 5567–5580. doi: 10.1016/j.electacta.2006.03.016.
11. Forsyth, Maria, Luca Porcarelli, Xiaoen Wang, Nicolas Goujon, and David Mecerreyes. "Innovative electrolytes based on ionic liquids and polymers for next-generation solid-state batteries." *Accounts of Chemical Research* 52 (2019): 686–694. doi: 10.1021/acs.accounts.8b00566.
12. Zhong, Cheng, Yida Deng, Wenbin Hu, Jinli Qiao, Lei Zhang, and Jiujun Zhang. "A review of electrolyte materials and compositions for electrochemical supercapacitors." *Chemical Society Reviews* 44 (2015): 7484–7539. doi: 10.1039/C5CS00303B.
13. Salanne, Mathieu. "Ionic liquids for supercapacitor applications." *Ionic Liquids* II (2018): 29–53. doi: 10.1007/978-3-319-89794-3_2.
14. Hayyan, Maan, Farouq S. Mjalli, Mohd Ali Hashim, Inas M. AlNashef, and Tan Xue Mei. "Investigating the electrochemical windows of ionic liquids." *Journal of Industrial and Engineering Chemistry* 19 (2013): 106–112. doi: 10.1016/j.jiec.2012.07.011.
15. Piatti, Erik, Luca Guglielmero, Giorgio Tofani, Andrea Mezzetta, Lorenzo Guazzelli, Felicia D'Andrea, Stefano Roddaro, and Christian Silvio Pomelli. "Ionic liquids for electrochemical applications: Correlation between molecular structure and electrochemical stability window." *Journal of Molecular Liquids* 364 (2022): 120001. doi: 10.1016/j.molliq.2022.120001.
16. Kim, Eunhwan, Juyeon Han, Seokgyu Ryu, Youngkyu Choi, and Jeeyoung Yoo. "Ionic liquid electrolytes for electrochemical energy storage devices." *Materials* 14 (2021): 4000. doi: 10.3390/ma14144000.
17. Miao, Ling, Ziyang Song, Dazhang Zhu, Liangchun Li, Lihua Gan, and Mingxian Liu. "Ionic liquids for supercapacitive energy storage: A mini-review." *Energy & Fuels* 35 (2021): 8443–8455. doi: 10.1021/acs.energyfuels.1c00321.
18. Karuppasamy Karuppasamy, Jayaraman Theerthagiri, Dhanasekaran Vikraman, Chang-Joo Yim, Sajjad Hussain, Ramakant Sharma, Thandavaryan Maiyalagan, Jiaqian Qin, and Hyun-Seok Kim. "Ionic liquid-based electrolytes for energy storage devices: A brief review on their limits and applications." *Polymers* 12 (2020): 918. doi: 10.3390/polym12040918.
19. Wang, Yu, and Wei-Hong Zhong. "Development of electrolytes towards achieving safe and high-performance energy-storage devices: A review." *ChemElectroChem* 2 (2015): 22–36. doi: 10.1002/celc.201402277.
20. Xu, Chenxuan, Guang Yang, Daxiong Wu, Meng Yao, Chunxian Xing, Jiahe Zhang, Haitao Zhang, et al. "Roadmap on ionic liquid electrolytes for energy storage devices." *Chemistry–An Asian Journal* 16 (2021): 549–562. doi: 10.1002/asia.202001414.
21. Sheikh, Setareh, and Alireza Haghpanah Jahromi. "Ionic liquids in green energy storage devices: Lithium-ion batteries, supercapacitors, and solar cells." *Monatshefte für Chemie-Chemical Monthly* 155 (2024): 383–399. doi: 10.1007/s00706-024-03178-4.
22. Matuszek, Karolina, Samantha L. Piper, Alina Brzęczek-Szafran, Binayak Roy, Saliha Saher, Jennifer M. Pringle, and Douglas R. MacFarlane. "Unexpected energy applications of ionic liquids." *Advanced Materials* 36 (2024): 2313023. doi: 10.1002/adma.202313023.
23. Choudhary, Gaurav, Jyoti Dhariwal, Moumita Saha, Shruti Trivedi, Manoj K. Banjare, Rahul Kanaoujiya, and Kamalakanta Behera. "Ionic liquids: Environmentally sustainable materials for energy conversion and storage applications." *Environmental Science and Pollution Research* 31 (2024): 10296–10316. doi: 10.1007/s11356-023-25468-w.

24. Zhang, Shiguo, Qinghua Zhang, Yan Zhang, Zhengjian Chen, Masayoshi Watanabe, and Youquan Deng. "Beyond solvents and electrolytes: Ionic liquids-based advanced functional materials." *Progress in Materials Science* 77 (2016): 80–124. 10.1016/j.pmatsci.2015.10.001.
25. MacFarlane, Douglas R., Maria Forsyth, Patrick C. Howlett, Mega Kar, Stefano Passerini, Jennifer M. Pringle, Hiroyuki Ohno, et al. "Ionic liquids and their solid-state analogues as materials for energy generation and storage." *Nature Reviews Materials* 1 (2016): 1–15. doi: 10.1038/natrevmats.2015.5.
26. Tang, Shaokun, Gary A. Baker, and Hua Zhao. "Ether-and alcohol-functionalized task-specific ionic liquids: Attractive properties and applications." *Chemical Society Reviews* 41 (2012): 4030–4066. doi: 10.1039/C2CS15362A.
27. Armand, Michel, Frank Endres, Douglas R. MacFarlane, Hiroyuki Ohno, and Bruno Scrosati. "Ionic-liquid materials for the electrochemical challenges of the future." *Nature Materials* 8 (2009): 621–629. doi: 10.1038/nmat2448.
28. Stettner, Timo, and Andrea Balducci. "Protic ionic liquids in energy storage devices: Past, present and future perspective." *Energy Storage Materials* 40 (2021): 402–414. doi: 10.1016/j.ensm.2021.04.036.
29. Ray, Apurba, and Bilge Saruhan. "Application of ionic liquids for batteries and supercapacitors." *Materials* 14 (2021): 2942. doi: 10.3390/ma14112942.
30. Zhou, Teng, Chengmin Gui, Longgang Sun, Yongxin Hu, Hao Lyu, Zihao Wang, Zhen Song, and Gangqiang Yu. "Energy applications of ionic liquids: Recent developments and future prospects." *Chemical Reviews* 123 (2023): 12170–12253. doi: 10.1021/acs.chemrev.3c00391.
31. Boujibar, Ouassim, Fouad Ghamouss, Arunabh Ghosh, Ouafae Achak, and Tarik Chafik. "Activated carbon with exceptionally high surface area and tailored nanoporosity obtained from natural anthracite and its use in supercapacitors." *Journal of Power Sources* 436 (2019): 226882. doi: 10.1016/j.jpowsour.2019.226882.
32. Pham, Duy Tho, Tae Hoon Lee, Dinh Hoa Luong, Fei Yao, Arunabha Ghosh, Viet Thong Le, Tae Hyung Kim, Bing Li, Jian Chang, and Young Hee Lee. "Carbon nanotube-bridged graphene 3D building blocks for ultrafast compact supercapacitors." *ACS Nano* 9 (2015): 2018–2027. doi: 10.1021/nn507079x.
33. Saha, Moumita, Ambrish Kumar, Rahul Kanaoujiya, Kamalakanta Behera, and Shruti Trivedi. "A comprehensive review of novel emerging electrolytes for supercapacitors: Aqueous and organic electrolytes versus ionic liquid-based electrolytes." *Energy & Fuels* 38 (2024): 8528–8552. doi: 10.1021/acs.energyfuels.4c00685.
34. Chen, Wei, Ying Wei, Xinyi Ge, Shuangqing Li, Xinyuan Zhang, Zhong-Xia Wang, Zheng Xing, Qingguo Zhang, and Xiang Liu. "Insight of alkyl imidazolium tetrafluoroborate ionic gels as supercapacitors and motion sensors: Effects of alkyl chain length and intermolecular interactions." *Journal of Power Sources* 599 (2024): 234224. doi: 10.1016/j.jpowsour.2024.234224.
35. Seki, Shiro, Yasutaka Ohno, Hajime Miyashiro, Yo Kobayashi, Akira Usami, Yuichi Mita, Nobuyuki Terada, Kikuko Hayamizu, Seiji Tsuzuki, and Masayoshi Watanabe. "Quaternary ammonium room-temperature ionic liquid/lithium salt binary electrolytes: Electrochemical study." *Journal of the Electrochemical Society* 155 (2008): A421. doi: 10.1149/1.2899014.
36. Kerner, Manfred, and Patrik Johansson. "Pyrrolidinium FSI and TFSI-based polymerized ionic liquids as electrolytes for high-temperature lithium-ion batteries." *Batteries* 4 (2018): 10. doi: 10.3390/batteries4010010.
37. Kalinova, Radostina, Ivaylo Dimitrov, Christo Novakov, Svetlana Veleva, and Antonia Stoyanova. "Modular platform for synthesis of poly (ionic liquid) electrolytes for electrochemical applications in supercapacitors." *ChemistrySelect* 6 (2021): 3795–3801. doi: 10.1002/slct.202004240.
38. Kundu, Kaushik, Bidyut K. Paul, Soumik Bardhan, and Swapan K Saha. "Recent advances in bioionic liquids." *Ionic Liquid-Based Surfactant Science: Formulation, Characterization, and Applications* (2015): 397.

39. Molahalli, Vandana, Pemika Hirankittiwong, Aman Sharma, Huddad Laeim, Apoorva Shetty, Nattaporn Chattham, and Gurumurthy Hegde. "Roadmap on ionic liquid crystal electrolytes for energy storage devices." *Materials Science and Engineering: B* 305 (2024): 117369. doi: 10.1016/j.mseb.2024.117369.
40. Karuppasamy, K., Dhanasekaran Vikraman, In-Tae Hwang, Hyun-Jung Kim, A. Nichelson, Ranjith Bose, and Hyun-Seok Kim. "Nonaqueous liquid electrolytes based on novel 1-ethyl-3-methylimidazolium bis (nonafluorobutane-1-sulfonyl imidate) ionic liquid for energy storage devices." *Journal of Materials Research and Technology* 9 (2020): 1251–1260. doi: 10.1016/j.jmrt.2019.11.052.
41. Nordness, Oscar, and Joan F. Brennecke. "Ion dissociation in ionic liquids and ionic liquid solutions." *Chemical Reviews* 120 (2020): 12873–12902. doi: 10.1021/acs.chemrev.0c00373.
42. Sato, Takaya, Gen Masuda, and Kentaro Takagi. "Electrochemical properties of novel ionic liquids for electric double layer capacitor applications." *Electrochimica Acta* 49 (2004): 3603–3611. doi: 10.1016/j.electacta.2004.03.030.
43. Li, Qi, Juner Chen, Lei Fan, Xueqian Kong, and Yingying Lu. "Progress in electrolytes for rechargeable Li-based batteries and beyond." *Green Energy & Environment* 1 (2016): 18–42. doi: 10.1016/j.gee.2016.04.006.
44. Gunawan, Christian A., Bryan H.R. Suryanto, and Chuan Zhao. "Electrochemical study of copper in room temperature protic ionic liquids ethylammonium nitrate and triethylammonium methylsulfonate." *Journal of the Electrochemical Society* 159 (2012): D611. doi: 10.1149/2.054210jes.
45. Wu, Chongchong, Alex De Visscher, and Ian D. Gates. "Comparison of electronic and physicochemical properties between imidazolium-based and pyridinium-based ionic liquids." *The Journal of Physical Chemistry B* 122 (2018): 6771–6780. doi: 10.1021/acs.jpcb.8b00764.
46. Aparicio, Santiago, and Mert Atilhan. "Mixed ionic liquids: The case of pyridinium-based fluids." *The Journal of Physical Chemistry B* 116 (2012): 2526–2537. doi: 10.1021/jp212258b.
47. Voroshylova, Iuliia V., Sergei R. Smaga, Elena V. Lukinova, Vitaly V. Chaban, and Oleg N. Kalugin. "Conductivity and association of imidazolium and pyridinium based ionic liquids in methanol." *Journal of Molecular Liquids* 203 (2015): 7–15. doi: 10.1016/j.molliq.2014.12.028.
48. Bandrés, Isabel, Diego F. Montano, Ignacio Gascón, Pilar Cea, and Carlos Lafuente. "Study of the conductivity behavior of pyridinium-based ionic liquids." *Electrochimica Acta* 55 (2010): 2252–2257. doi: 10.1016/j.electacta.2009.11.073.
49. Yan, Runyu, Markus Antonietti, and Martin Oschatz. "Toward the experimental understanding of the energy storage mechanism and ion dynamics in ionic liquid based supercapacitors." *Advanced Energy Materials* 8 (2018): 1800026. doi: 10.1002/aenm.201800026.
50. Silvester, Debbie S., Rabia Jamil, Simon Doblinger, Yunxiao Zhang, Rob Atkin, and Hua Li. "Electrical double layer structure in ionic liquids and its importance for supercapacitor, battery, sensing, and lubrication applications." *The Journal of Physical Chemistry C* 125 (2021): 13707–13720. doi: 10.1021/acs.jpcc.1c03253.
51. Fedorov, Maxim V., and Alexei A. Kornyshev. "Towards understanding the structure and capacitance of electrical double layer in ionic liquids." *Electrochimica Acta* 53 (2008): 6835–6840. doi: 10.1016/j.electacta.2008.02.065.
52. Wang, Yan, Kaiyuan Xue, Changzeng Yan, Yuehui Li, Xingyun Zhang, Kailimai Su, Pengjun Ma, Shanhong Wan, and Junwei Lang. "Tuning of ionic liquid–solvent electrolytes for high-voltage electrochemical double layer capacitors: A review." *Batteries* 10 (2024): 54. doi: 10.3390/batteries10020054.
53. Lian, Cheng., Kun Liu, Katherine L. Van Aken, Yury Gogotsi, David J. Wesolowski, H. L. Liu, De-en. Jiang, and Jianzhong Wu. "Enhancing the capacitive performance of electric double-layer capacitors with ionic liquid mixtures." *ACS Energy Letters* 1 (2016): 21–26. doi: 10.1021/acsenergylett.6b00010.

54. Guan, Li, George Z. Chen, Anna K. Croft, and David M. Grant. "Perspective—Redox ionic liquid electrolytes for supercapattery." *Journal of The Electrochemical Society* 169 (2022): 030529. doi: 10.1149/1945-7111/ac5ba8.
55. Sun, Li, Kelei Zhuo, Yujuan Chen, Quanzhou Du, Suojiang Zhang, and Jianji Wang. "Ionic liquid-based redox active electrolytes for supercapacitors." *Advanced Functional Materials* 32 (2022): 2203611. doi: 10.1002/adfm.202203611.
56. Zdolšek, Nikola, Raquel P. Rocha, Jugoslav Krstić, Tatjana Trtić-Petrović, Biljana Šljukić, José L. Figueiredo, and Milica J. Vujković. "Electrochemical investigation of ionic liquid-derived porous carbon materials for supercapacitors: Pseudocapacitance versus electrical double layer." *Electrochimica Acta* 298 (2019): 541–551. doi: 10.1016/j.electacta.2018.12.129.
57. Sathyamoorthi, Sethuraman, Vembu Suryanarayanan, and David Velayutham. "Organo-redox shuttle promoted protic ionic liquid electrolyte for supercapacitor." *Journal of Power Sources* 274 (2015): 1135–1139. doi: 10.1016/j.jpowsour.2014.10.166.
58. Liang, Zhun, Chang Zhao, Wei Zhao, Yuan Zhang, Pattarachai Srimuk, Volker Presser, and Guang Feng. "Molecular understanding of charge storage in MoS_2 supercapacitors with ionic liquids." *Energy & Environmental Materials* 4 (2021): 631–637. doi: 10.1002/eem2.12147.
59. Pan, Shanshan, Meng Yao, Jiahe Zhang, Bosen Li, Chunxian Xing, Xianli Song, Peipei Su, and Haitao Zhang. "Recognition of ionic liquids as high-voltage electrolytes for supercapacitors." *Frontiers in Chemistry* 8 (2020): 261. doi: 10.3389/fchem.2020.00261.
60. Liu, Huan, and Haijun Yu. "Ionic liquids for electrochemical energy storage devices applications." *Journal of Materials Science & Technology* 35 (2019): 674–686. doi: 10.1016/j.jmst.2018.10.007.
61. Arbizzani, Catia, Maurizio Biso, Dario Cericola, Mariachiara Lazzari, Francesca Soavi, and Marina Mastragostino. "Safe, high-energy supercapacitors based on solvent-free ionic liquid electrolytes." *Journal of Power Sources* 185 (2008): 1575–1579. doi: 10.1016/j.jpowsour.2008.09.016.
62. Shahzad, Suniya, Afzal Shah, Elaheh Kowsari, Faiza Jan Iftikhar, Anum Nawab, Benoit Piro, Mohammad Salim Akhter, Usman Ali Rana, and Yongjin Zou. "Ionic liquids as environmentally benign electrolytes for high-performance supercapacitors." *Global Challenges* 3 (2019): 1800023. doi: 10.1002/gch2.201800023.
63. Liew, Chiam-Wen, Ramesh T Subramaniam, and Abdul K. Arof. "Good prospect of ionic liquid based-poly (vinyl alcohol) polymer electrolytes for supercapacitors with excellent electrical, electrochemical and thermal properties." *International Journal of Hydrogen Energy* 39 (2014): 2953–2963. doi: 10.1016/j.ijhydene.2013.06.061.
64. Thangavel, Ranjith, Aravindaraj G. Kannan, Rubha Ponraj, Vigneysh Thangavel, Dong-Won Kim, and Yun-Sung Lee. "High-energy green supercapacitor driven by ionic liquid electrolytes as an ultra-high stable next-generation energy storage device." *Journal of Power Sources* 383 (2018): 102–109. doi: 10.1016/j.jpowsour.2018.02.037.
65. Wu, Xin, Yao Dai, Nian Wu Li, Xiao Chun Chen, and Le Yu. "Recent progress in ionic liquid-based electrolytes for nonaqueous and aqueous metal batteries." *eScience* 4 (2024): 100173. doi: 10.1016/j.esci.2023.100173.
66. Aetizaz, Muhammad, Sehrish Sarfaraz, and Khurshid Ayub. "Interaction of imidazolium based ionic liquid electrolytes with carbon nitride electrodes in supercapacitors; A step forward for understanding electrode–electrolyte interaction." *Journal of Molecular Liquids* 369 (2023): 120955. doi: 10.1016/j.molliq.2022.120955.
67. Jiang, Xinglin, Xinzheng Wu, Yanting Xie, Zixing Wang, Junfeng Huang, Yuanxiao Qu, Dali Mu, Xiong Zhang, Weiqing Yang, and Haitao Zhang. "Additive engineering enables ionic-liquid electrolyte-based supercapacitors to deliver simultaneously high energy and power density." *ACS Sustainable Chemistry & Engineering* 11 (2023): 5685–5695. doi: 10.1021/acssuschemeng.3c00213.

68. Greaves, Tamar L., and Calum J. Drummond. "Protic ionic liquids: Properties and applications." *Chemical Reviews* 108 (2008): 206–237. doi: 10.1021/cr068040u.
69. Dong, Kun, Suojiang Zhang, and Jianji Wang. "Understanding the hydrogen bonds in ionic liquids and their roles in properties and reactions." *Chemical Communications* 52 (2016): 6744–6764. doi: 10.1039/C5CC10120D.
70. Hwang, Jinkwang, Kazuhiko Matsumoto, Chih-Yao Chen, and Rika Hagiwara. "Pseudo-solid-state electrolytes utilizing the ionic liquid family for rechargeable batteries." *Energy & Environmental Science* 14 (2021): 5834–5863. doi: 10.1039/D1EE02567H.
71. Deb, Debalina. "Flexible polymerized ionic liquids gel polymer electrolytes for supercapacitor application." In *Ionic Liquids-Recent Advances: Recent Advances* (2025): 89–114. doi: 10.5772/intechopen.1002097.
72. Manoharan, Sindhuja, Karthikeyan Krishnamoorthy, Arunprasath Sathyaseelan, and Sang-Jae Kim. "High-power graphene supercapacitors for the effective storage of regenerative energy during the braking and deceleration process in electric vehicles." *Materials Chemistry Frontiers* 5 (2021): 6200–6211. doi: 10.1039/D1QM00465D.
73. Alam, Mahamad Nabab. "Regenerative Braking and Rapid Acceleration System for Electric Vehicles Using Integrated Battery and Supercapacitor." In *2024 IEEE 4th International Conference on Sustainable Energy and Future Electric Transportation (SEFET)* (2024): 1–6. doi: 10.1109/SEFET61574.2024.10718265.
74. Njema, George G., Russel Ben O. Ouma, and Joshua K. Kibet. "A review on the recent advances in battery development and energy storage technologies." *Journal of Renewable Energy* 2024 (2024): 2329261. https://doi.org/10.1155/2024/2329261.
75. Fagiolari, Lucia, Matteo Sampò, Andrea Lamberti, Julia Amici, Carlotta Francia, Silvia Bodoardo, and Federico Bella. "Integrated energy conversion and storage devices: Interfacing solar cells, batteries and supercapacitors." *Energy Storage Materials* 51 (2022): 400–434. https://doi.org/10.1016/j.ensm.2022.06.051.
76. Arora, Swati, and Nagendra Verma. "A review: Advancing organic electronics through the lens of ionic liquids and polymerized ionic liquids." *RSC Applied Polymers* 2 (2024): 317–355. doi: 10.1039/D3LP00269A.
77. Azaïs, Philippe. "Manufacturing of industrial supercapacitors." *Supercapacitors: Materials, Systems, and Applications* (2013): 307–371. doi:10.1002/9783527646661.
78. Iwan, Agnieszka, Krzysztof A. Bogdanowicz, Robert Pich, Agnieszka Gonciarz, Jacek Miedziak, Ireneusz Plebankiewicz, and Wojciech Przybyl. "The safety engineering of sodium-ion batteries used as an energy storage system for the military." *Energies* 18 (2025): 978. doi: 10.3390/en18040978.
79. Pekias, Antonios, George S. Maraslidis, Markos G. Tsipouras, Fotis N. Koumboulis, and George F. Fragulis. "Power supply technologies for drones and machine vision applications: A comparative analysis and future trends." *Telecom* 4 (2023): 459–476. doi: 10.3390/telecom4030024.
80. Gao, Ziyan, Yang Zhou, Jin Zhang, Javad Foroughi, Shuhua Peng, Ray H. Baughman, Zhong Lin Wang, and Chun H. Wang. "Advanced energy harvesters and energy storage for powering wearable and implantable medical devices." *Advanced Materials* 36 (2024): 2404492. doi: 10.1002/adma.202404492.
81. Ghandi, Khashayar. "A review of ionic liquids, their limits and applications." *Green and Sustainable Chemistry* 4 (2014): 44–53. doi: 10.4236/gsc.2014.41008.
82. Küpers, Verena, Jan Frederik Dohmann, Peter Bieker, Martin Winter, Tobias Placke, and Martin Kolek. "Opportunities and limitations of ionic liquid-and organic carbonate solvent-based electrolytes for mg-ion-based dual-ion batteries." *ChemSusChem* 14 (2021): 4480–4498. doi: 10.1002/cssc.202101227.
83. Greer, Adam J., Johan Jacquemin, and Christopher Hardacre. "Industrial applications of ionic liquids." *Molecules* 25 (2020): 5207. doi: 10.3390/molecules25215207.

84. Wang, Xuehang, Maryam Salari, De-en Jiang, Jennifer Chapman Varela, Babak Anasori, David J. Wesolowski, Sheng Dai, Mark W. Grinstaff, and Yury Gogotsi. "Electrode material–ionic liquid coupling for electrochemical energy storage." *Nature Reviews Materials* 5 (2020): 787–808. doi: 10.1038/s41578-020-0218-9.
85. Costa, Susana P.F., Ana M.O. Azevedo, Paula C.A.G. Pinto, and M. Lúcia M.F.S. Saraiva. "Environmental impact of ionic liquids: Recent advances in (eco) toxicology and (bio) degradability." *ChemSusChem* 10 (2017): 2321–2347. doi: 10.1002/cssc.201700261.
86. Stark, Annegret, Peter Behrend, Oliver Braun, Anja Müller, Johannes Ranke, Bernd Ondruschka, and Bernd Jastorff. "Purity specification methods for ionic liquids." *Green Chemistry* 10 (2008): 1152–1161. doi: 10.1039/B808532C.
87. Doblinger, Simon, Taylor J. Donati, and Debbie S. Silvester. "Effect of humidity and impurities on the electrochemical window of ionic liquids and its implications for electroanalysis." *The Journal of Physical Chemistry C* 124 (2020): 20309–20319. doi: 10.1021/acs.jpcc.0c07012.
88. Kaur, Harpreet, Abhinay Thakur, Ramesh Chand Thakur, and Ashish Kumar. "A review on multifaceted role of ionic liquids in modern energy storage systems: From electrochemical performance to environmental sustainability." *Energy & Fuels* 39 (2025): 3703–3734. doi: 10.1021/acs.energyfuels.4c05274.
89. Yao, Nan, Xiang Chen, Zhong-Heng Fu, and Qiang Zhang. "Applying classical, Ab initio, and machine-learning molecular dynamics simulations to the liquid electrolyte for rechargeable batteries." *Chemical Reviews* 122 (2022): 10970–11021. doi: 10.1021/acs.chemrev.1c00904.
90. Venâncio, Raíssa, Débora A.C. da Silva, Ludmila O. Borges, Isabel A. Silva, Leonardo Morais da Silva, Gustavo Doubek, Leonardo Siqueira, and Hudson Zanin. "High performance of supercapacitors containing an ionic liquid electrolyte by means of electrochemical and computational studies." Available at SSRN 5195893. doi: 10.2139/ssrn.5195893.
91. Vatamanu, Jenel, Zongzhi Hu, Dmitry Bedrov, Carlos Perez, and Yury Gogotsi. "Increasing energy storage in electrochemical capacitors with ionic liquid electrolytes and nanostructured carbon electrodes." *The Journal of Physical Chemistry Letters* 4 (2013): 2829–2837. doi: 10.1021/jz401472c.
92. Pereira, José, Reinaldo Souza, and Ana Moita. "A review of ionic liquids and their composites with nanoparticles for electrochemical applications." *Inorganics* 12 (2024): 186. doi: 10.3390/inorganics12070186.
93. Abo-Hamad, Ali, Mohammed Abdul Hakim AlSaadi, Maan Hayyan, Ibrahim Juneidi, and Mohd Ali Hashim. "Ionic liquid-carbon nanomaterial hybrids for electrochemical sensor applications: A review." *Electrochimica Acta* 193 (2016): 321–343. doi: 10.1016/j.electacta.2016.02.044.
94. Lu, Chao, and Xi Chen. "Latest advances in flexible symmetric supercapacitors: From material engineering to wearable applications." *Accounts of Chemical Research* 53 (2020): 1468–1477. doi: 10.1021/acs.accounts.0c00205.
95. Liu, Qiu-Ping, Xue-Dan Hou, Ning Li, and Min-Hua Zong. "Ionic liquids from renewable biomaterials: Synthesis, characterization and application in the pretreatment of biomass." *Green Chemistry* 14 (2012): 304–307. doi: 10.1039/C2GC16128A.
96. Zotova, Julija, Zaneta Wojnarowska, Brendan Twamley, Marian Paluch, and Lidia Tajber. "Green synthesis of lidocaine ionic liquids and salts: Mechanisms of formation and interactions in the crystalline and supercooled states." *ACS Sustainable Chemistry & Engineering* 8 (2020): 18266–18276. doi: 10.1021/acssuschemeng.0c06811.
97. de Jesus, Sérgio S., and Rubens Maciel Filho. "Are ionic liquids eco-friendly?" *Renewable and Sustainable Energy Reviews* 157 (2022): 112039. doi: 10.1016/j.rser.2021.112039.

Index

For Product Safety Concerns and Information please contact our EU representative GPSR@taylorandfrancis.com Taylor & Francis Verlag GmbH, Kaufingerstraße 24, 80331 München, Germany